Garment Manufacturing Technology

The Textile Institute and Woodhead Publishing

The Textile Institute is a unique organisation in textiles, clothing and footwear. Incorporated in England by a Royal Charter granted in 1925, the Institute has individual and corporate members in over 90 countries. The aim of the Institute is to facilitate learning, recognise achievement, reward excellence and disseminate information within the global textiles, clothing and footwear industries.

Historically, The Textile Institute has published books of interest to its members and the textile industry. To maintain this policy, the Institute has entered into partnership with Woodhead Publishing Limited to ensure that Institute members and the textile industry continue to have access to high calibre titles on textile science and technology.

Most Woodhead titles on textiles are now published in collaboration with The Textile Institute. Through this arrangement, the Institute provides an Editorial Board which advises Woodhead on appropriate titles for future publication and suggests possible editors and authors for these books. Each book published under this arrangement carries the Institute's logo.

Woodhead books published in collaboration with The Textile Institute are offered to Textile Institute members at a substantial discount. These books, together with those published by The Textile Institute that are still in print, are offered on the Elsevier website at: http://store.elsevier.com/. Textile Institute books still in print are also available directly from the Institute's website at: www.textileinstitutebooks.com.

A list of Woodhead books on textile science and technology, most of which have been published in collaboration with The Textile Institute, can be found towards the end of the contents pages.

Related titles

Gupta and Zakaria, *Anthropometry, Apparel Sizing and Design*, Woodhead, 2014, 9780857096814

Wong and Guo, *Fashion Supply Chain Management Using Radio Frequency Identification (RFID) Technologies*, Woodhead, 2014, 9780857098054

Sinclair, *Textiles and Fashion*, Woodhead, 2014, 9781845699314

Woodhead Publishing Series in Textiles:
Number 168

Garment Manufacturing Technology

Edited by

Rajkishore Nayak and Rajiv Padhye

The Textile Institute

ELSEVIER

AMSTERDAM • BOSTON • CAMBRIDGE • HEIDELBERG
LONDON • NEW YORK • OXFORD • PARIS • SAN DIEGO
SAN FRANCISCO • SINGAPORE • SYDNEY • TOKYO
Woodhead Publishing is an imprint of Elsevier

Published by Woodhead Publishing in association with The Textile Institute
Woodhead Publishing is an imprint of Elsevier
80 High Street, Sawston, Cambridge, CB22 3HJ, UK
225 Wyman Street, Waltham, MA 02451, USA
Langford Lane, Kidlington, OX5 1GB, UK

Copyright © 2015 Elsevier Ltd. All rights reserved.

No part of this publication may be reproduced, stored in a retrieval system or transmitted in any form or by any means electronic, mechanical, photocopying, recording or otherwise without the prior written permission of the publisher.

Permissions may be sought directly from Elsevier's Science & Technology Rights Department in Oxford, UK: phone (+44) (0) 1865 843830; fax (+44) (0) 1865 853333; email: permissions@elsevier.com. Alternatively you can submit your request online by visiting the Elsevier website at http://elsevier.com/locate/permissions, and selecting Obtaining permission to use Elsevier material.

Notice
No responsibility is assumed by the publisher for any injury and/or damage to persons or property as a matter of products liability, negligence or otherwise, or from any use or operation of any methods, products, instructions or ideas contained in the material herein. Because of rapid advances in the medical sciences, in particular, independent verification of diagnoses and drug dosages should be made.

British Library Cataloguing in Publication Data
A catalogue record for this book is available from the British Library

Library of Congress Control Number: 2015932737

ISBN 978-1-78242-232-7 (print)
ISBN 978-1-78242-239-6 (online)

For information on all Woodhead Publishing publications visit our website at http://store.elsevier.com/

www.elsevier.com • www.bookaid.org

Contents

List of contributors		xi
Woodhead Publishing Series in Textiles		xiii

1 Introduction: the apparel industry   1
R. Nayak, R. Padhye
    1.1 Introduction   1
    1.2 Global scenario of apparel manufacturing   4
    1.3 Challenges in apparel production   10
    1.4 Role of various organisations   12
    1.5 Future trends   13
    1.6 Conclusions   15
    References   15

Part One   Product development, production planning and selection of materials   19

2 Product development in the apparel industry   21
M. Senanayake
    2.1 Introduction   21
    2.2 Product-development models and product-development process   22
    2.3 Variations in apparel product development: demand-led product development   29
    2.4 Apparel product-development technologies   32
    2.5 Apparel product standards, specifications, quality assurance and product technical package   33
    2.6 Apparel product life-cycle management (PLM) and supply-chain relationships   41
    2.7 Measures for apparel product development   43
    2.8 Future trends in apparel product development   44
    2.9 Case studies: PD tools and technologies   49
    2.10 Conclusions   55
    2.11 Sources of further information and advice   55
    References   56

3	**Role of fabric properties in the clothing-manufacturing process**	**59**
	B.K. Behera	
	3.1 Introduction	59
	3.2 Fabric properties and performance	59
	3.3 Garment make-up process and fabric properties	61
	3.4 Low-stress mechanical properties and make-up process	63
	3.5 Control system	67
	3.6 Fabric tailorability, buckling and formability	69
	3.7 Sewability	71
	3.8 Conclusions	79
	References	80
4	**Production planning in the apparel industry**	**81**
	S. Das, A. Patnaik	
	4.1 Introduction	81
	4.2 Production planning	81
	4.3 Production systems	83
	4.4 Production planning and control in the apparel industry	88
	4.5 Supply chain management in the apparel industry	94
	4.6 Inventory management	97
	4.7 Manufacturing performance improvement through lean production	99
	4.8 Waste management	103
	4.9 Human resource management	104
	4.10 New tools developed in production planning	105
	Acknowledgments	105
	References	105
5	**Fabric sourcing and selection**	**109**
	A. Vijayan, A. Jadhav	
	5.1 Introduction	109
	5.2 Fabric sourcing	110
	5.3 Fabric inspection	115
	5.4 Future trends	121
	5.5 Conclusions	127
	References	127
6	**Selecting garment accessories, trims, and closures**	**129**
	K.N. Chatterjee, Y. Jhanji, T. Grover, N. Bansal, S. Bhattacharyya	
	6.1 Part 1: introduction to garment accessories	129
	6.2 Part 2: selecting garment accessories	129
	6.3 Part 3: selecting supporting materials	136
	6.4 Part 4: selecting closures	138
	6.5 Part 5: accessories for children's wear	145
	6.6 Part 6: evaluation of quality of trims and accessories	148

	6.7	Part 7: fashion accessories	150
	6.8	Conclusions	183
		Acknowledgments	183
		Bibliography	183

Part Two Garment design and production — 185

7 Garment sizing and fit — 187
R. Pandarum, W. Yu

7.1	Introduction	187
7.2	Geometry of the human form	188
7.3	The human figure divided into body proportions	189
7.4	Garment size charts	193
7.5	Development of garment size charts	194
7.6	Sizing and fit systems	197
7.7	Three-dimensional (3D) body scanning – current and potential future applications in clothing manufacture and retailing	200
7.8	Conclusions	201
7.9	Sources of further information	201
	Acknowledgements	202
	References	202

8 Pattern construction — 205
K. Kennedy

8.1	Introduction	205
8.2	Pattern construction modes	206
8.3	Body, material and design	208
8.4	Pattern construction tools	213
8.5	Conclusions	218
8.6	2D and 3D CAD Web sites	219
	Acknowledgements	219
	References	219

9 Fabric spreading and cutting — 221
I. Vilumsone-Nemes

9.1	Introduction	221
9.2	Cut process planning	221
9.3	Spreading of textile materials	223
9.4	Cutting of textile materials	230
9.5	Fusing of cut textile components	240
9.6	Final work operations of the cutting process	242
9.7	Future trends	244
9.8	Conclusions	244
9.9	Sources of further information	245

10	Sewing, stitches and seams	247
	G. Colovic	
	10.1 Introduction	247
	10.2 Stitch classes	248
	10.3 Seam types	258
	10.4 Seam-neatening	269
	10.5 Future trends	270
	10.6 Conclusions	271
	10.7 Sources of further information and advice	272
	References	273
11	Sewing equipment and work aids	275
	P. Jana	
	11.1 Introduction	275
	11.2 Different bed types in industrial sewing machines	275
	11.3 Different feed types in industrial sewing machines	280
	11.4 Cyclic sewing machines	289
	11.5 Computerised sewing machines	291
	11.6 Work aids	295
	11.7 Sewing automats	302
	11.8 Sewing needles	306
	11.9 Sewing threads	310
	11.10 Future trends and conclusions	312
	11.11 Sources of further information and advice	313
	References	314
12	Sewing-room problems and solutions	317
	M. Carvalho, H. Carvalho, L.F. Silva, F. Ferreira	
	12.1 Introduction	317
	12.2 Seam pucker and other surface distortions	317
	12.3 Sewing defects caused by needles	322
	12.4 Material feeding and associated problems	324
	12.5 Problems in stitch formation	325
	12.6 Thread breakage	328
	12.7 Future trends	328
	12.8 Conclusions	334
	12.9 Sources of further information and advice	334
	References	334
13	Alternative fabric-joining technologies	337
	E.M. Petrie	
	13.1 Alternatives to sewing	337
	13.2 Adhesive bonding	341
	13.3 Conventional thermal welding	356
	13.4 Advanced thermal-welding processes	363
	13.5 Conclusions	370
	References	371

14	**Seamless garments**	373
	N. Nawaz, R. Nayak	
	14.1 Introduction	373
	14.2 Seamless technique	373
	14.3 Common seamless products	376
	14.4 Raw materials	377
	14.5 Seamless knitting machines	377
	14.6 Advantages of seamless garments	379
	14.7 Disadvantages of seamless garments	380
	14.8 Applications of seamless garments	381
	14.9 Future developments	382
	14.10 Conclusions	382
	References	383

Part Three Garment finishing, quality control, care labelling and costing 385

15	**Garment-finishing techniques**	387
	S. MacA. Fergusson	
	15.1 Introduction	387
	15.2 Garment finishing for functionality	388
	15.3 Knitwear finishing	393
	15.4 Denim garment finishing	396
	15.5 Pressing (factors and equipment)	397
	15.6 Future trends	401
	15.7 Conclusions	402
	References	402

16	**Quality control and quality assurance in the apparel industry**	405
	C.N. Keist	
	16.1 Introduction	405
	16.2 Quality control in the apparel industry	406
	16.3 Future trends	423
	16.4 Conclusions	424
	16.5 Sources of further information and advice	425
	References	425

17	**Care labelling of clothing**	427
	R. Nayak, R. Padhye	
	17.1 Introduction	427
	17.2 Requirements of care labelling	427
	17.3 Definition of care label	429
	17.4 Care labelling systems	431
	17.5 Future trends	444
	17.6 Conclusions	445
	References	446

18 Garment costing **447**
A. Singh, K. Nijhar
    18.1 Introduction **447**
    18.2 Costing need **447**
    18.3 Cost classification **449**
    18.4 Cost elements **450**
    18.5 Measures of efficiency **451**
    18.6 Profitability **452**
    18.7 Garment sales element analysis **454**
    18.8 Mark-downs **461**
    18.9 Managing cost through inventory control **463**
    18.10 Apparel costing sheet analysis **464**
    18.11 Conclusions **467**
           References **467**

Index **469**

List of contributors

N. Bansal   The Technological Institute of Textile and Sciences, Bhiwani, India

B.K. Behera   Indian Institute of Technology, New Delhi, India

S. Bhattacharyya   The Technological Institute of Textile and Sciences, Bhiwani, India

H. Carvalho   University of Minho, Guimarães, Portugal

M. Carvalho   University of Minho, Guimarães, Portugal

K.N. Chatterjee   The Technological Institute of Textile and Sciences, Bhiwani, India

G. Colovic   The College of Textile – Design, Technology and Management, Belgrade, Serbia

S. Das   Faculty of Science, Nelson Mandela Metropolitan University, Port Elizabeth, South Africa

S. MacA. Fergusson   RMIT University, Melbourne, VIC, Australia

F. Ferreira   University of Minho, Guimarães, Portugal

T. Grover   The Technological Institute of Textile and Sciences, Bhiwani, India

A. Jadhav   RMIT University, Melbourne, VIC, Australia

P. Jana   National Institute of Fashion Technology, New Delhi, India

Y. Jhanji   The Technological Institute of Textile and Sciences, Bhiwani, India

C.N. Keist   Western Illinois University, Macomb, IL, USA

K. Kennedy   RMIT University, Melbourne, VIC, Australia

N. Nawaz   School of Fashion and Textiles, RMIT University, Melbourne, VIC, Australia

R. Nayak   School of Fashion and Textiles, RMIT University, Melbourne, VIC, Australia

K. Nijhar   Yum Productions Pty. Ltd, Melbourne, VIC, Australia

R. Padhye   School of Fashion and Textiles, RMIT University, Melbourne, VIC, Australia

R. Pandarum   University of South Africa, Florida, South Africa

A. Patnaik   Faculty of Science, Nelson Mandela Metropolitan University, Port Elizabeth, South Africa; CSIR Materials Science and Manufacturing, Polymers and Composites Competence Area, Nonwovens and Composite Group, Port Elizabeth, South Africa

E.M. Petrie   Independent Consultant, Cary, NC, USA

M. Senanayake   California State Polytechnic University, Pomona, CA, USA

L.F. Silva   University of Minho, Guimarães, Portugal

A. Singh   RMIT University, Melbourne, VIC, Australia

A. Vijayan   RMIT University, Melbourne, VIC, Australia

I. Vilumsone-Nemes   Technical Faculty 'Michael Pupin', University of Novi Sad, Novi Sad, Serbia

W. Yu   Hong Kong Polytechnic University, Hong Kong SAR, China

Woodhead Publishing Series in Textiles

1 **Watson's textile design and colour Seventh edition**
   Edited by Z. Grosicki
2 **Watson's advanced textile design**
   Edited by Z. Grosicki
3 **Weaving Second edition**
   P. R. Lord and M. H. Mohamed
4 **Handbook of textile fibres Volume 1: Natural fibres**
   J. Gordon Cook
5 **Handbook of textile fibres Volume 2: Man-made fibres**
   J. Gordon Cook
6 **Recycling textile and plastic waste**
   Edited by A. R. Horrocks
7 **New fibers Second edition**
   T. Hongu and G. O. Phillips
8 **Atlas of fibre fracture and damage to textiles Second edition**
   J. W. S. Hearle, B. Lomas and W. D. Cooke
9 **Ecotextile '98**
   Edited by A. R. Horrocks
10 **Physical testing of textiles**
    B. P. Saville
11 **Geometric symmetry in patterns and tilings**
    C. E. Horne
12 **Handbook of technical textiles**
    Edited by A. R. Horrocks and S. C. Anand
13 **Textiles in automotive engineering**
    W. Fung and J. M. Hardcastle
14 **Handbook of textile design**
    J. Wilson
15 **High-performance fibres**
    Edited by J. W. S. Hearle
16 **Knitting technology Third edition**
    D. J. Spencer
17 **Medical textiles**
    Edited by S. C. Anand
18 **Regenerated cellulose fibres**
    Edited by C. Woodings
19 **Silk, mohair, cashmere and other luxury fibres**
    Edited by R. R. Franck

20	**Smart fibres, fabrics and clothing**	
	Edited by X. M. Tao	
21	**Yarn texturing technology**	
	J. W. S. Hearle, L. Hollick and D. K. Wilson	
22	**Encyclopedia of textile finishing**	
	H-K. Rouette	
23	**Coated and laminated textiles**	
	W. Fung	
24	**Fancy yarns**	
	R. H. Gong and R. M. Wright	
25	**Wool: Science and technology**	
	Edited by W. S. Simpson and G. Crawshaw	
26	**Dictionary of textile finishing**	
	H-K. Rouette	
27	**Environmental impact of textiles**	
	K. Slater	
28	**Handbook of yarn production**	
	P. R. Lord	
29	**Textile processing with enzymes**	
	Edited by A. Cavaco-Paulo and G. Gübitz	
30	**The China and Hong Kong denim industry**	
	Y. Li, L. Yao and K. W. Yeung	
31	**The World Trade Organization and international denim trading**	
	Y. Li, Y. Shen, L. Yao and E. Newton	
32	**Chemical finishing of textiles**	
	W. D. Schindler and P. J. Hauser	
33	**Clothing appearance and fit**	
	J. Fan, W. Yu and L. Hunter	
34	**Handbook of fibre rope technology**	
	H. A. McKenna, J. W. S. Hearle and N. O'Hear	
35	**Structure and mechanics of woven fabrics**	
	J. Hu	
36	**Synthetic fibres: Nylon, polyester, acrylic, polyolefin**	
	Edited by J. E. McIntyre	
37	**Woollen and worsted woven fabric design**	
	E. G. Gilligan	
38	**Analytical electrochemistry in textiles**	
	P. Westbroek, G. Priniotakis and P. Kiekens	
39	**Bast and other plant fibres**	
	R. R. Franck	
40	**Chemical testing of textiles**	
	Edited by Q. Fan	
41	**Design and manufacture of textile composites**	
	Edited by A. C. Long	
42	**Effect of mechanical and physical properties on fabric hand**	
	Edited by H. M. Behery	
43	**New millennium fibers**	
	T. Hongu, M. Takigami and G. O. Phillips	

44 **Textiles for protection**
   Edited by R. A. Scott
45 **Textiles in sport**
   Edited by R. Shishoo
46 **Wearable electronics and photonics**
   Edited by X. M. Tao
47 **Biodegradable and sustainable fibres**
   Edited by R. S. Blackburn
48 **Medical textiles and biomaterials for healthcare**
   Edited by S. C. Anand, M. Miraftab, S. Rajendran and J. F. Kennedy
49 **Total colour management in textiles**
   Edited by J. Xin
50 **Recycling in textiles**
   Edited by Y. Wang
51 **Clothing biosensory engineering**
   Y. Li and A. S. W. Wong
52 **Biomechanical engineering of textiles and clothing**
   Edited by Y. Li and D. X-Q. Dai
53 **Digital printing of textiles**
   Edited by H. Ujiie
54 **Intelligent textiles and clothing**
   Edited by H. R. Mattila
55 **Innovation and technology of women's intimate apparel**
   W. Yu, J. Fan, S. C. Harlock and S. P. Ng
56 **Thermal and moisture transport in fibrous materials**
   Edited by N. Pan and P. Gibson
57 **Geosynthetics in civil engineering**
   Edited by R. W. Sarsby
58 **Handbook of nonwovens**
   Edited by S. Russell
59 **Cotton: Science and technology**
   Edited by S. Gordon and Y-L. Hsieh
60 **Ecotextiles**
   Edited by M. Miraftab and A. R. Horrocks
61 **Composite forming technologies**
   Edited by A. C. Long
62 **Plasma technology for textiles**
   Edited by R. Shishoo
63 **Smart textiles for medicine and healthcare**
   Edited by L. Van Langenhove
64 **Sizing in clothing**
   Edited by S. Ashdown
65 **Shape memory polymers and textiles**
   J. Hu
66 **Environmental aspects of textile dyeing**
   Edited by R. Christie
67 **Nanofibers and nanotechnology in textiles**
   Edited by P. Brown and K. Stevens

68 **Physical properties of textile fibres Fourth edition**
    W. E. Morton and J. W. S. Hearle
69 **Advances in apparel production**
    Edited by C. Fairhurst
70 **Advances in fire retardant materials**
    Edited by A. R. Horrocks and D. Price
71 **Polyesters and polyamides**
    Edited by B. L. Deopura, R. Alagirusamy, M. Joshi and B. S. Gupta
72 **Advances in wool technology**
    Edited by N. A. G. Johnson and I. Russell
73 **Military textiles**
    Edited by E. Wilusz
74 **3D fibrous assemblies: Properties, applications and modelling of three-dimensional textile structures**
    J. Hu
75 **Medical and healthcare textiles**
    Edited by S. C. Anand, J. F. Kennedy, M. Miraftab and S. Rajendran
76 **Fabric testing**
    Edited by J. Hu
77 **Biologically inspired textiles**
    Edited by A. Abbott and M. Ellison
78 **Friction in textile materials**
    Edited by B. S. Gupta
79 **Textile advances in the automotive industry**
    Edited by R. Shishoo
80 **Structure and mechanics of textile fibre assemblies**
    Edited by P. Schwartz
81 **Engineering textiles: Integrating the design and manufacture of textile products**
    Edited by Y. E. El-Mogahzy
82 **Polyolefin fibres: Industrial and medical applications**
    Edited by S. C. O. Ugbolue
83 **Smart clothes and wearable technology**
    Edited by J. McCann and D. Bryson
84 **Identification of textile fibres**
    Edited by M. Houck
85 **Advanced textiles for wound care**
    Edited by S. Rajendran
86 **Fatigue failure of textile fibres**
    Edited by M. Miraftab
87 **Advances in carpet technology**
    Edited by K. Goswami
88 **Handbook of textile fibre structure Volume 1 and Volume 2**
    Edited by S. J. Eichhorn, J. W. S. Hearle, M. Jaffe and T. Kikutani
89 **Advances in knitting technology**
    Edited by K-F. Au
90 **Smart textile coatings and laminates**
    Edited by W. C. Smith
91 **Handbook of tensile properties of textile and technical fibres**
    Edited by A. R. Bunsell

92	**Interior textiles: Design and developments**
	Edited by T. Rowe
93	**Textiles for cold weather apparel**
	Edited by J. T. Williams
94	**Modelling and predicting textile behaviour**
	Edited by X. Chen
95	**Textiles, polymers and composites for buildings**
	Edited by G. Pohl
96	**Engineering apparel fabrics and garments**
	J. Fan and L. Hunter
97	**Surface modification of textiles**
	Edited by Q. Wei
98	**Sustainable textiles**
	Edited by R. S. Blackburn
99	**Advances in yarn spinning technology**
	Edited by C. A. Lawrence
100	**Handbook of medical textiles**
	Edited by V. T. Bartels
101	**Technical textile yarns**
	Edited by R. Alagirusamy and A. Das
102	**Applications of nonwovens in technical textiles**
	Edited by R. A. Chapman
103	**Colour measurement: Principles, advances and industrial applications**
	Edited by M. L. Gulrajani
104	**Fibrous and composite materials for civil engineering applications**
	Edited by R. Fangueiro
105	**New product development in textiles: Innovation and production**
	Edited by L. Horne
106	**Improving comfort in clothing**
	Edited by G. Song
107	**Advances in textile biotechnology**
	Edited by V. A. Nierstrasz and A. Cavaco-Paulo
108	**Textiles for hygiene and infection control**
	Edited by B. McCarthy
109	**Nanofunctional textiles**
	Edited by Y. Li
110	**Joining textiles: Principles and applications**
	Edited by I. Jones and G. Stylios
111	**Soft computing in textile engineering**
	Edited by A. Majumdar
112	**Textile design**
	Edited by A. Briggs-Goode and K. Townsend
113	**Biotextiles as medical implants**
	Edited by M. W. King, B. S. Gupta and R. Guidoin
114	**Textile thermal bioengineering**
	Edited by Y. Li
115	**Woven textile structure**
	B. K. Behera and P. K. Hari

116 **Handbook of textile and industrial dyeing. Volume 1: Principles, processes and types of dyes**
Edited by M. Clark
117 **Handbook of textile and industrial dyeing. Volume 2: Applications of dyes**
Edited by M. Clark
118 **Handbook of natural fibres. Volume 1: Types, properties and factors affecting breeding and cultivation**
Edited by R. Kozłowski
119 **Handbook of natural fibres. Volume 2: Processing and applications**
Edited by R. Kozłowski
120 **Functional textiles for improved performance, protection and health**
Edited by N. Pan and G. Sun
121 **Computer technology for textiles and apparel**
Edited by J. Hu
122 **Advances in military textiles and personal equipment**
Edited by E. Sparks
123 **Specialist yarn and fabric structures**
Edited by R. H. Gong
124 **Handbook of sustainable textile production**
M. I. Tobler-Rohr
125 **Woven textiles: Principles, developments and applications**
Edited by K. Gandhi
126 **Textiles and fashion: Materials design and technology**
Edited by R. Sinclair
127 **Industrial cutting of textile materials**
I. Viļumsone-Nemes
128 **Colour design: Theories and applications**
Edited by J. Best
129 **False twist textured yarns**
C. Atkinson
130 **Modelling, simulation and control of the dyeing process**
R. Shamey and X. Zhao
131 **Process control in textile manufacturing**
Edited by A. Majumdar, A. Das, R. Alagirusamy and V. K. Kothari
132 **Understanding and improving the durability of textiles**
Edited by P. A. Annis
133 **Smart textiles for protection**
Edited by R. A. Chapman
134 **Functional nanofibers and applications**
Edited by Q. Wei
135 **The global textile and clothing industry: Technological advances and future challenges**
Edited by R. Shishoo
136 **Simulation in textile technology: Theory and applications**
Edited by D. Veit
137 **Pattern cutting for clothing using CAD: How to use Lectra Modaris pattern cutting software**
M. Stott

| 138 | Advances in the dyeing and finishing of technical textiles
M. L. Gulrajani
| 139 | Multidisciplinary know-how for smart textiles developers
Edited by T. Kirstein
| 140 | Handbook of fire resistant textiles
Edited by F. Selcen Kilinc
| 141 | Handbook of footwear design and manufacture
Edited by A. Luximon
| 142 | Textile-led design for the active ageing population
Edited by J. McCann and D. Bryson
| 143 | Optimizing decision making in the apparel supply chain using artificial intelligence (AI): From production to retail
Edited by W. K. Wong, Z. X. Guo and S. Y. S. Leung
| 144 | Mechanisms of flat weaving technology
V. V. Choogin, P. Bandara and E. V. Chepelyuk
| 145 | Innovative jacquard textile design using digital technologies
F. Ng and J. Zhou
| 146 | Advances in shape memory polymers
J. Hu
| 147 | Design of clothing manufacturing processes: A systematic approach to planning, scheduling and control
J. Gersak
| 148 | Anthropometry, apparel sizing and design
D. Gupta and N. Zakaria
| 149 | Silk: Processing, properties and applications
Edited by K. Murugesh Babu
| 150 | Advances in filament yarn spinning of textiles and polymers
Edited by D. Zhang
| 151 | Designing apparel for consumers: The impact of body shape and size
Edited by M.-E. Faust and S. Carrier
| 152 | Fashion supply chain management using radio frequency identification (RFID) technologies
Edited by W. K. Wong and Z. X. Guo
| 153 | High performance textiles and their applications
Edited by C. A. Lawrence
| 154 | Protective clothing: Managing thermal stress
Edited by F. Wang and C. Gao
| 155 | Composite nonwoven materials
Edited by D. Das and B. Pourdeyhimi
| 156 | Functional finishes for textiles: Improving comfort, performance and protection
Edited by R. Paul
| 157 | Assessing the environmental impact of textiles and the clothing supply chain
S. S. Muthu
| 158 | Braiding technology for textiles
Y. Kyosev
| 159 | Principles of colour appearance and measurement
Volume 1: Object appearance, colour perception and instrumental measurement
A. K. R. Choudhury

160	**Principles of colour appearance and measurement**
Volume 2: Visual measurement of colour, colour comparison and management
A. K. R. Choudhury
161	**Ink jet textile printing**
C. Cie
162	**Textiles for sportswear**
Edited by R. Shishoo
163	**Advances in silk science and technology**
Edited by A. Basu
164	**Denim: Manufacture, finishing and applications**
Edited by R. Paul
165	**Fabric structures in architecture**
Edited by J. Ignasi de Llorens
166	**Electronic textiles: Smart fabrics and wearable technology**
Edited by T. Dias
167	**Advances in 3D textiles**
Edited by X. Chen
168	**Garment manufacturing technology**
Edited by R. Nayak and R. Padhye

Introduction: the apparel industry

R. Nayak, R. Padhye
School of Fashion and Textiles, RMIT University, Melbourne, VIC, Australia

1.1 Introduction

Apparel manufacturing is labour intensive, which is characterised by low fixed capital investment; a wide range of product designs and hence input materials; variable production volumes; high competitiveness and often high demand on product quality (Scott, 2006; Hassler, 2003; Forza and Vinelli, 2000). Although the manufacturing process is associated mainly with apparel and household linens, it is also used in a variety of industries and crafts such as upholstery, shoe-making, sail-making, bookbinding and the production of varieties of sporting goods. Sewing is the fundamental process, with ramifications into a variety of textile arts and crafts, including tapestry, quilting, embroidery, appliqué and patchwork.

The apparel-manufacturing process evolved as an art and underwent several technical changes. The technological advancements in the apparel industry include the use of computerised equipment (especially in design, pattern-making and cutting), 3D scanning technology, automation and robotics, integration of wearable technology and advanced material transport systems (Bailey, 1993; Forza and Vinelli, 2000). Another important development involves the increasing use of robotics to transport components and materials within the plant, which helps in improving production efficiency. However, the apparel industry — especially sewing technology — has remained significantly less automated compared to many other manufacturing industries.

There are wide varieties of clothing types that the apparel manufacturers have to handle, which can be broadly divided into two categories: outer clothing and inner clothing. Outer clothing includes work-wear and uniforms, leisure wear and sportswear (e.g. suits, pants, dresses, ladies' suits, blouses, blazers, jackets, cardigans, pullovers, coats, sports jackets, skirts, shirts short- or long-sleeved, ties, jeans, shorts, T-shirts, polo shirts, sports shirts, tracksuits, bathing shorts, bathing suits and bikinis). The underclothing (underwear) includes jersey goods and lingerie (e.g. underpants, undershirts, briefs, socks, stockings and pantyhose). These products are manufactured in a wide range of design and style variations, which increases the complexity of the manufacturing process.

1.1.1 Sewing: the art of joining materials

Sewing can be defined as the art and craft of fastening objects by stitches produced with a needle and a thread. It is considered one of the oldest textile arts, and started in the Palaeolithic Age before the invention of spinning yarn or weaving fabric (Angier, 1999).

During the Palaeolithic Age, animal hides were stitched together by sewing and used for clothing and shelter. Tools such as needles, pins and pin-cushions were used as sewing aids in the bridal wear of many European brides from the Middle Ages to the seventeenth century. It is also believed that Stone Age people across Asia and Europe used to produce clothing items from fur and skin by joining them with needles such as bone, antler or ivory and threads prepared from various animal body parts including sinew, catgut and veins.

In the American plains and Canadian prairies, the indigenous peoples had adopted sophisticated sewing practices to assemble tipi shelters. In Africa, sewn baskets were produced by the combination of sewing and weaving of plant leaves. For example, the baskets prepared by Zulu weavers used thin strips of palm leaves as 'thread'. The technology of weaving fabric from natural fibres dates back to the Middle East around 4000 BCE (before the common era) or perhaps earlier during the Neolithic Age, which also leads to the development of sewing.

Sewing was mainly considered a women's occupation, and women had an important role in extending the longevity of clothing items before the nineteenth century (Quataert, 1985). For most people, clothing was an expensive investment. Faded clothing generally continued to be worn by turning it inside-out. In many instances, the components of a piece of cloth would be separated and reassembled by sewing to suit other applications or made into quilts or other products. As civilisation developed, ramification of sewing started in the form of pattern-making, cutting, pressing and alterations. Hence, the skilled women often had to barter their expertise with one another.

Decorative needlework such as embroidery was considered a much valued skill. Young women with affordability and time preferred to acquire these skills, as in many cultures decorative embroidery was much valued. Although the origin of stitches used for embroidery works in the Western repertoire were traditionally British, Irish or Western European, the stitches used throughout the world today originated in different cultures. Romanian couching or Oriental couching, the Cretan open filling stitch and the Japanese stitch are some of the examples. During the Middle Ages, the embroidery stitches spread by way of the trade routes that were active. Chinese embroidery techniques were brought to the Western Asia and Eastern Europe by the Silk Road (Watt, 1997; Wood, 2004). Similarly, the embroidery practices that originated in the Middle East were spread to Southern and Western Europe through Morocco and Spain. Embroidery and sewing techniques were also spread around the globe by European imperial settlements. However, similar techniques of sewing indigenous to cultures in distant locations were also developed, which are unlikely to have been developed by cross-cultural communication. For example, a method of reverse appliqué known to the areas of South America was also known to Southeast Asia.

Sewing was done only by hand and survived for thousands of years before the invention of the sewing machine in the nineteenth century. Prior to the introduction of the machines, hand-sewing was considered to be an art performed by skilled people. Although technological developments have helped mass production of sewn objects in the twentieth century, hand-sewing is still in use for producing some unique features in clothing around the world. Fine hand-sewing pursued by both textile artists and hobbyists as a means of creative expression is a characteristic feature of high-quality tailoring, haute couture fashion and custom dressmaking.

1.1.2 Sewing: the technology of making clothes

The textile production was shifted from the household to industries for mass production by the Industrial Revolution. In 1790, Thomas Saint patented the world's first sewing machine, which consisted of an overhanging arm, a feed mechanism (adequate for the short lengths of leather), a vertical needle bar and a looper (Head, 1982). Until 1874, Saint's contribution was not made public. In the later part of 1874, William Newton Wilson built a machine from the drawings of Saint's patent in the London Patent Office, which worked following some adjustments to the looper (Cooper, 1968). The model of the machine that Wilson built from Saint's drawings is now in display in the London Science Museum.

French inventor Barthélemy Thimonnier invented a sewing machine in 1829 and signed a contract in 1830 with Auguste Ferrand, a mining engineer. Auguste prepared the requisite drawings and filed a patent application supported by the French government. The patent was issued on 17th July 1830 to both of them. In the same year, Barthélemy, with some of his friends, opened the first machine-based apparel-manufacturing company in the world to create army uniforms. A mob of tailors, believing that the machines would put them out of work, broke into Barthélemy's shop and threw the machines out of the windows and burned down the shop.

In the 1850s, an American inventor, actor and entrepreneur, Isaac Merritt Singer, developed the first sewing machine that could operate faster and with accuracy to surpass the productivity of a tailor sewing by hand (Mossoff, 2011; Head, 1982). The machine was made of wood and used a barbed needle, which works in the principle of today's lock-stitch machines. The needle passes downwards through the cloth, grabs the thread and pulls it upwards to form a loop to be locked by the next loop. He revolutionised the sewing world and established the Sewing Machine Company named 'Singer'. Although several other inventions were patented before Singer, his machine was widely adopted due to the ease of use at home and its availability on an instalment payment basis.

The technology of sewing underwent further developments during the twentieth century. There was substantial growth in the use of sewn products, and sewing machines became more affordable to the working class. Information on the latest fashion trends became available in periodicals during the late nineteenth and early twentieth centuries, which led to higher demand for the stylish fashion items. During this period, average women liked to stay up-to-date with the fashion trends by altering their previously styled clothing. Middle- and lower-class women adjusted their clothing by adding new neck collars, shortening skirts or cinching shirt waists, whereas women with higher income preferred to go for the new styles to fit changes in fashion.

Ready-made garments (RMGs) paved their way during this period as women (with sewing skill) in larger numbers joined the paid workforce, leaving them with less time to sew. The first RMGs were military uniforms for men, mass-produced during the War of 1812. Subsequently, good-quality RMGs became available for women as well. Although the concept of RMG existed at this time, they were not widely available until the beginning of the twentieth century. Since then, the RMG sector both for men's and women's clothing has seen tremendous growth, and today it has almost replaced the customised production of clothing items.

1.2 Global scenario of apparel manufacturing

The textile and apparel industries contribute significantly towards the national economy of many countries (Jones, 2006; Dicken, 2003). Although the apparel industry is global in nature, the manufacturing facilities from developed countries are shifting to developing countries to reduce the labour cost (Bheda et al., 2003). Even in these developing countries, the garment industries are facing the greatest challenges in spite of the cheap labour cost, due to the short production life-cycle, high volatility, low predictability, high level of impulse purchase and the quick market response (Bruce et al., 2004). To reduce the cost of production, the garment industries in developing countries are rather focussing on sourcing of cheaper raw materials and minimising delivery cost than labour productivity due to the availability of cheap labour.

Although some of the apparel manufacturers still survive in the developed countries, they are struggling for survival due to low profitability. In the global competitive scenario, the advantage of manufacturing clothing items locally over manufacturing abroad includes closeness to the market and the ability to react to fashion changes faster than the foreign competitors. However, the local manufacturers are gradually reducing the production and focussing on performing only the entrepreneurial functions involved in apparel manufacturing, which include buying raw materials, designing clothes and accessories, preparing samples and arranging for the production, distribution and marketing of the finished product. Global clothing production has now shifted to, and is gradually centralising in, countries such as China, Bangladesh, India, Korea, Cambodia, Pakistan and Vietnam (Tewari, 2005).

The global trade of apparel and textile products is no longer governed by quotas since 1 January 2005, when the agreement on textiles and clothing was terminated (Diao and Somwaru, 2001). Recently, the global trade has been governed by legislation within the multilateral trading system, which helped in the steady increase of imports from countries with low labour costs. When customer satisfaction is considered, the speed of replenishment comes into play. Thus, the countries with proximity are more competitive for those goods where replenishment is important, and these economic factors will intensify (Abernathy et al., 2004). Therefore, cost and locations are the two drivers underpinning the sourcing decisions. Developed countries like the US, Australia and the UK are facing a steady decline of indigenous textile and apparel production (Abernathy et al., 2004).

The countries in Asia are the leading manufacturers of clothing items around the globe. Among the Asian countries, China continues to be the leading exporter of textiles and clothing items. Its share in world exports increased to 33% for textiles

Introduction: the apparel industry

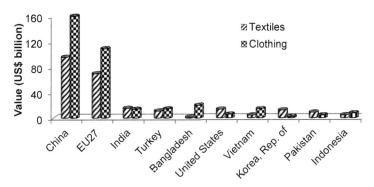

Figure 1.1 Global exporters of textiles and clothing in 2012.
From WTO (2013).

(up from 32% in 2011) and to 38% for clothing (up from 37%). The European Union (EU) and the USA remain the major markets for clothing, accounting for 38% and 20%, respectively, of world imports in 2012. Figure 1.1 shows the leading global exporters of textiles and clothing.

Tables 1.1 and 1.2 show the leading global exporters and importers of clothing for the year 2012. It can be observed that China and the EU are the leading exporters and importers of clothing, respectively. Furthermore, a majority of the Asian countries are among the leading exporters of clothing items.

It was anticipated that the post-multi-fibre arrangement (MFA) era is likely to witness stiff competition among the apparel manufacturers in developing countries. Hence, the improvement in the productivity, quality and technology fronts will be the deciding factors for success (Bheda et al., 2003). After the phase-out of the MFA and the end of the 'China safeguard' (international trade to protect communities from development aggression or home industries from foreign competition such as China, which was effective till December 11, 2013), international buyers got the freedom of sourcing textile and apparel products from any country, subjected to the system of tariffs (Cammett, 2006). Hence, the success of the apparel business in the post-MFA era depended mostly on the cost competitiveness. As procurement practices are no longer constrained by country-specific quotas, buyers can demand many more attributes such as product variety, quality and timely delivery in addition to the price. As there is no limit on the volume of clothing import from a given country, the volume factor could take care of the export competitiveness purely based on cost competitiveness. Hence, the global apparel production facilities would most likely be concentrated only among the most competitive firms.

In the post-MFA regime, it was expected that the global apparel market will expand largely in low-cost countries (such as China, Bangladesh and India) and countries in close proximity to the major markets (such as Mexico, Turkey and some Eastern European countries) (Kaplinsky and Morris, 2009). The countries where the apparel sector had grown under quota protection will lose substantial amount of their share.

Table 1.1 Leading exporters of clothing in 2012

Exporters (Billion dollars and percentage)	Value	Share in world exports				Annual percentage change			
	2012	1980	1990	2000	2012	2005–12	2010	2011	2012
China[a]	160	4.0	8.9	18.2	37.8	12	21	18	4
European Union (27)	109	–	–	28.4	25.8	3	2	17	–7
Extra-EU (27) exports	29	–	–	6.6	6.8	6	2	26	2
Hong Kong, China	23	–	–	–	–	–3	5	2	–8
Domestic exports	0	11.5	8.6	5.0	0.1	–38	–28	–14	–29
Re-exports	22	–	–	–	–	2	6	2	–8
Bangladesh[b]	20	0.0	0.6	2.6	4.7	16	25	29	4
Turkey	14	0.3	3.1	3.3	3.4	3	10	9	2
Vietnam[b]	14	–	–	0.9	3.3	17	22	27	7
India	14	2	2	3.0	3.3	7	–6	31	–6
Indonesia	8	0.2	1.5	2.4	1.8	6	15	18	–6
USA	6	3.1	2.4	4.4	1.3	2	12	11	7
Malaysia[a]	5	0.4	1.2	1.1	1.1	9	24	18	0
Mexico[a]	4	0.0	0.5	4.4	1.1	–7	6	6	–4
Cambodia	4	–	–	0.5	1.0	10	25	31	8
Thailand	4	0.7	2.6	1.9	1.0	1	15	6	–6
Pakistan	4	0.3	0.9	1.1	1.0	2	17	16	–7
Sri Lanka[b]	4	0.3	0.6	1.4	0.9	5	7	21	–5
Above 15	370	–	–	78.7	87.5	–	–	–	–

[a]Includes significant shipments through processing zones.
[b]Includes Secretariat estimates.
Courtesy of WTO (International Trade Statistics, 2013).

Table 1.2 Leading importers of clothing in 2012

Importers (billion dollars and percentage)	Value 2012	Share in world imports					Annual percentage change			
		1980	1990	2000	2012	2005–12	2010	2011	2012
European Union (27)	170	–	–	41.1	38.5	4	3	14	−11
Extra-EU (27) imports	90	–	–	19.8	20.3	5	4	15	−12
United States	88	16.4	24.0	33.0	19.9	1	14	8	−1
Japan	34	3.6	7.8	9.7	7.7	6	5	23	3
Hong Kong, China	16	–	–	–	–	−2	7	4	−5
Retained imports	–	–	–	–	–	–	–	–	–
Canada[c]	9	1.7	2.1	1.8	2.1	7	10	15	−2
Russian Federation[b,c]	9	–	–	0.1	2.1	39	85	23	0
Korea, Republic of	6	0.0	0.1	0.6	1.4	12	31	38	3
Australia[c]	6	0.8	0.6	0.9	1.4	10	19	21	4
Switzerland	6	3.4	3.1	1.6	1.3	4	1	16	−7
China[a]	5	0.1	0.0	0.6	1.0	16	37	59	13
United Arab Emirates[b]	4	0.6	0.5	0.4	0.8	13	2	21	13
Saudi Arabia, Kingdom of[b]	3	1.6	0.7	0.4	0.8	13	278	28	18
Mexico[a,c]	3	0.3	0.5	1.8	0.7	2	9	20	8
Chile	3	0.2	0.0	0.2	0.6	17	35	36	7
Turkey	3	0.0	0.0	0.1	0.6	19	32	15	−18
Above 15[d]	348	–	–	92.4	78.9	–	–	–	–

[a]Includes significant shipments through processing zones.
[b]Includes Secretariat estimates.
[c]Imports are valued f.o.b. (free on board).
[d]Excludes retained imports of Hong Kong, China.
Courtesy of WTO (International Trade Statistics, 2013).

As expected, apparel manufacturers in countries such as Africa and Latin America whose 'export competitiveness' heavily relied on the quota system have lost market shares, whereas apparel manufacturers in countries such as India and China have gained appreciation. A number of countries such as Bangladesh, Cambodia, Vietnam, Indonesia, Sri Lanka, Pakistan and Honduras have maintained or increased their market shares (Frederick and Gereffi, 2011). Apparel sourcing by retailers in the US and EU from regional suppliers has decreased due to the rapid growth in the procurement from low-cost Asian countries.

In addition to the free-trade policies, the apparel industries are also driven by technical advancements. The technical advancements in the equipment used in cutting and sewing are generic over the period; however, the other noticeable advancements are in 3D printing and digital printing (Manners-Bell and Lyon, 2012). The 3D printing technology, which was used for decades by architects and scientists to create 3D models, is not far from the fashion world (Rachel, 2013). The 3D printing technology in the past was only used by large companies. The technological advancements have enabled the use of this pioneering technology in clothing and fashion, as the cost of the technology has plummeted in recent years. In the past, when designers planned to manufacture a dress, they were in a need of selling hundreds to make the cost of production worthwhile. However, 3D printing has reduced the designers' manufacturing costs to zero until an order has been received. This has facilitated the designers' ability to experiment in small batches and sell the apparel in limited editions. Famous designers such as Catherine Wales (British), Michael Schmidt (German) and Iris Van Herpen (Dutch) have started using this technique to fabricate personalised clothing.

Companies such as **Hot Pop Factory** and retailers such as **New Balance** are now using this technology to print jewellery and shoes, respectively. A San Francisco-based start-up, Continuum, allows its customers to design their own apparel online, which is manufactured for them using 3D printers (Nate, 2013). Customers design bikinis on Continuum's website, specifying their body shapes and measurements. The company then uses nylon to print out each unique order. Founder Mary Huang believes that this intersection of fashion and technology will be the future because it 'gives everyone access to creativity'. The emerging 3D technology is exciting, but what is the purpose of a 3D-printed necklace without a matching shirt or dress? A couple of years ago 3D printing and fashion just met, but now their friendship is off to a promising start. Figure 1.2 shows the clothing and fashion accessories produced by the use of 3D printing technology.

The use of the Internet has facilitated the buying and selling of clothing and fashion accessories. However, online retailers deal with a high percentage of returns due to poor fit, material quality and customer satisfaction (Kartsounis et al., 2003; Cordier et al., 2003). Smart online shopping tools are being developed recently that can dramatically reduce returns and minimise shipping processes. For example, MyShape has developed a patented technology that matches shoppers with items that correspond to their personal measurements and preferences. Online programmes allow the consumers to create their style variations and save their body measurements (see Figure 1.3).

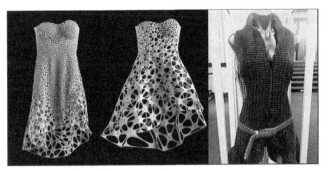

Figure 1.2 3D-printed fashion clothing materials.

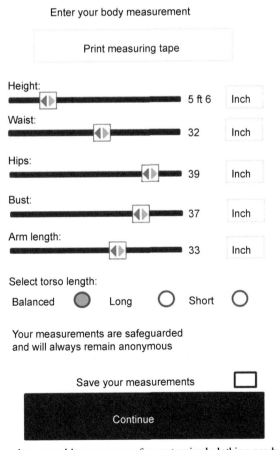

Figure 1.3 Online scheme used by consumers for customised clothing production.

1.3 Challenges in apparel production

1.3.1 Consumer choice and demographic variability

Apparel manufacturers have to produce a diverse product mix as consumers are difficult to understand and predict (Sutton, 1998). Their choice is unstable and unpredictable, and there is wide variation in their demographics and physiographics. In addition, consumer's age, household income, education level, occupation and ethnicity also affect their choice. Consumers of recent times are becoming more selective, multidimensional and complex (Forward, 2003) due to increased awareness and readily available information on the products and designs. Consumer's choice is shifting from traditional designs towards luxury high-fashion items. A majority of consumers prefer to shop at mass merchandisers such as Target, Walmart, Sears or JC Penny.

Consumers' choices and expectations widely vary within a target market. Almost all demand both moderate pricing and frequent style changes, but some prefer to follow the latest fashion trend whereas others desire to purchase investment clothing (Brown and Rice, 2001). However, a manufacturer cannot directly transfer consumers' choice and preferences in different product styles to satisfy their needs. Consumer demand for fashionable items and frequent style changes has necessitated reduction in the design and product cycle time. Although the traditional production cycle time is 18–24 months, apparel companies such as ZARA (Ghemawat et al., 2003) and PRADA have introduced new styles every 4–5 weeks to cater to these needs.

1.3.2 Challenges related to supply chain

A complete garment is fabricated from various components. For example, a man's casual shirt consists of cotton or polyester/cotton fabric, nonwoven interlining, woven or printed care labels, polyester embroidered brand labels, buttons or snap fasteners and sewing thread. As each component is manufactured by different suppliers, the non-arrival of any single component can cause delay in the production. On average, the major manufacturers of consumer apparel handle about 2000–40,000 products annually (Ghemawat et al., 2003). However, trend-setting companies such as ZARA handle about 11,000 different products in the same time. About 292 activities are involved in each apparel development cycle for each seasonal product. Hence, the apparel manufacturers' design-change activities may range from a low value of 7300 to as high as 300,000 (i.e. products multiplied by activities).

The apparel supply chain is complex and demanding due to global outsourcing, longer lead times and shorter seasons. It is extremely difficult to forecast the changing styles, and the supply chain is complicated by hard-to-model constraints, costs and lead times. The major aspect of the supply chain is responsible sourcing, which takes into account factors such as sustainability, compliance, chemical safety and product safety. In modern clothing manufacture, the emphasis on quality is increasing and more intelligence and agility in the supply chain is expected. Retailers are managing with smaller inventories and rapidly reacting to consumer needs. There are always efforts by the retailers not to increase the retail prices in proportion with the rise in the cost of inputs. Hence, the recent pressure on the profit margin is greater than ever.

The apparel retailers are facing challenges from Internet or online sales, which continues to upend the apparel sector. Brick-and-mortar companies are brainstorming to harness the power of Internet sales. As the proliferation of Internet-only companies continues, there will be stiff competition among them. In addition, Internet selling always involves the risk of incorrect fit and the wrong material, colour and style.

1.3.3 Challenges in design and production

An apparel item such as a pair of jeans looks simple compared to a vehicle, which may contain about 15,000 parts. However, the former is involved with enormous design changes in a season whereas the later undergoes few changes (e.g. two major and eight minor; Tassey et al., 1999) even for many years. Therefore, the apparel manufacturers have to handle an extraordinary amount of design changes. Traditionally, the apparel-manufacturing process has been performed by production workers who perform the cutting, sewing and other operations in an assembly line. In spite of the technical advancements, the apparel industry still remains as a labour-intensive unit. One of the major global concerns in apparel production is rising labour cost.

1.3.4 Stiff competition

Apparel manufacturers and retailers always state that apparel manufacturing is highly competitive. An apparel competitor sells similar clothing items to the same market segment, consumers or store types. The globalisation and post-MFA conditions have accentuated the competition. Merchandisers perform *comparison shopping* to analyse the price, quality and service of their competitors.

1.3.5 Ecological challenges

The major environmental threat involved in the textile manufacturing process is contributed by wet processing, which pollutes water (Correia et al., 1994). The toxic effluents that are harmful to aquatic lives as well as to humans are directly released into rivers. Due to the action of dissolved oxygen in water, the pollutants are broken down chemically and biologically. Hence the amount of dissolved oxygen in the water is reduced, creating difficulties for the survival of aquatic life.

Generally, a piece of clothing is rejected when it fails to provide its intended service or goes out of fashion. At the end of their life, clothing items are either recycled or thrown out as general waste, which often goes into landfill sites. For a number of reasons, landfill is recognised as the least attractive option for disposing of the waste. Landfill materials have the potential to pollute the air or enter the water, causing environmental threats. Hence, to reduce the threats, environmentalists' watch words 'reduce, reuse and recycle' should be used in clothing production.

The global trend in apparel manufacturing is moving towards the use of recyclable materials and reduced chemical finishes (Fletcher, 2013). The demand is shifting to develop eco-friendly products from materials derived from sources other than petroleum, which is a limited resource. However, several clothing items are manufactured as blends of natural and synthetic fibres (such as polyester and nylon). Although the natural fibres are biodegradable, the major concern is caused by the synthetic material.

The concept of zero discharge of hazardous chemicals requires manufacturers and retailers to take responsibility for the environmental concerns. Strict legislation by federal and state governments can help in achieving this target. Although this may be hard to achieve, it can help to substantially reduce the waste level and hence environmental pollution.

Consumers around the globe are becoming more aware of environmental concerns and the potentially harmful effects on human health from the toxic chemicals and ingredients in clothing and food items. The textile industry, especially wet processing, uses many chemical pollutants, allergens and carcinogens. Strict legislation is required for the manufacturers to make their products safer by minimising the use of hazardous chemicals and heavy metals. The chemicals that are ecologically safe should only be permitted. The textile plants should also address the problems of direct disposal of wastewater containing toxic chemicals into waterways to reduce water pollution.

1.4 Role of various organisations

1.4.1 Role of academic institutions

The academic institutions should provide industry-oriented courses to students and prepare them for situations they will face once they have entered the job market. In addition, the interaction between academic institutions and the industry needs to be improved by proper coordination. The students should regularly visit the manufacturing units, and complete some projects or training in-house to acquire knowledge about real situations on the production floor. The students should be aware of the recent trends in design, production and quality control. Furthermore, they also should understand the key challenges faced by the apparel industry and the ways to meet these challenges. The academic institutions should organise visit of the students to trade fairs, fashion shows, seminars and conferences in related fields to acquire knowledge on the current state of the art.

1.4.2 Role of governing bodies

After the phase-out of the MFA in 1 January 2005, apparel manufacturers obtained the freedom of free trade. But even in the free-trade regime, government policies often affect the apparel-manufacturing units. Stricter import and export regulations create difficulties in the import of raw materials and export of finished goods. Friendly policies for import and export can help to achieve the business objectives.

Import tariffs are used to raise the landed cost of a good produced elsewhere and imported into the domestic market. Import tariffs serve as a means of collecting revenue in addition to raising the cost of imports and thereby improving the competitiveness of locally produced competing goods. In many developing countries this form of revenue collection is of critical importance and supersedes any objective of local industry protection. Often, the tariffs offer little protection to local manufacturers, where the landed cost (including a tariff) is still substantially lower than the products that can be produced locally. Hence, the policies of the governing bodies should be favourable for domestic producers, importers as well as exporters.

1.4.3 Role of industries

The key parameters to be successful in the competitive global market depend on the buzzwords such as low product cost, better quality, durability, customer satisfaction, durability, comfort and style. In spite of technical advancements, the clothing industries still remain as labour-intensive production systems, which results in low productivity and often delays. Industries should emphasise on automation, training to the workers, modernisation, waste reduction, ecological considerations and adhering to specific quality systems, which can lead to an increase in the productivity and quality and a decrease in the amount of waste, stopovers and production delays. Each of the employees should work to achieve the organisational goal, and the industry should also look after the employees with fair remuneration, incentives, leaves, other perks and facilities.

1.5 Future trends

The global apparel industries have undergone dramatic changes in the past four to five decades both in terms of technological changes and volume expansion. The textile and clothing production facilities today are in a transition zone between traditional production methods and realisation of highly focused design and production of value-added clothing items. A large number of smart and intelligent textiles are becoming increasingly popular and commercially successful, as they provide additional functionality in the clothing. These textiles can sense and respond to external stimuli and are able to perform special functions, which an average fabric cannot do. Smart textiles are manufactured by the application of smart materials or finishes, such as shape-memory polymers or phase-changing materials, or electronic or e-textiles.

Smart and intelligent textiles are being used in various applications such as physiological monitoring (such as heart rate, respiration rate, temperature, activity and posture), sports training data acquisition, monitoring potential external hazards or hazards involved in handling of hazardous materials and tracking the position and status of soldiers in action (Nayak et al., 2015). To evaluate these smart textiles and clothing, new instrumental analysis techniques and new test methods should be devised. The fashion and textile designers should be adequately trained in the areas of advanced fabrics and state-of-the-art technologies used in smart textiles.

Digital printing is also a revolutionising technical development, which allows small-quantity production and sought-after designers to create high-demand clothing without much waste (Bal et al., 2014). Numerous opportunities exist with digital printing, which reduces the water usage and energy consumption and minimises textile effluents. Printing of photo-quality images directly onto fabric and reproductions of hand-painted art are the other advantages of this technique. In future, digital printing will be widely adopted by manufacturers to precisely print customised designs in fabric as well as in garments (Gupta, 2001).

The use of whole-garment or seamless technology is gaining impetus in the apparel industry due to smooth fit, comfort, invisibility and easy-care properties (Nayak and Mahish, 2006). The seamless technique helps in waste reduction, cost-saving, lower lead time and flexibility to incorporate certain design features. The global fashion trends suggest that seamless garments are becoming popular among consumers, especially the youth. There is a potential for seamless garments to hold 50% of the industry's sales within the next 10 years by providing educational and training facilities and adding versatile design features.

The Internet is influencing consumers' choice of clothing as well as apparel manufacturers' and retailers' style of marketing approach. For example, Brazilian retailer C & A has initiated an online campaign known as 'Fashion Like' that influences consumers' decision-making. Whenever a customer 'likes' a clothing item online at the C & A website, that 'like' gets totalled on a screen embedded in a clothes hanger on display. The in-store consumers can then decide whether to select the most-liked item or go for the less-liked pieces. In future, the Internet will influence the fashion market in ways that never happened before.

The use of 3D printing technology in clothing and fashion is rapidly growing. As the technology continues to grow, its limitations and costs will gradually diminish. The idea of mass-customised design can become a reality with the application of 3D printing, which can reduce the problem of size and fit. Who does not want to wear a one-of-a-kind, perfectly tailored piece? It can be economically used to prepare a dress to match each individual's liking, or shoes designed for the exact shape of one's feet. The new printing technology can open up the possibilities for personalised styles.

Several apparel manufacturers are moving towards a quick response and just-in-time approach, where the sold apparel items are quickly replaced by the manufacturers rather than from a large inventory kept with the retailer. The electronic data interchange and the use of barcode systems have helped in rapid communication between the manufacturer and the retailer, providing information on the consumer requirements, inventory, production progress, technical issues and delivery status. Social media such as Twitter and Facebook are helping consumers in understanding the fashion trends and getting closer to brands and the opportunity for new brands to emerge more readily.

The recent care labels used in garments are either printed or woven labels storing limited information. The electronic care labels are paving their way, which use radio frequency identification (RFID) technology for storing information electronically on a garment. However, the feasibility of this technology is a challenge due to its high cost, health risks and other technical challenges (Nayak et al., 2007).

Although there is some automation, the apparel industries are still far behind the other sectors and rely on manual intervention. The automation of various instruments in spreading, cutting, sewing and material handling can reduce the production cost and minimise faults. Systems like Kaizen and lean manufacturing are increasingly applied to garment manufacturing. Kaizen is a system that merges work and process improvement without increasing the cost. Kaizen involves every employee — from upper management to the jobber — and each individual is encouraged to come up with small improvement suggestions on a regular basis. Lean manufacturing focuses on achieving the shortest possible cycle time by eliminating process waste in a systematic approach and through continuous improvement (Hodge et al., 2011; Paneru, 2011). Lean manufacturing emphasises on producing the same goods utilising fewer resources.

Design and product development will make use of 3D tools that allow for garments to be created in 3D and converted automatically to 2D for traditional manufacturing methods. Digital body models can be created by 3D body-scanning systems, which will allow the digital products to be draped over them. The simulation of these digital products on the body can be obtained from the fibre and fabric characteristics. These data can be shared digitally and monitored through the product development process without the need for the development of a physical fabric or garment sample. Various organisations will benefit from this, including equipment suppliers, logistics companies, transportation providers, advanced technology centres, educational institutions and training organisations.

1.6 Conclusions

Technological advancements have dramatically influenced clothing and fashion in the modern age, which transformed them from ancient times into the modern world. Industrialisation brought many changes into the manufacturing of clothing items. In many nations, clothing crafted by hand has largely been replaced by goods from industries. The diversity of clothing styles and designs is hardly found in any other industry. Consumer choice is always unpredictable and now shifting from traditional designs towards luxury high-fashion items. Furthermore, the clothing industries around the globe are facing stiff competition every now and then. Hence, the apparel manufacturers should establish standard operating rules, minimise waste, reduce material wastage and optimise logistics to become successful.

References

Abernathy, F.H., Volpe, A., Weil, D., 2004. The Apparel and Textile Industries after 2005: Prospects and Choices. Harvard Center for Textile and Apparel Research, Cambridge.

Angier, N., 1999. Furs for evening, but cloth was the stone age standby. New York Times 2, 1.

Bailey, T., 1993. Organizational innovation in the apparel industry. Ind. Relat.: J. Econ. Soc. 32, 30–48.

Bal, N., Houshyar, S., Gao, Y., Kyratzis, I.L., Padhye, R., Nayak, R., 2014. Digital printing of enzymes on textile substrates as functional materials. J. Fiber Bioeng. Inform. 7 (4), 595–602.

Bheda, R., Narag, A., Singla, M., 2003. Apparel manufacturing: a strategy for productivity improvement. J. Fash. Mark. Manage. 7, 12–22.

Brown, P.K., Rice, J., 2001. Ready-To-Wear Apparel Analysis. Prentice Hall, Upper Saddle River, NJ.

Bruce, M., Daly, L., Towers, N., 2004. Lean or agile: a solution for supply chain management in the textiles and clothing industry? Int. J. Oper. Prod. Manage. 24, 151–170.

Cammett, M., 2006. Development and the changing dynamics of global production: global value chains and local clusters in apparel manufacturing. Compet. Change 10, 23–48.

Cooper, G.R., 1968. The Invention of the Sewing Machine. Smithsonian Institute, Washington, DC.

Cordier, F., Seo, H., Magnenat-Thalmann, N., 2003. Made-to-measure technologies for an online clothing store. IEEE Comput. Graphics Appl. 23, 38–48.

Correia, V.M., Stephenson, T., Judd, S.J., 1994. Characterisation of textile wastewaters—a review. Environ. Technol. 15, 917–929.

Diao, X., Somwaru, A., 2001. Impact of the MFA Phase-out on the World Economy an Intertemporal, Global General Equilibrium Analysis. IFPRI, Trade and Macroeconomic Division.

Dicken, P., 2003. Global Shift: Reshaping the Global Economic Map in the 21st Century. Sage, London.

Fletcher, K., 2013. Sustainable Fashion and Textiles: Design Journeys. Earthscan, London.

Forward, R., April 2003. Twenty Trends for 2010: Retailing in an Age of Uncertainty. Retail Forward, Columbus, OH.

Forza, C., Vinelli, A., 2000. Time compression in production and distribution within the textile-apparel chain. Integr. Manuf. Syst. 11, 138–146.

Frederick, S., Gereffi, G., 2011. Upgrading and restructuring in the global apparel value chain: why China and Asia are outperforming Mexico and Central America. Int. J. Technol. Learn. Innovation Dev. 4, 67–95.

Ghemawat, P., Nueno, J.L., Dailey, M., 2003. Zara: Fast Fashion. Harvard Business School, Boston, MA.

Gupta, S., 2001. Inkjet printing—a revolutionary ecofriendly technique for textile printing. Indian J. Fibre Text. Res. 26, 156–161.

Hassler, M., 2003. The global clothing production system: commodity chains and business networks. Global Netw. 3, 513–531.

Head, C., 1982. Old Sewing Machines. Osprey Publishing, Oxford.

Hodge, G.L., Goforth Ross, K., Joines, J.A., Thoney, K., 2011. Adapting lean manufacturing principles to the textile industry. Prod. Plann. Control 22, 237–247.

Jones, R., 2006. The Apparel Industry. Wiley-Blackwell, New York.

Kaplinsky, R., Morris, M., 2009. The Asian drivers and SSA: is there a future for export-oriented African industrialisation? World Econ. 32, 1638–1655.

Kartsounis, G., Magnenat-Thalmann, N., Rodrian, H.-C., 2003. E-TAILOR: Integration of 3D Scanners, CAD and Virtual-Try-On Technologies for Online Retailing of Made-To-Measure Garments. Springer, Berlin/Heidelberg.

Manners-Bell, J., Lyon, K., 2012. The implications of 3D printing for the global logistics industry. Transp. Intell. 1–5.

Mossoff, A., 2011. Rise and fall of the first American patent thicket: the sewing machine war of the 1850s. Ariz. L. Rev. 53, 165.

Nate, C., 2013. Continuum's 3-D Printed Clothing Offers a Glimpse into the Future of Fashion (Online). Available from: http://www.huffingtonpost.com/2013/04/16/continuum-3-d-printed-clothing_n_3093541.html (accessed 10.09.14).

Nayak, R., Wang, L., Padhye, R., 2015. Electronic Textiles for Military Personnel. Woodhead Publishing Ltd., Cambridge, UK, pp. 241–258.

Nayak, R., Mahish, S.S., 2006. Seamless garment: An overview. Asian Text. J. 15 (4), 77–80.

Nayak, R., Chatterjee, K.N., Khurana, G.K., Khandual, A., 2007. RFID: Tagging the new era. Man-Made Text. India 50 (5), 174–177.

Paneru, N., 2011. Implementation of Lean Manufacturing Tools in Garment Manufacturing Process Focusing Sewing Section of Men's Shirt. Degree Programme in Industrial Management. Oulu University of Applied Sciences, Oulu, Finland.

Quataert, J.H., 1985. The shaping of women's work in manufacturing: guilds, households, and the state in central Europe, 1648–1870. Am. Hist. Rev. 90, 1122–1148.

Rachel, H., 2013. 3D Printing Hits the Fashion World (Online). Available from: http://www.forbes.com/sites/rachelhennessey/2013/08/07/3-d-printed-clothes-could-be-the-next-big-thing-to-hit-fashion/ (accessed 15.09.14).

Scott, A.J., 2006. The changing global geography of low-technology, labor-intensive industry: clothing, footwear, and furniture. World Dev. 34, 1517–1536.

Sutton, S., 1998. Predicting and explaining intentions and behavior: how well are we doing? J. Appl. Soc. Psychol. 28, 1317–1338.

Tassey, G., Brunnermeier, S.B., Martin, S.A., 1999. Interoperability cost analysis of the US automotive supply chain. Planning report 99–101. The National Institute for Standards and Technology (NIST).

Tewari, M., 2005. The role of price and cost competitiveness in apparel exports, post-MFA: a review. Indian Counc. Res. Int. Econ. Relat. Work. Pap. 1–62.

Watt, J.C., 1997. When Silk Was Gold: Central Asian and Chinese Textiles. Metropolitan Museum of Art, New York, NY.

Wood, F., 2004. The Silk Road: Two Thousand Years in the Heart of Asia. University of California Press, Berkeley and Los Angeles, CA.

WTO, 2013. International Trade Statistics. World Trade Organization, Geneva.

Part One

Product development, production planning and selection of materials

Product development in the apparel industry

M. Senanayake
California State Polytechnic University, Pomona, CA, USA

2.1 Introduction

Guess how many different models of an automotive brand are introduced per year? Will that be a couple of two- and four-door sedan models, a few cross-over models, a few SUV models, a few convertible models, and a couple of sports models with a few colors? Now compare this with an apparel brand, which produces many product lines each with many styles in multiple colors/prints, in multiple sizes, for a number of seasons per year. In most situations the brand may also have different product lines for different markets with different price points. The number of stock-keeping units (SKUs) produced by an apparel brand per year cannot be meaningfully compared to the SKUs developed by an automotive brand. Now if you think of the product-development (PD) efforts for these two industries, it is quite clear that the PD for an apparel brand is quite complex. The PD gets further complicated with the time pressure to meet market demand for fashion items. Therefore, it is important and beneficial to understand the apparel PD process for anyone interested in careers in the apparel industry.

PD is defined as the design and engineering of products that are serviceable for the target consumer, marketable, manufacturable, and profitable (Glock and Kunz, 2005). Therefore, it is the process of setting up the goals for the season, carrying out market and trend research, conceptualizing and finalizing a line of products for the season, developing product styles using the technical design process, making samples of the products to finalize product styles with standards, selling the product line, and sourcing and manufacturing the line. Even though one might not see the relationship, this process is also involved in functions such as material requirement planning, inventory control, production planning and scheduling, quality control, logistics, and finance. (Senanayake and Little, 2001). "Product development is used by both wholesale manufacturers, who develop products for signature brands and retailers, who use it for private-label development for their own stores" (Stone, 2006).

Even though the historical approach was to have the PD and production in house for many manufacturers and brands, today it has quite become decentralized. The globalization of garment production has resulted in a tendency to split elements of the PD and production processes between different countries (Goworek, 2010). The rapid development of PD technologies and communication technologies has enabled people at multiple locations around the globe to seamlessly integrate, collaborate, communicate, and manage the PD process. Also, it is apparent that not only apparel manufacturers but also apparel contractors, subcontractors, and specialty contractors have launched design and development units within their organizations for the retailers and brands to select designs from without spending time and effort in PD.

2.2 Product-development models and product-development process

2.2.1 Product-development models

The models of apparel PD are helpful to look at in detail to understand the stages and activities in the PD process. Most previously published PD models for apparel are of the sequential type even though the elements in the PD process are of a concurrent nature and take place simultaneously. Some of these models define the process with general stages and others use lists of activities. An apparel product line contains multiple products or styles, with companies developing several product lines per year. At a given time there can be multiple products or styles in the line at different stages of completion. This causes the PD process to be complex, requiring close coordination and monitoring.

2.2.1.1 No-interval coherently phased product-development (NICPPD) model

The author found the no-interval coherently phased product-development (NICPPD) model (May-Plumlee and Little, 1998) for apparel a detailed and easy-to-understand process model. The model shows responsible departments with activities in each phase of the PD process, and was specifically developed for apparel. The manuscript that discusses this model also summarizes other previous PD models, which were considered in developing NICPPD model. This is a six-phase apparel PD model that shows the involvement of four functional areas: marketing, PD, merchandising, and production planning and control. Though the model only discusses these four functional areas, one can relate it to other functional areas depending on how the departments are organized in an apparel company. This is because different companies have different departmental structures. The model includes functional overlaps and recycling ideas through previous development phases for further refinement.

Major corporate decisions are shown as fuzzy gates in the model, implying that teams involved in the PD must collectively decide on the next phases for a product line or product style. The system constraints may vary depending on the type of product line and business model. Some of the examples for these constraints are vendor reliability, raw material availability, customer constraints such as personal consumption expenditures, consumer wants, marketing channels, and available technology. The reader is encouraged to refer to this model to have an in-depth examination of each phase of the development process to clearly understand the six phases. These six phases are visually presented in the manuscript for better comprehension. The summary of these six phases is discussed in Table 2.1.

2.2.2 Product-development process

Based on our understanding of the PD model discussed above, we can now discuss a more general PD process considering the functions from designer's idea or the initial design concept to make the product ready for final production. These stages are discussed considering the apparel-manufacturing and scope of this book. When developing products, it is important to know the business model (such as wholesale brand, private label, store brand, customized product, etc.) and if the product is a new design or knock-off, so that the functions in these stages can be tailored to the business model's product situation. Here we will discuss the PD process considering five stages: (1) line planning and research, (2) design concepts:line concept through research, (3) design development:line development, (4) line presentation and marketing, and (5) production planning:pre-production and optimization.

2.2.2.1 Line planning and research

The PD process is a team effort by a number of people involved from a number of departments in an apparel organization. The initial step of this process is planning a line of products for a particular season or particular time period, depending on the company's selling seasons or selling practices. The PD team is generally comprised of several people from designing, merchandising, research and development (R&D) and/or raw material development, technical design or product engineering, sales and marketing, finance, graphic design, sourcing, operations, planning, and quality assurance. The PD team uses the information from research on trends, colors, materials, previous successes or failures, past sales records, experience from previous lines, and mark-down reports, etc. to brainstorm a plan for the new line. The information from this effort will assist the designers and the PD team to formulate a plan for the new line with a positive (brand) image to influence the consumer segment that the company is targeting for its sales (Stone, 2006).

2.2.2.2 Design concepts: line concept through research

After planning, the PD process begins with the design, which is a critical component in the development of fashion products. The design process begins with a line concept, which explains the mood, theme, and other key elements that contribute to the identity of the line (Keiser and Garner, 2012). To develop the line concept, the designers obtain their inspirations for designs by conducting research. This involves market research and fashion research, from which they interpret findings into styles considering the brand and the target consumer. The market research provides information that helps the company to understand consumer demands. This is done by investigating the target consumers and their behavior (through consumer research), their preferred product

Table 2.1 **The six phases of the NICPPD model by May-Plumlee and Little (1998)**

Stages	Functions or activities
Phase 1: Line planning and research	Research efforts from marketing division such as market research, target consumer research, and previous sales data to develop marketing plan that triggers sales forecast. Merchandising and PD teams to carry out fabric and trim research, color research, and trend research. These efforts will lead to line plan that includes financial and sales goals.
Phase 2: Design/concept development	Initiating the development of specific products. PD team uses information from Phase 1 to develop concepts. Marketing division carries out concept testing using techniques such as mall intercepts. Decisions are made at the color and concept meeting by marketing, merchandising, and PD teams. Fabric, design, trim, and color are selected and approved at design review meeting by PD team. The output from this phase is a preliminary line represented by style sketches and design specifications for each style.
Phase 3: Design development and style selection	The activities involved in this phase are raw material development, testing and approval, color testing and approval, acquisition of sample yardage, pattern-making and fit approval, style evaluation and approval, wear testing, and preliminary costing, leading to finalizing sample specifications and translating the line from sketches to actual product line. Merchandising, marketing, and PD teams will review samples to make decisions on final line adoption. Merchandising team identifies the assortment, makes volume decisions, and establishes pricing and gross margins.
Phase 4: Marketing the line	PD team will order raw materials for more duplicates. Production planning and control division will carry out detailed costing and develop product specs. Marketing and merchandising teams will develop promotional materials for sales reps. Line is presented at markets to retail channels using sales samples by marketing team. Marketing and merchandising teams will review retail orders, compare with sales forecast, and add/drop styles, colors, and sizes to come up with the final modified line.
Phase 5: Pre-production	Translating the decided prototypes and first patterns into complete size range for final production. PD team will finalize quality, production, and process standards for manufacturing. Production planning and control division will manage grading, marker making, planning, and sourcing of both material and production. Quality, material, and engineering specifications will be finalized by merchandising and production planning and control division.
Phase 6: Line optimization	Merchandising team will review the final line against orders, drop styles with inadequate orders, and replace with new styles, colors, and sizes to optimize the line for profitability. If changes are made, the model will direct to previous phases as appropriate. This phase will continue while production is in progress.

designs and characteristics (through product research), and the general market trends (through market analysis). The fashion research provides trends for the season as silhouettes, design details, colors, fabrics, and trims (Burns et al., 2011).

Textile and graphic designers play a vital role in bringing the textile design ideas. This is because the fabric choice can play a substantial role in the aesthetics and performance of the designer's fashion ideas (McKelvey and Munslow, 2003). These textile designers get their inspirations from mood boards from the design team, catwalk reports, print suppliers, and observing other fashion manufacturers or retailers (Goworek, 2010).

The companies that conduct trend research evaluate what has occurred in the past and project what may happen in the next season, tracking economic trends, social and cultural trends, technological advances, and political influences, which may all have an impact on the product design as well as on consumer behavior and spending. Knowledge of the brand position and target market is vital for successful positioning of a product line (Bubonia, 2012). Johnson and Moore (2001) discussed a number of sources that can be helpful in research. In addition, there are new sources such as the Sourcing Journal (https://www.sourcingjournalonline.com).

Today, the design process is challenged with analyzing infinite information generated from sources such as social media, e-commerce, and other electronic data sources such as blogs. Experts with the knowledge of data analysis must work with the designers for successful PD. These analytics will lead to better trend forecasts of fibers, fabrics and trims, color, silhouettes, design details, inspirations, and other aspects for better product designs. For example, the fast-fashion company Zara employs cross-functional teams comprised of specialists from fashion, commercial, and retail within the design department to pre-approve designs (Christopher, 2000).

Collaborating with the designer, the merchandiser will review the trends, analyze previous sales, consider the budget allocated and projected sales for the department or account, and come up with the line plan (Lee and Steen, 2010). With the product designs for the line finalized, the design-development process begins.

2.2.2.3 Design development: line development

At this stage the designers interpret the research findings and translate the line concepts into styles considering the sales potential, appropriateness for the brand, target consumer, and product line. The fabric, color, print, trim, and silhouette decisions can be made by the designers while working with the merchandisers to ensure that they are working within controlled parameters for the assortment plan. Designers will sketch and develop muslins for 25—50% more garments for the line, considering possible drop-offs during concept presentations due to variables such as unfeasible target price points, schedule conflicts, and design aesthetics (Keiser and Garner, 2012). The PD team will view the work and weight the designs from two perspectives: (1) based on their merit and (2) based on the suitability for the line as a whole (Stone, 2006). Storyboards and mood boards are used to discuss the themes to understand the line relationship to the brand, target consumer, price point, etc. Pattern-makers and sample-sewers will be involved in creating garment patterns and sewing prototypes with available fabrics in the sample size for the styles, which are considered to be feasible to incorporate in the line. These prototypes provide a good understanding of manufacturing feasibility and allow evaluating fit. The technical designers will start developing a technical package (tech-pack) for each style, which comprises standards and specifications to produce the style.

Another ongoing process during design development is the function of raw-material development with the anticipation of using the materials in the product line. More time needs to be spent if new raw materials such as fabric are to be developed. This involves working with the R&D teams from raw-material suppliers to develop materials and test them to achieve expected standards. As this is a time-consuming process, it must be well planned in advance and monitored as scheduled in the PD calendar. Preliminary cost estimates are developed for the styles, which are required to make financial decisions for the existence of the style in the line.

2.2.2.4 Line presentation and marketing

The styles to be adopted in the line are reviewed in the line review meeting attended by executive decision-makers from sales, sourcing, finance, operations, planning, and manufacturing, in addition to the PD teams who will make every attempt to convince the executives to adopt their styles in the line. In general, a fitting session with a live fit model is conducted during this decision-making process. Designs are dropped, changes and modifications to materials and designs are suggested, and even at times, alternative designs are suggested in addition to accepting some during this review process. This process may continue for a number of meetings back and forth. Accepted designs win a style number to move officially into manufacturer's line (Stone, 2006). The tech-pack will be further refined at this stage by the technical designers for each style approved with the collaboration of the PD team. This tech-pack is some times used as a "bid package" by the sourcing department to select contractors to get the orders placed for manufacturing.

2.2.2.5 Production planning: pre-production and optimization

During the production planning stage of the PD process, sourcing decisions are made to identify which production facility (own manufacturing, cut, make (sew) and trim (CMT) contractors (a firm that is contracted to cut, make, and trim a product from fabric, findings and cutting marker), full-package contractors, or specialty contractors) will produce the approved styles in the line. The functions at this stage of the PD process will depend on what type of contractor will be used for production of the styles, such as CMT versus a full-package contractor (a contractor that carries out all steps of production for a style from fabric purchase to cutting, sewing, trimming, packaging, and distribution). For example, if a full-package contractor is used, there is not much involvement of the PD team on purchasing the developed fabric from the fabric supplier, but must monitor the bulk fabric quality in collaboration with the contractor. Therefore, it is important to understand how each contractor carries out its functions, which will have an effect on the PD process. During the sourcing decision process, a sample along with the corresponding tech-pack is sent to the contractor (bid package). The contractor will use this to negotiate the production cost and other style-specific information including delivery, quality, subcontracting needs, etc., to come to an agreement. Once the bid packages are reviewed from multiple contractors, the sourcing decision will be made considering a number of factors, such as cost, quality, lead time, reputation, compliance, and previous relationship. The production capacities will then be reserved based on the assortment plan.

As scheduled in the PD calendar, material requirement planning will follow the production plan, where the fabrics will be ordered along with the orders for the other trims and components for the styles. The garment costs will be finalized and finances are cleared to move forward with producing the line.

The PD team will work with the contractor to make a production sample—"the final sewn prototype that has been corrected, perfected and tested for fit, function and aesthetic appeal" (Bubonia, 2012)—to make sure that the cut-and-sew contractors are capable of meeting expected quality standards and specifications and product manufacturability. Further, a production sample size-set is produced by the contractor, which will help the PD team to verify pattern-grading accuracy as well as the style's fit, function, and aesthetic performance of the size range. Once the production size set is approved, bulk production can be started for private-label PD. In a wholesale PD situation, salesmen's samples are produced, which will be presented to retail buyers at the manufacturers' seasonal shows (such as Men's Apparel Garment Industry Convention (MAGIC)). The manufacturer will negotiate with the retail buyers the order minimums, lead times, and wholesale cost for the styles to accept the retail buyer's order (Stone, 2006). Based on the orders received, production quantities are fine-tuned with the contractors to start the production process. The style that does not get attention from retail buyers may have a chance to get dropped at this stage, but this is rare in most situations.

2.2.3 Product-development calendar

As a complex process, the activities in the PD process must be well planned and scheduled, identifying who will initiate each process and who is responsible to achieve tasks to accomplish process goals in a timely manner. It is important to understand that even though many functions in the PD process are sequential, some functional deadlines must be met simultaneously or concurrently. The coordinating and monitoring tool that is used by companies to plan the PD process is called the PD calendar or merchandising calendar.

A PD calendar can be developed by using a Gantt chart format, as shown in Figure 2.1. This is a basic example of a PD calendar, which comprises 40 development activities and requires about 35 weeks to accomplish (AAMA, 1991). Each task is allotted a set amount of time and must be completed in this time-frame. The tasks are spread out over weeks, giving targets so that the future processes will be able to continue without being delayed by earlier processes (Schertel, 1998). More detailed calendars may have additional columns highlighting the department requesting each activity and the department that is responsible for accomplishing them. Instead of showing one column for each week, some calendars may have additional columns for more detailed start and completion times.

The timeline and activities in the PD calendar may vary from company to company depending on factors such as the product type, fashion versus basic, original versus

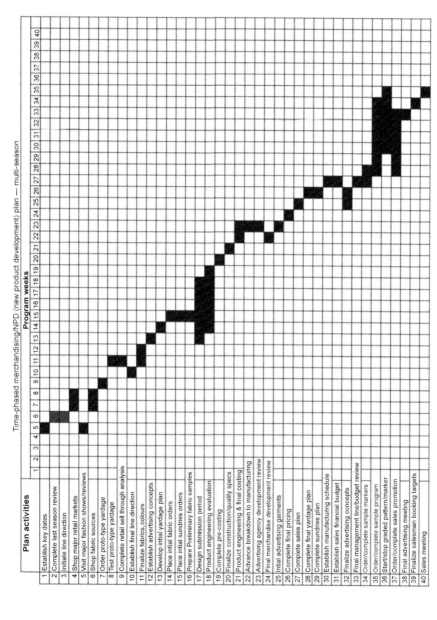

Figure 2.1 Multi-season PD calendar (AAMA, 1991).

knock-off, and the company's business practice such as wholesale, retail private label, fast-fashion, or mass customization. The traditional timetable for completion of a PD cycle is 4–6 months. If production is sourced offshore or if materials come from locales far from the production site, the lead time might be longer (Keiser and Garner, 2012). Many companies have spent a great deal of money to shrink the PD cycle time using technology and management systems developed over time. Companies such as Zara have shrunk the PD cycle to less than 2 weeks (Mihm, 2010).

Companies may use the calendar either as a stand-alone software solution or it may be integrated with the company-wide product data management (PDM) or product life-cycle management (PLM) solutions. In some systems there is a functionality to notify the responsible party with flagging when certain activities are not met according to the scheduled time.

2.3 Variations in apparel product development: demand-led product development

Fast-fashion companies are involved in made-to-demand production processes today. It is important to understand the PD strategies that are demanded by such business models. As an effort to understand the functions for such a demand-led PD process, the author conducted an interview with Ken Watson (Figure 2.2), who is a supply-chain/fast-fashion consultant and the managing director for Industry Forum Services, UK. In addition, the PD information for the following section was gathered through secondary research about fast-fashion companies.

2.3.1 Fast-fashion product development

The definition of fast fashion is neither very clear nor straightforward. As Watson (2014) describes, it is the company's ability to consistently bring the "on-trend"

Figure 2.2 Ken Watson, a fashion supply-chain consultant and the Managing Director of Industry Forum Services, UK, explained his views on demand-led product development.

fashion products to the consumer in season. In order to be able to get products on-trend within the season, the PD team must be able to innovate the product line fast at the beginning of the season and must have the ability to replenish the winners following the demand signals. With the fast-fashion PD and production strategies, companies such as Zara bring products to the market within 14 days responding to changes in the demand signals. These companies consider the process from concept to customer, avoiding all possible delays in PD and production process. To achieve this speed, a high degree of planning and flexibility is required. For that, there are some fundamental changes that need to be made to the company's apparel PD process and its supply-chain model (Watson, August 21, 2014, pers. comm.).

2.3.1.1 *Design and development for small production runs of multiple styles with short lead times*

To cater to the fragmented demand and to be responsive and turn fashion products quickly, an apparel company must be able to produce small runs of multiple options. The success of companies like Zara is based on short lead times with more fashionable clothes and smaller quantities, providing shopping excitement for consumers, which leads to an increase in sales (Watson, August 21, 2014, pers. comm.). With the ability to bring a new style within 2—4 weeks, Zara produces about 12,000 styles per year (Dutta, 2002). These multiple style options are preapproved and the production capacities are prebooked.

The cross-functional teams, consisting of specialists from fashion, commercial, and retail, work within the design department to preapprove designs reflecting the latest international fashion trends. These designs are from inspirations developed through visits to fashion shows, campuses, competitors' stores, pubs, cafes, clubs, etc. relevant to lifestyle of the target market. The design decisions are further enhanced by the data that flow regularly from the point of sale of retail stores around the world (Christopher, 2000). In addition, the store managers track selling trends in real time and send both trend information and new products that customers are looking for to more than 300 in-house designers at Zara's head quarters. Zara also uses advanced inventory optimization models and demand-forecasting models that are used in making design-concept approval decisions (Mihm, 2010).

2.3.1.2 *Design approval*

Having the designs preapproved with certain number of styles, it is then test marketed by introducing varying styles in the line in small quantities to the market to identify the winners to respond to the retailer's needs very fast. The result is deciding what to produce to maximize potential sales and to get the product-line concepts or the styles approved quickly. This will be a very expensive process unless it is carried out effectively. The company must understand the risk involved in this kind of a business model and must be able to enjoy the success as well as accept the risk. Standardized garment patterns, standardized sampling methods, use of virtual digital technologies, virtual prototyping, and single-person decisions (to eliminate waiting for decisions or waiting for action) are some strategies to get the style approval quickly (Watson, August 21, 2014, pers. comm.). Strategies such as postponement and delayed product differentiation assist demand-led PD (Senanayake and Little, 2010). Garment dyeing is also used, allowing the manufacturer to produce best-selling colors quickly (Mihm, 2010).

Allocation of the right amount and kind of products to the individual stores enhances profitability and provides information about winners so that PD can be more effectively done for future seasons. This is a pull strategy, rather than the conventional push strategy where retailers push products to the market, expecting and pushing consumers to buy. Another strategy to manage an international demand-led PD process, such as Zara's, is by standardizing the major portion (85%) of the product lines from country to country, leaving about 15% varying for the local taste (Mihm, 2010).

2.3.1.3 Raw material readiness

To achieve demand-led PD, it is necessary to have a flexible flow of fabrics within the season with the on-trend color and print applied onto it. Designers, fabric suppliers, and manufacturers must constantly communicate to transform trends into reality. To achieve this goal, the made-to-demand company works with preapproved fabrics and preapproved colors and prints. Suppliers keep neutral or greige fabrics that can be finished as needed based on trend colors. It is not possible to be responsive unless the capacities are prebooked for both materials and apparel production (Watson, August 21, 2014, pers. comm.).

2.3.1.4 Supply-chain relationships

The apparel company must have a good relationship with the business partners vertically through the fast-fashion supply chain, particularly textile suppliers and garment manufacturers and those where historically queues have led to long lead times. Also, it is imperative that the suppliers and manufacturers understand the market dynamics and are ready to respond to the retailer, considering the dynamic nature of the market (Watson, August 21, 2014, pers. comm.). Sophisticated logistic technology with end-to-end services can provide fast-moving of merchandise without delays. However, it is important to understand that the fundamental concept is that it should not compromise the quality of the product due to the speed (Watson, August 21, 2014, pers. comm.). Fast-fashion company Zara tries to carry out the operations that enhance cost efficiency through economies of scale (labeling, packing, etc.) in house and sourcing other manufacturing operations from a network of specialized contractors. These contractors work very closely, receiving technological, financial, and logistical support to achieve stringent quality and time targets (Christopher, 2000).

Most experts suggest that vertical integration is a key to demand-led PD. However, it is also important to understand that most vertically integrated apparel companies are performance- and capacity-driven rather than flexibility-driven, which is a bad recipe for a business model of this kind. Vertical integration helps if it is organized for the demand-led business needs. For that, the performance matrices for the management need to be reorganized for flexibility. To achieve success in demand-led PD, a company must expect a significant cultural change in the organization, with exceptional employees who are challenged with a significant change in the performance matrices compared to the conventional PD process (Watson, August 21, 2014, pers. comm.).

2.4 Apparel product-development technologies

Apparel PD is heavily supported today by computer systems, where team members in each phase of the development process can access and use data on a real-time basis. Technologies such as automatic body-measurement extraction (body scanning); avatar generation with animation capabilities; computer-aided design (CAD), which provides two-dimensional (2D) and three-dimensional (3D) garment pattern solutions; 3D garment draping simulation, visualization, and virtual fitting; CAD systems for textile design; digital printing systems for 2D and 3D printing; costing solutions; optical-scanning technologies; full-fashion garment-knitting solutions; and radio-frequency identification (Nayak et al., 2007) are a few of the many technologies that have assisted companies to improve their PD process. When orders are processed, technologies such as electronic data interchange; and web- and cloud-based communication systems have enabled the PD partners to act upon the same demand data from one stage to another to avoid the bullwhip effect (Christopher, 2000).

2.4.1 Technology solutions

As a strategy to bring information about the latest technology solutions commercially used for apparel PD today, the author interviewed executives from apparel technology companies and apparel industry consultants. The objective of this effort was to make the reader understand the useful PD tools and technology solutions available for apparel companies today. These case studies are available at the end of the chapter in Section 2.9. Further, these interviews contributed to gather information regarding future PD technology trends, are discussed in Section 2.8.

2.4.2 Digital color communication

Color is one of the main intrinsic quality attributes in an apparel product. The designers generally used a physical color swatch in the past to communicate color among the members of the design team or with the suppliers. This physical sample swatch has been lately replaced by a digital color sample. The digital color sample can be presented and communicated electronically in the form of reflectance data measured by a spectrophotometer. This facilitates color communication without needing to mail sample swatches back and forth, which is costly and time-consuming. A software solution attached to the supplier's spectrophotometer can import the designer's digital color sample (standard), calculate the recipe, carry out the lab dip (color fabric), and compare with the standard color. This process can be repeated until the designer's color standard is met (Che and Li, 2010).

As the appearance is more than the color, it is advantageous that color combined with fabric substrate could be viewed accurately on a calibrated device screen and communicated precisely along the textile supply chain. There are still many textile samples, for example, yarn-dyed fabric, which cannot be accurately measured by a spectrophotometer. Therefore, color-and-image systems have been developed for more reliable digital color presentation on various textile substrates. However, these systems need further development. When developed, these color-and-image systems will assist efficient color communication throughout the supply chain (Che and Li, 2010).

The information from the discussion above and the case studies in Section 2.9 provide a good synopsis of the PD technologies used by apparel companies today. These technology solutions will allow retailers and consumers to take a proactive role in the PD process, where integration of pre-production, production, and postproduction processes will be facilitated. The technologies assisting the PD function are changing fast as researchers from these technology providers are looking to improve the technology to be competitive in their marketplace. The result is improving the efficiency of the PD functions, which in turn will reduce the development time and cost. Therefore, PD process has a direct impact on business performance and profitability in the apparel industry. Readers can obtain an overview of new technology trends from Apparel Magazine and Gartner Inc., a leading IT and supply-chain research company in the USA (http://apparel.edgl.com/home).

2.5 Apparel product standards, specifications, quality assurance and product technical package

2.5.1 Apparel product standards and specifications

The apparel company must make sure that the apparel items produced are exact reproductions of the final sample approved in the PD process. This will provide product consistency irrespective of where they are sourced. To achieve this goal, apparel companies establish standards for design, product, and performance during the PD process. Standards are set for characteristics and procedures, which are the basis for resources and production decisions (Glock and Kunz, 2005). Companies use standards they have established for a product to develop specifications, which are graphical representations, written guidelines, or descriptions of materials and their performances, procedures, dimensions, placements, aesthetics, etc. When these specifications are followed and incorporated in an apparel product, one can achieve the set standards for the product. For example, the seam strength can be a company standard for a side seam of a pair of jeans, which can be achieved by specifications such as seam type, stitch type, sewing thread, and stitches per inch (SPI).

2.5.2 Quality assurance through standards and specifications

Standards define the quality level and quality characteristics of a product for the target market that must be established during PD, which must be followed throughout the supply chain. This allows following the quality assurance plan for all the partners in the supply chain and leads to quality control during production and delivery. The retailers or brands establish these standards and specifications based on predetermined preference of target consumers, order volume, price points, and capabilities of production and sourcing. While compliance to some of the standards is mandatory, companies use voluntary standards for the apparel product to describe their quality level. For example, the Consumer Product and Safety Commission enforces flammability standards for children's wear sold in the USA, which are mandatory standards to follow by children's wear apparel companies. Companies can use international standards, national standards, industry standards, or company-specific standards in defining the quality level of the product. Examples of sources providing such standards are the International Organization for Standardization, the American Society for Testing and Materials, and the American Association of Textile Chemists and Colorists. In addition to providing set standards for the apparel industry, these organizations provide standard test methods to evaluate performance of material during quality assurance and quality control. These test methods can become part of the operating procedures to achieve material performance specifications used by the manufacturers.

2.5.3 Product technical package

During the design development stage of the PD process, the design ideas are transformed into finished products. While the patterns are developed to achieve fit and the samples are sewn with the available or planned materials and trims, technical designers in PD must develop a packet of specifications considering every aspect of the product they develop for manufacturing. This specification packet is called the technical package, or tech-pack, for a given style. Tech-packs vary considerably from company to company depending on the product, company size, sourcing regions, contractor profiles, etc. Some large companies have developed sourcing manuals incorporating standards and specifications that they share with sourcing partners. When specifications are well developed, shared and implemented with stringent quality control, the final product will guarantee the concept features that the designer envisioned during PD. These not only help in the product quality but also act as a legal and financial bargaining tool when it comes to chargebacks due to unacceptable products from contractors.

Considering the nature of the global apparel industry, these specifications must be developed using graphic and written descriptions to be easily understood by supply-chain and other partners. Software applications such as PLM have tech-packs as part of their integrated information management system, providing a platform to share these specifications in real time. For example, Lectra (a technology solution company for the apparel industry) provides its Fashion PLM application to capture and transform information in a single tech-pack from design and PD to production.

A technical package consists of all the specifications for each style being considered for adoption. However the specifications can be organized so that, there should be sufficient information in a tech-pack for the supply-chain partners to follow in producing the product. The next section discusses specifications in the form of a sample tech-pack developed by senior PD class students instructed by the author at California State Polytechnic University, Pomona.

2.5.3.1 Design specifications

These are specifications related to the aesthetic appeal of the apparel product, such as fashion details, style details, and characteristics related to the product concept. Some style information can be transferred from the line sheet. This information can be shown on a "style summary" of a tech-pack, as shown in Figure 2.3. In addition to the general style information, the front and back flats (technical drawing) and special instructions (such as a graphic placement, label placement, a reinforcement tape placement inside the garment, etc.) are included in the style summary sheet. (Technical facts provided by Ms. Alejandra Parise, adjunct faculty at CPP (Cal Poly Pomona).)

2.5.3.2 Product specifications

Product specifications provide the intrinsic quality expectations of the garment and its components as well as standard procedures in producing and completion of the product such as size and fit specifications, garment assembly specifications, finishing and packing specifications, and labeling specifications. Such specifications can be shown using the bill of materials (BOM), branding and packaging information, fit evaluation, and garment assembly worksheets of a tech-pack.

An example of a BOM is shown in Figure 2.4. The BOM includes information pertaining to all the materials needed to complete construction of the style, such as self-fabric, contrast fabrics, linings, interlinings, trims, thread, etc., together with the point of placement of materials in the garment, the supplier information, color name/code, shrinkage (if any), and a component sketch and swatch section showing the garment components with physical or digital swatches.

The standard branding and packaging information shown in Figure 2.5 can be a part of the BOM that shows label, hang tag, packaging material specifications including where these items are placed, their dimensions, images, color, etc. In some instances, instructions can be included on how to fold, pack, and present the final garment for delivery.

An example of a fit specification page of the tech-pack, as shown in Figure 2.6, provides standard graded measurements for all the sizes (also called the size chart). This also includes the point of measurements (POM) indicating how to measure each finished garment measurement accurately. A tolerance is provided for each measurement, allowing production flexibility and fit quality control during production.

An example of a garment assembly page of a tech-pack is shown in Figure 2.7. This page includes the sequence of assembly operations along with the stitch count (stitches per inch), seam and stitch notations, machine type for each operation, and any special instructions assisting the apparel production. The component sketch area shows the components that are available for assembly.

Style summary					
					Cal Poly Pomona
Style #:	AM21401	Description:	Open back scoop neck T	Status:	In progress
Design/proto #:	#3	Fabric group:	Knit	Current stage:	PD
Division:	Womens	Season:	Fall 2014	Approved date:	Pending
Brand:	AM2	Base size:	S	Date created:	01/21/14
Product class:	Tops	Size range:	XS,S,M,L	Date revised:	05/10/14
Factory:	Cal poly pomona	Designer:	Megan S		
Country of origin:	USA	Product development:	Savannah C		
Fabric type:	60% rayon 40% cotton jersey knit	Tech designer:	Andrew N		
Front & back flats			Special instruction		
			Callouts, comments, instructions		

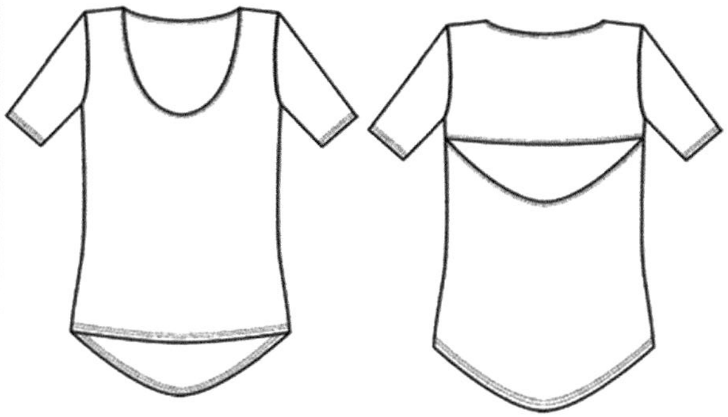

Figure 2.3 Tech-pack: style summary.

BOM								Cal Poly Pomona	
Style #:	AM21401	Description:	Open back scoop neck T			Status:	In progress		
Design/proto #:	#3	Group:	Knit			Current stage:	PD		
Division:	Womens	Season:	Fall 2014			Approved date:	Pending		
Brand:	AM2	Base size:	S			Date created:	01/21/14		
Product class:	Tops	Size range:	XS,S,M,L			Date revised:	05/10/14		
			Colour name/colour code						
Name/code/description	Use	Supplier information	Colour1	Colour2	Shrinkage		Component sketch & swatches		
Fabric1: 60% rayon 40% cotton jersey knit	All over	Alibaba	Black and white space dyed		3%				
Fabric2:									
Lining:									
Interlining:									
Thread: TEX 27, 100% spun polyester	All over	Fashion supplies	Black: T27S2						
Trim1									
Trim2									
Trim3									

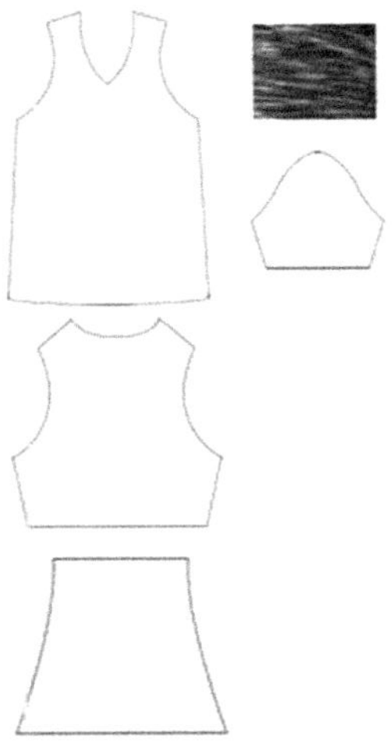

Figure 2.4 Tech-pack: bill of materials.

Standard branding & packaging information				
Branding				
Main label White satin main label/ black text	Attach brand label at center back neck	Pacific coast beach label company	1" × 1.5"	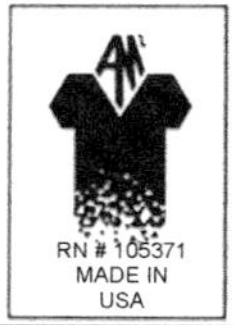
Care label White satin care label/ black text	Attach care label at side seam 3" above hem	Pacific coast bach label company	1" × 1.5"	Machine wash cold, delicate cycle. Wash with like colours. Do not use bleach. Lay flat to dry. Iron with low heat when necessary. 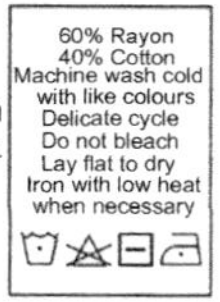
Hang tag 8.5" X 11" transparent	Place 2" under left armhole	Kinkos		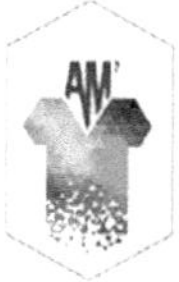
Packaging **None**				

Figure 2.5 Tech-pack: branding and packaging information.

Fit specifications

Style #:	AM21401	Description:	Open back scoop neck T	Status:	In progress
Design/proto #:	#3	Group:	Knit	Current stage:	PD
Division:	Womens	Season:	Fall 2014	Approved date:	01/21/14
Brand:	AM2	Base size:	S	Date created:	
Product class:	Tops	Size range:	XS, S, M, L	Date revised:	05/10/14

Approval status		Sizes				Fit		
POM	Description	XS	S	M	L	Fit spec. (L)	Delta (+/−)	Revised spec.
1	Back bottom panel length center back to bottom hem	18 1/2	18 3/4	19	19 1/4	19	3/8	
2	Center back neck to shoulder	7 1/2	7 3/4	8 1/4	8 1/2	8	1/4	
3	Center back neck to yoke hem	12 1/2	12 3/4	13	13 1/4	13	3/8	
4	Center front neck to hem	21 1/4	21 3/4	22 1/4	22 3/4	22 1/4	1/2	
5	Front neck width	18	18 1/2	19	19 1/2	20	3/8	
6	Neck drop	5 1/2	5 1/2	5 1/2	5 1/2	6 3/4	1/4	
7	Sideseam	18	18 1/4	18 1/2	18 3/4	18	1/2	
8	Sleeve length top	9 1/2	9 3/4	10	10 1/4	9 1/2	1/4	
9	Sleeve opening	11 1/4	12	12 3/4	13 1/2	12 1/2	1/4	
10	Sweep	38	40	42	44	43	1/2	

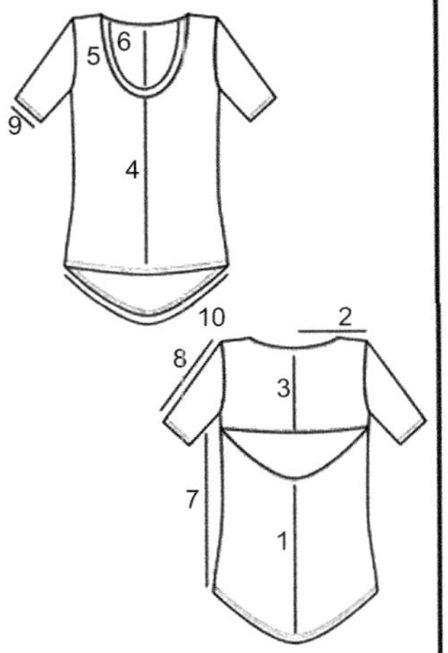

Comments
1)
2)
3)
4)

CB: Centre back

Pom comments—flats with measurements

Cal Poly Pomona

Figure 2.6 Tech-pack: fit specifications.

Garment assembly — Cal Poly Pomona

Style #:	AM21401	Description:	Open back scoop neck T	Status:	In progress
Design/proto #:	#3	Group:	Knit	Current stage:	PD
Division:	Womens	Season:	Fall 2014	Approved date:	Pending
Brand:	AM2	Base size:	S	Date created:	01/21/14
Product class:	Tops	Size range:	XS, S, M, L	Date revised:	05/10/14

Component sketch

Operation	SPI + OR - 2	Seam	Stitch	Machine	Special instructions
Overlock front and back neck (3/8")	13	Serging [Efd]	514	4-thread overlock	
Hem back yoke (1" double fold)	13	Double fold hem [Efd]	401	Single needle chainstich	
Hem top of bottom back (1" double fold)	13	Double fold hem [Efd]	401	Single needle chainstich	
Attach shoulders, right sides together (3/8")	13	Plain seam [Ssa]	514	4-thread overlock	Attach brand label center back neck
Hem neck (3/8")	13	Single hem [Efa]	401	Single needle chainstich	Shoulder seams placed towards back
Set sleeves into shoulder, right sides together (3/8")	13	Plain seam [Ssa]	514	4-thread overlock	Attach care label 3" from bottom raw hem (left side seam)
Sew side seams (3/8")	13	Plain seam [Ssa]	514	4-thread overlock	
Hem sleeves (3/4")	13	Single hem coverstitch [Efl]	406	2 needle coverstich	
Hem bottom (3/4")	13	Single hem coverstitch [Efl]	406	2 needle coverstich	

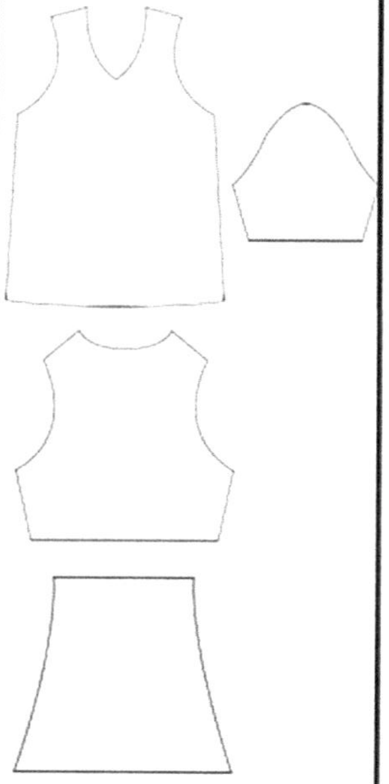

Figure 2.7 Tech-pack: garment assembly.
Note: The entries inside the square bracket [], are different types of seam classes.

2.5.3.3 Performance specifications

Specifications related to materials and their performance (fabric, thread, trims, and findings including branding and packaging information) are generally specified in the BOMs laid out for the style. These specifications can be fabric structure, weight, color, shrinkage, strength, flammability, etc. In addition, suppliers of the materials, minimum quantities, and price per unit can also be included.

2.5.3.4 General style information

General information related to the style—such as the allocated style number, style description, product division, fabric type, fabric group, season, base size, size range, country of origin, status of the style, etc.—are available in most worksheets of the tech-pack and may be automatically updated when once entered.

The tech-pack is a dynamic document, which will undergo many changes until the style is approved. Therefore, a more informative tech-pack can include a style history worksheet that shows the history of major changes to the style. An example of a style history sheet is shown in Figure 2.8. In addition, a tech-pack may include a pattern page that has a listing of all the pattern pieces that make up the pattern set needed to produce the style. This page may also include the drawing of each pattern shape, name of the piece, number of pieces to cut, and the yield for each fabric, lining, or interlining to produce a garment.

2.6 Apparel product life-cycle management (PLM) and supply-chain relationships

With the complexity of the apparel business and the need for fast turnaround time, companies have to improve the way they handle information at every phase of the business. There is continuous economic pressure on fashion companies coming from both ends of the supply chain, with consumers and retailers expecting fast fashion at lower prices while the cost of sourcing, manufacturing, and delivering tends to increase. The pressure further increases with the need to conduct business in a responsible manner with corporate responsibility guidelines such as ethical sourcing, concerns about conflict materials, sustainable PD and production, supply-chain transparency, and corporate governance. To comply with these needs, the company needs to manage information throughout the supply chain effectively. A good PLM application streamlines the task of keeping information to comply with these needs.

2.6.1 PLM systems

Many companies historically used spreadsheets to manage their PD information. However, spreadsheets lack the functionality and data structure needed to collaborate and communicate to conduct this fast-moving consumer-driven industry. Therefore, PLM solutions are becoming essential to deal with these complex business dynamics.

Style history						Cal Poly Pomona
Style #:	AM21401	Description:	Open back scoop neck T	Status:	In progress	
Design/proto #:	#3	Group:	Knit	Current stage:	PD	
Division:	Womens	Season:	Fall 2014	Approved date:	Pending	
Brand:	AM2	Base size:	S	Date created:	01/21/14	
Product class:	Tops	Size range:	XS,S,M,L	Date revised:	05/10/14	
Date		Comments				By
Apr 7, 2014	Standardize seam allowance on all garments. 3/8" neck, shoulder, sideseam. 3/4 hems					Savannah C
Apr 14, 2014	Revised neck measurements—see fit evaluation					Andrew N
Apr 28, 2014	Shrinkage added to the garment- 3%					Megan S

Figure 2.8 Tech-pack: style history.

The PLM applications are designed to concurrently support product design, development, technical design, merchandising, sourcing, sales, marketing, and other supply-chain partners with current information so that creative concepts can be turned into commercial products quickly. This will enable the company to respond to rapid changes in demand signals identified by powerful "analytics" applications today. To get the maximum benefits, a PLM application must offer rich integration with other PD technologies, which are discussed in Sections 2.4 and 2.9. The information can then seamlessly flow through PD and manufacturing ERP (Enterprise resource planning) systems and other supply-chain partners. The reader is encouraged to read more on various PLM applications for the apparel industry provided by companies such as Gerber, Lectra, and Human Solutions (e.g., Gerber Yunique PLM, Lectra Fashion PLM, and Human Solutions GoLive PLM). These systems help retailers, brand owners, and manufacturers to manage the information associated with their products from conception to store shelf and help them communicate with their suppliers around the world more effectively.

2.7 Measures for apparel product development

With the increase in competition, the apparel industry's focus on measures to monitor the PD process has increased. These measures can be used to benchmark with competitors or to adopt best practices. The Measures for Excellence report by the Quick Response Leadership Committee of the American Apparel and Footwear Association, formerly called the American Apparel and Manufacturer's Association (AAMA, 1991), discussed some of these measures. The number of measures used by companies over the time period has increased (Little and Heinje, 1998; Jang et al., 2005). The next sections outline a few first-level measures related to apparel PD that can be used to measure PD performance. The reader is encouraged to refer to the manuscript entitled "Measures for New Product Development" (Senanayake and Little, 2001), by the author, which discusses a wide range of measures in relation to apparel PD.

2.7.1 Sample adoption ratio

The sample adoption ratio is the percentage of PD samples that is actually adopted into a line. This measure is becoming increasingly important, particularly in high-style environments such as fast fashion. Sample adoption ratios in 1994 typically measured at 20–30% for fashion products and 40–75% for basic and fashion-basic products (Senanayake and Little, 2001). It is interesting to understand the general sample adoption ratios for fast-fashion and non-fast-fashion apparel companies today with the latest technologies used for business functions such as virtual prototyping.

2.7.2 Seasons per year

Selling seasons are the number of clearly differentiated (by styling, fabric weight, or other factors) selling seasons for a company in a year. This provides an indication of how often new lines are to be developed. Over time, the number of seasons per year has increased. PD cycle time will vary depending on the number of seasons, which can be from few weeks to few months. The concept of buying/selling seasons may be eliminated as manufacturers and retailers respond to individual demands.

2.7.3 Product-development cycle time

Product-development cycle time is the time between designer's concept and when the style is released for production. As the number of seasons increases, and as the diversity of most product lines expands, the need to shorten the time required to develop new products becomes more important. Companies with an efficient PD process may have cycle times in the 1- to 3-month range for fashion garments and in the three- to four-week range for basic garments (Senanayake and Little, 2001). However, fast-fashion companies today have PD cycle times that range from a few days to a few weeks. It is not only a PD measure but also a measure of merchandising excellence.

2.7.4 Initial forecast accuracy and forecast accuracy

The line plans can be evaluated according to the initial forecast accuracy, that is, wholesale orders placed as a percent of demand projected when a style or merchandise group is accepted in the line. The ratio of actual order demand to the forecasted order demand expressed as a percentage is the forecast accuracy.

2.7.5 SKU planning frequency

The SKU planning frequency relates to how frequently a firm plans production as a reaction to changes in forecast or in order demand. The frequency can be monthly, biweekly, weekly, or daily (AAMA, 1991).

The above measures are just a sample of many measures available to benchmark the PD process. Other measures can be in the areas of finance, inventory, logistics, process performance, etc. There is also a need to develop new measures in PD for areas such as electronic commerce, virtual prototyping and fast fashion.

2.8 Future trends in apparel product development

As an approach to inquire into future trends in apparel PD, the author interviewed Dr Mike Fralix (Figure 2.9), an industry consultant and the CEO/President of Textile/Clothing Technology Corporation or $[TC]^2$, based in Cary, North Carolina. In addition, the author questioned how technology leaders envisioned the future of apparel PD during the interview process to develop the case studies in Section 2.9.

Figure 2.9 Dr Mike Fralix, an apparel industry consultant and the CEO/President of $[TC]^2$, USA, explained his views on the future of apparel PD.

2.8.1 Future PD trends: a consultant's view

2.8.1.1 Digital product development

A new trend in apparel PD is apparent where more and more functions are expected to go digital, with fewer physical entities such as garment patterns and samples, whether they are in 2D or in 3D. This trend will not just apply to apparel PD but will carry along the preproduction, production, and postproduction functions in the apparel supply chain. Fralix believes that body-scanning technology, which $[TC]^2$ has pioneered in developing, is ready to be commercially used for apparel PD today. This technology combined with other technologies also have the capability of developing a digital fit model, which can be draped using electronically sewn, digitally decorated (from digital libraries) garment patterns to create digital samples for virtual-fit visualization and fit decisions. This virtual-prototyping capability has been enhanced with technology that can create a digital fit model or the fit model avatar to be animated to simulate various realistic activities and poses, such as cycling when developing bike shorts and sitting when developing dress pants. There are a number of apparel companies that have reduced the number of physical samples by adopting virtual prototyping technologies, where on average companies today make more than five samples during the apparel PD process otherwise (Fralix, August 20, 2014, pers. comm.).

2.8.1.2 Digital fabric performance

When asked about the technological readiness of digital fabric libraries with fabric parameters to simulate true-to-life fabric drape simulations, Fralix explained that, as product developers have a good idea about how most fabrics that have been around for some time behave when you wear them, one can make fit decisions quite accurately. However, if it is a new fabric, many tests need to be done, and the test information needs to be incorporated into the libraries before that fabric can realistically be digitally draped for virtual prototyping (Fralix, August 20, 2014, pers. comm.).

2.8.1.3 Digital printing

Digital fabric printing applications for apparel PD are becoming popular. From the PD point of view, companies could use digitally printed sample yardage to replace the expensive sample yardage development process, which has a long lead time. For example, if a yarn-dyed fabric needs to be developed, rather than going through an expensive physical yarn and fabric development process, one can digitally print multiple yarn-dyed fabric effects on white fabric and use them for the sampling process. Once the exact yarn-dyed fabric effect is approved, the physical sample development process can be started. This strategy will reduce the time and cost of developing multiple arrays of fabrics to arrive at the decision (Fralix, August 20, 2014, pers. comm.).

2.8.1.4 Other PD tools

It is apparent that most widely discussed new tools and technologies for PD are around the technical design function. However, there are other tools that assist in more efficient apparel PD processes, such as the costing solution Quick True Cost by Methods Workshop, a US-based company, which provides the cost of the product in advance during PD, which is claimed to be within 95% accuracy when compared to the actual engineered product cost. Considering the trends in technology applications, the new digital technologies will integrate with 3D-printing technologies to improve the apparel PD process. It is expected that new materials for 3D printing will develop that have the characteristics to replicate textile materials. Research is underway to achieve these innovations. The new PDM or PLM systems that provide seamless information-sharing for decision-making in the apparel supply chain, in addition to accurate reporting and version control, also have a great impact on the apparel PD process. Further, companies such as TEXbase, which provide PLM system for fabrics, would enhance the PD process for apparel in the future (Fralix, August 20, 2014, pers. comm.).

When the US-apparel industry is considered, an industry trend for apparel is re-shoring. Some companies that moved to off-shore locations for their PD and production are moving back to the USA. Knowing where the production will take place, and the cost associated with it, has a positive influence on the effectiveness of the apparel PD process (Fralix, August 20, 2014, pers. comm.).

2.8.2 Future PD trends: technology companies' view

2.8.2.1 Compressing the PD cycle time

Almost all the executives from the technology companies who were interviewed believe that shortening the PD cycle is a trend that is very critical, as there is a lot of pressure for apparel companies to cater for types of customers with limited resources. To achieve this goal, apparel companies must implement virtual and digital 3D product-development technologies using virtual fitting, prototyping, realistic simulation and visualization, and using standard effective communication tools such as PLM systems across the supply chain. It is important to have PLM systems allowing the PD and supply-chain partners to share, document, store, and retrieve information to make quick decisions across time zones and cultures. Companies using advanced technology tools to radically reduce the need for physical samples will be at a comparative advantage over other companies.

2.8.2.2 Fit customization and virtual human modeling

The fashion, textiles, and apparel industries are moving towards global retailing or remote shopping, with the concept of enabling customers to purchase garments by conducting "virtual try-on" using their own body size and shape. Therefore, the virtual human modeling concept has become very important, and companies are conducting research on systems to quickly develop 3D avatars for virtual try-on for global retailing. The ability to create color body scans, and the capability of 4D body scanning, will have useful applications for apparel PD in the future. With tech-companies becoming more transparent, tools from one company will be able to communicate with the tools from another tech-company, providing better technological collaboration for apparel PD.

2.8.2.3 Mobile product development

There will be many applications developed to be used on mobile devices, providing apparel companies as well as consumers the opportunity to design, develop, visualize, and purchase apparel products fast. The virtual representations will look real, with virtual collections, dressing rooms, show rooms, etc. The technologies are becoming more affordable, so that the so-called complex, expensive technologies used by high-tech industries such as aerospace, will be used for consumer products such as apparel. The new generation of consumers will not agree looking at a picture of an apparel product but will look for a virtual dressing room to experience virtual fitting before making a purchase decision. For this, the technological tools will become more available through the web and cloud-based applications on mobile devices.

2.8.3 Apparel Made for You (AM4U): PD for virtual inventory apparel manufacturing

The Cal Poly Pomona Apparel Merchandising and Management Department, in collaboration with a number of technology companies, initiated a project named Apparel Made for You (AM4U) to demonstrate the opportunity to produce customized or personalized garments on demand by integrating emerging business and technology solutions. In March 2012, the author, as one of the founding members of this project, with the AM4U group, demonstrated that custom-designed, custom-fitted clothing could be developed, purchased, manufactured, and shipped (or delivered) to the consumer in less than 4 hours from undyed fabric. The demonstration involved the linkage of retail e-commerce with body scanning, 2D and 3D CAD, digital dyeing and printing, computerized cutting, and computer-integrated garment-assembly technologies.

Since then, AM4U has built an integrated mini-factory (IMF) capable of printing, cutting, and sewing fabric on demand with one of two operational strategies: (1) digital demand replenishment (DDR), to keep retail supply pipelines flowing to replace on-trend fabrics and garments in-store; and (2) direct-to-consumer demand, called purchase-activated manufacturing (PAM), to supply customized or personalized products for individual consumer demand.

AM4U technology integration uses the following steps: (1) generating/adjusting garment patterns from consumer size data (including body scanning) and customer's style selection; (2) decorating the patterns from the customer's color, print, and logo selection; (3) digital printing the pattern pieces with colors, prints, and/or logos on transfer paper; (4) transferring the decorated patterns onto fabric with cutting lines using the new patent-pending Active Tunnel Infusion (ATI™) dyeing and printing process; (5) optical or digital identification of pattern pieces and cutting using optical or digital cutting technology; (6) computer-integrated flexible sewing system; and (7) final quality inspection, packing, and shipping to the customer.

ATI™ technology can permanently dye, print, or imprint on either one side or both sides of the fabric in one continuous passage, and has no minimum order quantity. As the whole operation can be confined into about 8000 ft^2 space, it is called an IMF, as shown in Figure 2.10. In addition to its small physical footprint, its environmental footprint is very small. This is because the fabric coloring and printing technology does not

Figure 2.10 Integrated mini-factory (IMF) of AM4U Inc.

use water and uses less energy without any post-treatment processes. An apparel company can develop an infinite number of styles, colors, prints, and garment patterns during the PD process, and keep them as virtual inventory on the cloud that can be market-tested online before retail distribution using the DDR or customized and personalized-to-consumer demand using the PAM.

2.9 Case studies: PD tools and technologies

Case study 1
Company: Human Solutions Assist AVM, Cary, NC, USA.
Interviewees: Tim Guenzel, Account Manager, Body Scanning; and Oliver Meier, Sales Director, Fashion Applications.

Human Solutions provide technology solutions for a number of areas in the apparel PD process. The company provides body-scanning solutions (both laser-based and depth-sensor-based), 2D and 3D CAD solutions, a web-based sizing and fit solution, a PLM solution, a cut-order planning solution, a marker solution, and a digital virtual try-on solution.

Body Scanning: In addition to measurement extraction, which is the primary function (to obtain appropriate fit by translating measurements into 2D patterns) of body scanning for apparel PD, the Anthroscan software developed by Human Solutions has the functionality to construct an avatar from the scan with a single click of a button without having the need to reconstruct with identification of feature points, etc., which was the case before and which took a considerable amount of time and skill. This scanned, reconstructed mesh (the mesh is used to construct the avatar) of the subject is considered to be watertight (ready for further processes with any surface gaps from the scan closed in a most realistic way using algorithms that use databases of actual shapes). This type of watertight mesh is required for further processes such as avatar creation (see Figure 2.11) and 3D printing. This avatar, which is created in less than a minute, can be imported to the 3D CAD program for further processing. In addition, the software has the capability to prepare the scan for draping visualization (for example, to simulate a pair of athletic shorts on an avatar with an athletic pose), as the avatar can be incorporated with a skeleton systematically placed by selecting joint locations within the body. The skeleton creation runs fully automatic without requiring any user interaction (Guenzel, August 14, 2014, pers. comm.). The body scanner price has come down tremendously, with a lower footprint and enhanced speed. As shown in Figure 2.12, the depth sensor scanners are more portable and will fit into the size of a golf bag. The calibration process is simple, with an easy-to-use menu-driven graphic user interface. Unlike some other scanners, the depth sensor scanners do not need to run in a dark environment.

Body-Scanning-Related Other Sizing Solutions: Human Solutions has conducted, or has partnered with institutions, for a number of sizing surveys in a number of countries. In addition, they have purchased body-measurement data from other size surveys. From this database of body measurements and generated avatars, the company has developed standard avatars for European, Asian, and American body sizes. These data are available in a web solution (iSize Portal), which can be used by apparel companies to compare and analyze their sizing standards' accuracy related to the respective population. Without having a body scanner, a company can use standard avatars from the web solution for PD considering their customer sizes. Once the company decides on an actual sizing system based on real data, it is also possible to create avatars for the decided size range, which can be brought into a 3D CAD program for PD such as virtual prototyping. iSize is a subscription-based web service to companies (Guenzel, August 14, 2014, pers. comm.).

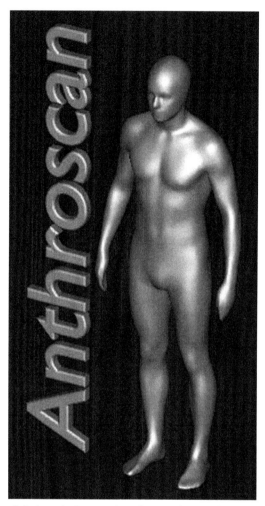

Figure 2.11 Human Solutions Anthroscan-based avatar that can be created in minutes from scanning.

2D and 3D CAD Solutions: With the appropriate 2D digital patterns imported, 3D digital prototypes can be electronically stitched on the avatar with up to 21 layers (inside and out side of the base layer) of different materials. Another functionality that is important for virtual prototyping is the capability to visually see how the digital garment fits on the virtual avatar considering the looseness, tightness, grain direction, weft and warp stretch, pattern alignment accuracy, etc. which most fashion 3D CAD suites have today. The true nature of virtual draping will depend on how well the fabric characteristics can simulate the garment drape using fabric mechanics defined by fiber type, yarn type, fabric structure, fabric finishes, etc. The 3D CAD software also provides the capability to visualize tight- and loose-fitting areas, warp or weft directional stretch, and pattern sewing accuracy with directional fabrics. The 3D CAD software, called Vidya, also provides the opportunity to integrate the individual body scan

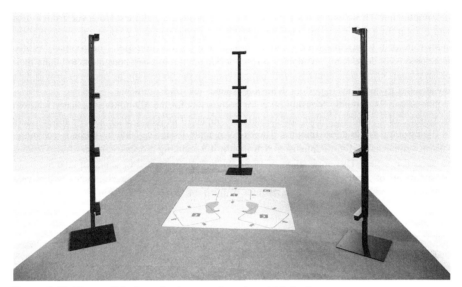

Figure 2.12 Human Solutions depth-sensor-based portable body scanner.

with the made-to-measure CAD suite. If a customer has his or her digital patterns for a garment style set up with made-to-measure features and his or her personal body scan has been obtained, the system can predict the best-fitting size and suggest alterations to the patterns. The CAD patterns, which are prepared in 2D, can then be automatically adjusted to the changes that are done in 3D. Further, the software provides the ability for the user to select and save multiple poses of an avatar created from the body scan, for a quick animation to test the draping. The avatar is also ready to be processed using other animation software for continuous live animations for fashion runways (Guenzel, August 14, 2014, pers. comm.).

Product Life-Cycle Management (PLM) Solution: Some functions of the Human Solution's PLM GoLive system can be used in the line development stage in addition to real-time information sharing through integration of the body measurements to 2D and 3D CAD and Automarker applications. A budget plan for the new line can be developed based on previous years' styles and sales information with the knowledge of development budget, manufacturing cost, margins, etc. This can be done using the cost suite for domestic and international production or sourcing considering currencies, importing costs, etc. Costing can be carried out for package sourcing (without detailed breakdown) or individual sourcing (with detailed breakdown). The PLM system can be configured to be either web-based or cloud-based (Meier, August 18, 2014, pers. comm.).

Case study 2
Company: Size Stream, Cary, NC, USA.
Interviewee: Dr. David Bruner, Vice President.

Size Stream, a company specialized in 3D depth-sensor-based body scanners, now can produce color body scans in addition to general scanning used for apparel PD such as body-measurement extraction, size standards development, and avatar creation. These body scanners also provide a great service to the custom clothing industry. Dr Bruner (August 14, 2014, pers. comm.) believes that the affordable price of scanners today will enable the retail sector and brands to better fit their population. These low-cost scanners cannot see through the clothes. However, from a scan with correct-fitting clothes, one may be able to advise the consumer about the right size for another garment where the scanner can be used as a size-prediction tool. The company is also in the process of developing an application for mobile phones to assist consumers to obtain their measurements using statistical data with either scanning the body with the phone or providing a few answers to questions regarding their body shapes and measurements.

Size Stream is currently focusing on color body scans for the 3D-printing industry while in the process of researching other applications for apparel PD. These scans can be made ready for 3D printing within a few minutes. Their new scanner is also capable of 4D body scanning, with movement as the fourth dimension. During the 4D scanning process, number of scan frames per second is saved. This application is useful when there is a need to scan and take the body measurements of a person in motion. It is expected that these latest developments will have a good use in apparel PD in the future (Bruner, August 14, 2014, pers. comm.).

Case study 3
Company: Optitex USA, New York, NY, USA.
Interviewee: Yoram Burg, President/Lead Sales.

Optitex provides PD, production, design and merchandizing software applications for the apparel industry. The company's major focus today is providing 2D and 3D CAD solutions and linking these for virtual PD with the enterprise worlds of design, merchandizing, and customer experience. The company claims that their PD applications can simulate and visualize a virtual prototype on a virtual-fit-model avatar that is so realistic that it cannot be distinguished from the real sample on the physical-fit model. The capability of virtually creating, visualizing, and editing the garment samples in one place (Figure 2.13) has allowed their clients to reduce the PD cycle time drastically. One of their US-based clients has reduced the number of physical samples by adopting virtual PD, so that 90% of the styles are being approved from one virtual sample today. They believe that this ability to develop 3D virtual design concepts fast and

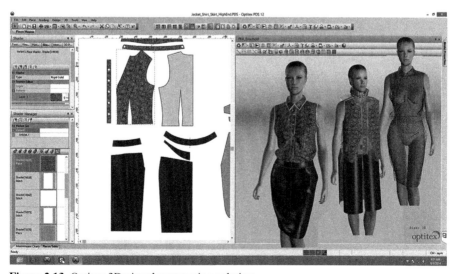

Figure 2.13 Optitex 3D virtual-prototyping solution.

visualize them digitally will open up enormous opportunities in many areas including PD in the apparel industry (Burg, August 27, 2014, pers. comm.).

Even though most companies have not looked at investments in PD technology before, many big retailers in the USA today are in advanced stages of deploying 3D PD and fast-fashion technologies. The major focus of these companies is to reduce PD cost and time. There is considerable corporate sponsorship for these technology investments by apparel companies today. The companies cannot forget about these efficient PD technologies, as technology developments move very fast and the companies that deploy them are at a comparative advantage over the other companies in terms of time and operation cost. Due to its XML (extensible markup language), API (application programing interface), and DLL (dynamic link library)-based applications, the Optitex applications today interface with any other CAD, CAM, PLM, or ERP application platforms providing flexibility of usage for apparel companies (Burg, August 27, 2014, pers. comm.).

Case study 4
Company: Tukatech Inc., Los Angeles, CA, USA.
Interviewee: Ram Sareen, CEO/Founder.

The technology focus of the company today is facilitating virtual PD in addition to linking PD to production applications such as grading and marker making. The company also provides hardware solutions such as plotters.

The "e-fit" technology, a virtual PD tool, along with other CAD solutions, provides the opportunity to develop a 3D virtual sample with actual size measurements and visualize it on a 3D virtual-fit model. This is done by (1) importing the artwork (planned prints and colors as per fabrics) developed on a computer-aided textile design application onto the garment patterns developed on the 2D pattern-making application; (2) developing Slopers in 3D, adding style lines, and changing fabric properties and texture; and (3) sewing and draping electronically on a 3D "e-fit" application. Draping can be done on a 3D fit-model avatar. Further, with the built-in motion simulator, one can see the actual drape of the garment on the fit-model avatar in motion, simulating the model on a runway and allowing designers to approve samples digitally. The ability of the designers to change their inspirations and visualize them virtually provides unlimited opportunities that were restricted before. This is because 99% of the time, designers change their original ideas after the physical prototypes are made, leading to an increased number of sample iterations before a product can be approved for a line. The block print placement trims such as laces and bindings, linings, etc. can also be simulated and rendered to show the final look with animation applied internally. This will also allow the development team to share the designs online without creating physical samples. Once approved, an unlimited number of virtual samples can be made very quickly using combinations of artwork, colors, features, etc. This tool can be used to create story boards, virtual runways, photos, and sample videos for further decision-making in line review and line presentation and marketing (Sareen, August 25, 2014, pers. comm.).

Retail clients have endorsed the Tukatech PD tools that assist in virtual PD due to the fact that there is a strong consistency between the digital samples on the "e-fit" model avatar and the physical samples on the live-fit model, as shown in Figure 2.14. Also some retailers have decided to work with digital sample approvals, entirely eliminating physical samples for children's clothing lines, and reducing the PD cycle time from months to days or even hours. While most eCommerce retailers take 6—10 months from concept to consumer, this technology can assist a company to develop products and go live with a new collection in less than 30 days. Using the tools wisely, one may be able to work without physical samples,

Figure 2.14 Tukatech realistic view of a virtual dress-drape simulation on a virtual-fit model.

photo shoots, and inventory where the digital images can be transferred to the eCommerce site and offered for sale. Once the initial product line is introduced to consumers, the company can rely on analytics to replenish high-demand products (Sareen, August 25, 2014, pers. comm.).

Case study 5
Company: Styku, Los Angeles, CA, USA.
Interviewee: Raj Sareen, CEO and Co-Founder.

The price of body scanners has come down by 50–75% with improved portability. Body scanning is not the full solution to the fit problem, but can assist in a way that can provide not only body measurements but also other solutions such as 3D digital body models for e-fitting and garment-size prediction for consumers. As soon as mobile devices include 3D sensors in the near future, there is expected to be an influx of body-scanning applications for consumers to use. Apparel companies will benefit from this "mobile body scanning" capability for size-prediction applications and supplying correct-fitting clothing to consumers. It is also expected that the scanning applications can be accessed through subscription systems (Sareen, August 27, 2014, pers. comm.).

2.10 Conclusions

The PD models are helpful to understand the complex apparel PD process. The PD calendar helps to coordinate, monitor, and control the many functions in apparel PD to design, develop, produce, and deliver the goods in a timely manner. Emphasis on improving the PD process in apparel firms is growing. Companies concentrate on improving the cost-effectiveness of the process by streamlining and shortening the PD cycle using the latest technologies and improving the market responsiveness. The digital PD process using virtual technologies provides the opportunity for companies to shrink the PD cycle time from months to days. Communicating the standards and specifications among supply-chain partners using tech-packs that were developed during PD, provides the opportunity for quality assurance and control in the apparel-manufacturing process. The effectiveness of sharing PD information among the PD teams in various geographical locations has improved with new PLM systems. Understanding the PD strategies used by fast-fashion companies is an approach to learn the demand-led PD process, which is important for a made-to-demand apparel business. Companies can compare their PD processes by benchmarking, with best-in-class companies using measures such as PD cycle time, sample adoption ratio, etc. to improve their business process. The future of digital PD will move to mobile technologies with more and more technology solutions developed for mobile devices such as phones and tablets. It is expected that the consumer will participate in the PD process while having the opportunity for more customization and personalization options. PD for virtual inventory manufacturing is developing with projects such as AM4U with green manufacturing strategies.

2.11 Sources of further information and advice

NICPPD model: May-Plumlee, T., Little, T.J. 1998. No-interval coherently phased product-development model for apparel. International Journal of Clothing Science and Technology 10(5), 342–364.
Measures for new product development: Senanayake, M., Little, T. 2001. Measures for new product development. Journal of Textile and Apparel Technology and Management 1(3), 1–14.
Sourcing Journal https://www.sourcingjournalonline.com/.
Apparel Magazine http://apparel.edgl.com/home.

The National Association for the Sewn Products Industry (SEAMS) http://www.seams.org/index.html.
Apparel Made for You (AM4U Inc.) http://www.sresearch.com/am4u/
Virtual inventory manufacturing alliance (VIMA) http://www.vimalliance.org/

References

AAMA, 1991. The Impact of Technology on Apparel: Part One (Technical Advisory Committee). Author, Arlington, Virginia.
Bruner, D., August 14, 2014. Interviewed by: Senanayake, Muditha.
Bubonia, J.E., 2012. Apparel Production Terms and Processes. Fairchild books, New York.
Burg, Y., August 27, 2014. Interviewed by: Senanayake, Muditha.
Burns, L., Bryant, N., Mullet, K., 2011. The Business of Fashion: Designing, Manufacturing, and Marketing (Distributor). Fairchild; Bloomsbury, New York: London.
Che, J., Li, R., 2010. Engineered colour appearance with digital approach. In: Proceedings of the 2010 ASEE Gulf-southwest Annual Conference, McNeese State, USA.
Christopher, M., 2000. The agile supply chain: competing in volatile markets. Ind. Mark. Manage. 29, 37–44.
Dutta, D., 2002. Retail @ the Speed of Fashion, Third Eyesight, Cast Study. Retrieved from: http://thirdeyesight.in/articles/ImagesFashion_Zara_Part_I.pdf.
Fralix, M., August 20, 2014. Interviewed by: Senanayake, Muditha.
Glock, R.E., Kunz, G.I., 2005. Apparel Manufacturing: Sewn Product Analysis. Pearson Education, Inc., Upper Saddle River, NJ.
Goworek, H., 2010. An investigation into product development processes for UK fashion retailers: A multiple case study. J. Fash. Mark. Manage. 14 (4), 648–662.
Guenzel, T., August 14, 2014. Interviewed by: Senanayake, Muditha.
Jang, N., Dickerson, K.G., Hawley, J.M., 2005. Apparel product development: measures of apparel product success and failure. J. Fash. Mark. Manage. 9 (2), 195–206.
Johnson, M., Moore, E., 2001. Apparel Product Development. Prentice Hall, Upper Saddle River, NJ.
Keiser, S., Garner, M., 2012. Beyond Design: The Synergy of Apparel Product Development. Fairchild Publications (Imprint of Bloomsbury publications Inc.), New York.
Lee, J., Steen, C., 2010. Technical Source Book for Designers. Fairchild Books, New York.
Little, T.J., Heinje, R.K., 1998. Does your quick response program measure up? Bobbin 39 (9), 42–47.
May-Plumlee, T., Little, T.J., 1998. No-interval coherently phased product development model for apparel. Int. J. Cloth. Sci. Technol. 10 (5), 342–364.
McKelvey, K., Munslow, J., 2003. Fashion Design: Process, Innovation and Practice. Blackwell Publishing, Oxford, London.
Meier, O., August 18, 2014. Interviewed by: Senanayake, Muditha.
Mihm, B., 2010. Fast fashion in a flat world: global sourcing strategies. Int. Bus. Econ. Res. J. 9 (6), 55–64.
Nayak, R., Chatterjee, K., Khurana, G., Khandual, A., 2007. RFID: Tagging the new era. Man-Made Tex. India 50 (5), 174–177.
Sareen, R., August 27, 2014. Interviewed by: Senanayake, Muditha.
Sareen, R., August 25, 2014. Interviewed by: Senanayake, Muditha.

Schertel, S., 1998. New Product Development: Planning and Scheduling of the Merchandising Calendar (Master Dissertation). North Carolina State University.

Senanayake, M., Little, T., 2001. Measures for new product development. J. Text. Apparel, Technol. Manage. 1 (3), 1—14.

Senanayake, M.M., Little, T.J., 2010. Mass customization: points and extent of apparel customization. J. Fash. Mark. Manage. 14 (2), 282—299.

Stone, E., 2006. In Fashion: Fun! Fame! Fortune! Fairchild, New York; Troika [distributor], London.

Watson, K., August 21, 2014. Interviewed by: Senanayake, Muditha.

Role of fabric properties in the clothing-manufacturing process

B.K. Behera
Indian Institute of Technology, New Delhi, India

3.1 Introduction

The clothing-manufacturing process can be defined as a method of conversion of a flat two-dimensional structure into a three-dimensional shell structure. Fabric is the basic raw material of the clothing industry. The quality of the fabric influences not only the quality of the garment but also the ease with which a shell structure out of flat fabric can be produced. The fabric specifications for different end-use requirements are different, and the selection of an appropriate fabric is one of the most difficult jobs for the clothing manufacturer. The specifications of fabrics for apparel manufacturing can be considered in terms of primary and secondary characteristics. The primary characteristics are static physical dimensions, and the secondary characteristics are the reactions of the fabric to an applied force. The apparel manufacturer is primarily interested in the secondary characteristics of the fabric. The consumer is mainly interested in appearance, comfort, and wearability of the fabric. The production of garments from high-quality fabrics not only gives comfort to the wearer but also helps in the smooth working of manufacturing processes and leads to defect-free garments (Behera and Hari, 2009, 1990; Behera, 1999). With the advent of high-speed automatic line production systems, the interrelationship between fabric mechanical properties and fabric processability in tailoring has become very important. The selection of manufacturing processes requires selection of machine and process variables based on the specific properties of the fabric being processed. Therefore, it is essential and logical to study various fabric characteristics influencing directly or indirectly the apparel making-up process and finished clothing product performance. This chapter deals with the relationship between some important fabric properties and the making-up process.

The importance of low-stress mechanical properties and fabric formability is also discussed in the context of tailorability. Determination of sewability as a measure of fabric quality assessment for clothing manufacture is discussed in detail.

3.2 Fabric properties and performance

3.2.1 Dimensional and physical properties

Usually fabric dimensional parameters are determined as primary requirements for fabric selection; these include thread linear density, ends and picks per cm in woven fabric or courses and wales per cm in knitted fabric, areal density (fabric mass in g/m^2), length and width, weave, fabric cover, and dimensional stability (to washing and dry cleaning). Other mandatory tests include color fastness to washing, dry cleaning, light, perspiration, and rubbing. The blend percentage of component fibers is determined for blended fabric. Sometimes flammability and the presence of certain hazardous chemicals are determined depending on the end-use and buyer's requirement. Fabric thickness, fabric density, crimp in warp-and-weft yarn, and moisture regain value are determined for detailed characterization of fabric quality depending on the requirement. Bow and skewness measurement are carried out, particularly in the case of check and stripe design fabrics.

3.2.2 Mechanical and other miscellaneous properties

The mechanical properties of apparel fabrics are important from the point of view of stresses applied to the fabrics in making-up as well as the physical changes in the fabric that result from application of forces in a garment during its use. Bending, tensile compression and shear properties are considered important from the point of view of garment making-up. These properties influence both sewability and the shape of the sewn fabrics. Other mechanical properties such as drape, tear strength, abrasion resistance, wrinkle and crease recovery, and pilling behavior of fabric are evaluated from a performance point of view. Some special properties are also evaluated, like bursting strength, elastic modulus, stretchability, drying speed, light reflectance, weather resistance, moth resistance, size content, resin content, oily and fatty matter, solvent extract, scouring loss, degumming loss, glossiness, color index, foreign matter, and nets. From a comfort point of view, water absorbency, air permeability, and thermal insulation are tested (Nayak et al., 2009). Under eco-testing, pH values of extracted liquid, barium activity number, and free formaldehyde content are examined.

3.2.3 Performance

Mechanical properties such as tensile, bending, shear, compression, and surface properties are considered important in deciding the utility and mechanical comfort performance of a fabric. However, analysis of the tailoring process reveals that these properties are equally important in the making-up process of the garment. In the tailoring process, an initially flat fabric is converted into a three-dimensional (3D) garment. This conversion requires complex mechanical deformation of the fabric at very low loads. Formability, the ability of a fabric to be converted into a 3D shape to fit a 3D surface, is dependent on the above mechanical properties. At the same time, it is also dependent on the techniques of garment manufacture, particularly on the amount of overfeeding adopted in certain operations. The fabric must provide the following requirements demanded by a garment (Kawabata and Niwa, 1989; Mori, 1994):

1. Utility performance to provide the individual with adequate physiological protection.
2. Comfort performance to ensure a sufficient degree of wearing comfort such as fitting to the human body.

3. Aesthetic performance to improve the aesthetic appearance of the wearer, highlighting certain anatomical particulars and concealing other.
4. Fabric performance for engineering of clothing manufacture as clothing material.

Textile fabrics as clothing materials provide protection for the human body from injury to skin caused by mechanical contact with outer bodies and protection from cold or hot environments. In addition, high-quality fabrics must be comfortable for the human body mechanically and thermally and give a good aesthetic appearance.

3.3 Garment make-up process and fabric properties

The main task for a clothing manufacturer is to produce shell structures out of flat fabrics to match the shape of human body. The overall scheme of the garment-manufacturing process is illustrated in Figure 3.1. In all shape-producing methods, there will be an interaction between the particular method used and various physical properties of the fabric. Some of the interactions between various major manufacturing processes and fabric properties are discussed in the following sections (Harlock, 1989; Postle, 1990).

3.3.1 Pattern grading

The first step in manufacturing a garment is the creation of design and construction of patterns for the components of design. This requires determination of geometrical shape of the body surface in order that appropriate shell structures can be produced. For pattern grading, anthropometric data should be available for the market in which the garment is to be sold. Pattern grading is a process of enlarging or diminishing a style pattern, making it possible to obtain proper fit for all sizes without changing

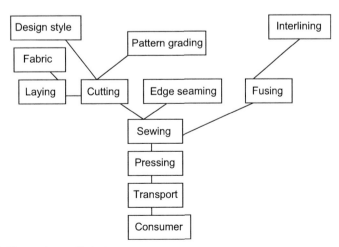

Figure 3.1 Garment-manufacturing process.

the title for a given compilation of anthropometric measurements that are suitable for a person whose body measurements lie within certain tolerance limits of the garment size measurements. The shape and size of the garment relative to the shape of the body, known as the fit, will be strongly influenced by physical and mechanical properties such as tendency of the fabric to stretch, shrink, distort, and drape due to stresses induced during use under static and dynamic situations, which are to be taken into account while drafting the pattern for a garment.

3.3.2 Laying-up and cutting

In laying and cutting procedures, layers of fabrics are superposed on a table to be cut simultaneously into garment components for further processing. During these operations, it is essential that each layer of fabric is laid in an unrestrained state in order that the dimensions to which various parts of the garments are cut are stable dimensions of the fabric. The fabric properties that govern this are:

Extensibility in laying direction decides the ease with which the fabric can be laid in an unrestrained state. It is essential that the fabric should not be too extensible at low loads.

The coefficient of friction helps to make multiple layers more or less stable. Stretch in fabric layers produced by in-plane or lateral stresses on superposed fabric layers will affect the stress-free dimensions of the garment patterns.

Building up multilayered structures in certain elements is often a necessary processing step in order to fulfill various functional requirements. Examples of such garment elements are jacket fronts, waistbands in trousers and skirts, and collar and cuffs in shirts. Multilayered structures should produce desirable bending stiffness, drape, and extension. The use of fusible interlinings for this purpose is an established practice.

Three-dimensional shapes in flat apparel fabrics can be achieved by either cutting and sewing processes or by shaping using steam-pressing and molding techniques. The former goal is met by appropriate pattern drafting of garment components, for example, the shoulder part of the sleeve; and by cutting drafts in the fabric components, for example, ladies dresses; which after application of stitches gives the desired shape or form. The resulting 3D form will be strongly influenced by certain fabric properties. During cutting of multiple layers of fabrics, which is normally done by the action of a reciprocating blade, a minimum level of stability of fabric dimension is necessary. To be more specific, a certain level of tensile rigidity both in warp and weft directions and some amount of surface friction are necessary for this process.

3.3.3 Seaming and sewing

Seam is the line where two or more layers of fabric are held together by stitches. It joins two or more pieces of fabric together, often leaving a seam allowance with raw edges inside the joining. During this operation, the fabric should be kept under the control of the operator and the sewing-machine feed dog. This requires a degree of in-plane fabric stability. Adequate tensile strength, shear rigidities, and fabric-to-metal coefficient of friction are basic requirements for seaming.

Sewing is the method of joining fabrics using stitches made with a needle and a thread. Sewing is one of the oldest of the textile arts dating back to the Paleolithic era. The incorporation of sewing thread into the fabric structure while maintaining a flat fabric surface requires some rearrangement of the yarns within the fabric. In order to avoid puckering of the seam, it is necessary that shear rigidity be quite low to permit interlaced yarns in the fabrics to rotate and accommodate the sewing thread; on the other hand, bending rigidity should be high enough to counteract internal forces necessary to bend the fabric out of its plane.

3.3.4 Pressing

The pressing or molding process produces a varied shape by application of one or more of the following: heat, moisture, and pressure. Again, the ease and stability of the resulting shape will be strongly influenced by certain fiber and fabric properties. The aim of fabric pressing is to remove distortions in the fabric surface, for example, wrinkles in the zone of the seam, which depend on the buckling behavior partially related to bending rigidity. The maximum pressure that can be exerted during pressing is limited by the need to maintain the appearance of the fabric surface. The relevant mechanical properties in this case are fabric surface compressibility and resilience.

3.3.5 Other processes

Fusing of interlinings in garment manufacturing is a very important process. Interlinings are the accessories used between two layers of fabric to keep the different components of apparel in a desired shape or to improve the aesthetics and/or performance. Generally, interlinings are soft, thick, and flexible fabric made of cotton, nylon, polyester, wool and viscose or their blends, which may be coated with some resins. There are two types of interlinings in use in the garment production: fusible and non fusible. The interlinings are carefully selected so that they can withstand the conditions during the fabric care and maintenance without any damage during the useful life of a garment. Once the garments are finished and inspected, they are packaged and transported to the retailers or the point of sale to the consumers.

3.4 Low-stress mechanical properties and make-up process

Low-stress mechanical properties of the fabrics have established themselves as an objective measure of the quality and performance of a garment. There are two major reasons why low-stress fabric mechanical properties are important in tailoring (Kawabata et al., 1992; Stylios and Liods, 1990; Kawabata and Niwa, 1994; Shishoo, 1989; Behera and Chand, 1997; Potluri et al., 1996). The first reason is that fabrics are more extensible in the low-load region. The property in this region is closely related to the tailoring process and the comfort of a wearer. In the tailoring process, an initially flat fabric is formed into stable, complex three-dimensional garment shapes. The conformation of a flat fabric to any three-dimensional surface requires complex mechanical deformation of the fabric such as bending, extension, longitudinal compression, and shearing in the fabric at very low loads. Garment patterns often require different lengths of fabric to be sewn together by overfeeding the longer of the two fabrics to form a seam of intermediate length as a means of imparting a three-dimensional character to the garment; the seams are inserted either in the warp or weft direction or at some angle. In this case, the shorter length is extended and the longer length is compressed longitudinally.

The second reason is that fabric extensibility at low loads causes difficulty in the handling of fabrics during the cutting and sewing processes. Fabrics having high extensibility cause dimensional distortion. Thus the fabric tensile, longitudinal compressive, and shear properties are the main mechanical properties relevant to the tailoring performance. Line production of garments and high-quality requirements are now the major issues faced in the apparel industry. Objective measurement of fabric mechanical properties is being used as an aid to the process and buying control of fabrics in the apparel industry. Fabric mechanical properties given in Table 3.1 are measured under low loads, so that conditions similar to actual fabric deformation in use are considered. The hysteresis behavior in tensile, shear, bending, and compression is measured to determine the fabric resilience or springiness. The fabric surface properties are also measured to detect the roughness by human senses.

3.4.1 Tensile properties

1. LT, the linearity of load elongation, affects fabric extensibility in the initial strain range; low values of LT give high extensibility but fabric dimensional stability is reduced.
2. RT, the tensile resilience; high values make the fabric more elastic.
3. EM, the tensile strain; larger values of EM in warp cause many problems in tailoring due to distortion of fabric during sewing; but in weft they are important for comfort in wearing and easier tailoring.

3.4.2 Shear properties

1. G, the shear rigidity; high values cause difficulty in tailoring and discomfort in wearing.
2. 2HG5, the hysteresis of shear force at higher shear angle (5°); high values give distortion in tailoring and wrinkling during wear.

3.4.3 Bending properties

Bending properties influence the formability a product of fabric bending rigidity and the longitudinal compressibility of fabric in its own plane prior to buckling. Bending rigidity is one of the basic fabric mechanical properties, and it also affects puckering and fabric cutting.

Table 3.1 Low-stress mechanical properties of fabric

Symbols	Property	Unit	Remarks
Tensile property			
LT	Linearity of load elongation curve	None	LT = 1, completely linear; LT = 0, extremely nonlinear
WT	Tensile energy	gf cm/cm²	Higher value of WT corresponds to higher extensibility. (Note: This is not a general rule, as WT must be interpreted in conjunction with WT)
RT	Tensile resilience	%	RT = 100%, completely elastic; and RT = 0%, completely inelastic
EM	Extensibility	%	Strain at maximum load (500 gf/cm)
Bending property			
B	Bending rigidity	gf cm²/cm	Bending rigidity per unit width of fabric
2HB	Hysteresis of bending moment	gf cm/cm	Hysteresis of bending moment observed in the bending moment–curvature relationship. A larger value of 2HB means a greater fabric inelasticity
Shear property			
G	Shear stiffness	gf/cm degree	Resistance to rotational movement of the warp and weft threads when subjected to low shear deformation. The lower the value of G, the more readily the fabric will conform to three-dimensional curvatures.
2HG	Hysteresis of shear force at 0.5° of shear angle	gf/cm	The higher the value of 2HG, the lesser will be the recovery from shear deformation. This will create more trouble in tailoring and formation of wrinkle during the wearing of the fabric.

Continued

Table 3.1 Continued

Symbols	Property	Unit	Remarks
2HG5	Hysteresis of shear force at 5° of shear angle	gf/cm	The higher the value of 2HG5, the lesser will be the recovery from shear deformation. This will create more trouble in tailoring and formation of wrinkle during the wearing of the fabric.
Compression property			
LC	Linearity of compression–thickness curve	None	LC = 1, completely linear; LC = 0, extremely nonlinear
WC	Compressional energy	gf cm/cm^2	Higher value of WC corresponds to higher compressibility
RC	Compressional resilience	%	RC = 100%, completely elastic; and RC = 0%, completely inelastic
Surface property			
MIU	Coefficient of friction	None	Higher value corresponds to higher friction
MMD	Mean deviation of MIU	None	Higher value corresponds to larger variation of friction
SMD	Geometrical roughness	μm	Higher value corresponds to geometrically rough surface
Weight and thickness			
W	Fabric weight	mg/cm^2	Indicates lighter or heavier fabric
T_o	Fabric thickness	mm	Thickness at a pressure of 0.5 gf/cm
T_m	Fabric thickness	mm	Thickness at a pressure of 50 gf/cm

3.4.4 Compression and surface properties

Fabric compression is measured as the degree of compressional force a fabric can sustain in certain direction before the fabric buckling occurs. The lower is the compressibility, the fabric is unable to accommodate the compressional load and the higher is the chance of seam pucker. Tension is developed during sewing by the sewing thread, which tends to relax by shortening the thread path in a stitch. This is achieved by lateral and longitudinal compression.

Fabric surface property, weight and thickness also influence cutting and sewing operations.

3.5 Control system

A tailoring control chart developed in Japan is shown in Figure 3.2 (Kawabata et al., 1992; Hari and Sundaresan, 1993). If all the properties of a fabric fall inside the "non-control" zone, the tailoring of this fabric is easy and it will not have defects in its

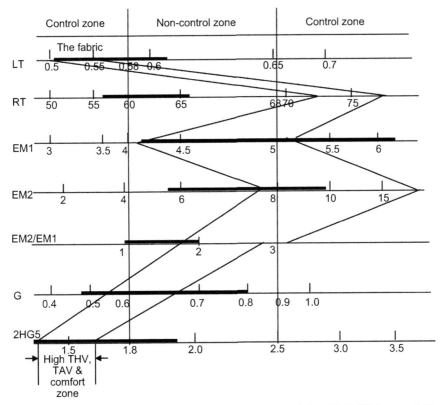

Figure 3.2 Tailoring process control chart. EM1, Warp extensibility; EM2, Weft extensibility; THV, Total hand value; TAV, Total appearance value.

appearance. However, if it was found that the properties of the fabrics having good handle are not necessarily in the "non-control" zone, the fabric might not be suitable for tailoring. This means good suiting cannot be processed in tailoring. The dots and horizontal lines show the mean value and the range of distribution of the fabric properties.

The control chart has been introduced for tailoring of higher grade fabrics, with special instructions for the tailoring process based on those mechanical properties that fall outside of the "non-control" zone.

The difference in the smoothness of the seam line comes from the mechanical properties of the constituent fabrics. The mechanical property related to this problem is mainly bending property in the weft direction. Simple criteria for the tailoring of suiting are given in Table 3.2. Another important property of fabrics for processing is fabric shrinkage caused by the steam-pressing process in the production-line system of suits.

Inspection of fabric properties before tailoring is done to determine the fabric's acceptance. The ranges of rejection values (Kawabata and Niwa, 1989; Hari and

Table 3.2 Ranges of rejection for tailoring

Mechanical property and its rejection range	Remarks
EM1 > 8	Not suitable for business suits from the shape retention. Unstable in processing
EM1 > 10	Not suitable for all types of suits
EM2 < 4	Difficulty in overfeed (overaction of overfeed)
2HG5 < 0.8	Too deformable and causes difficulty in tailoring suit, shape retention problem
2HG5 > 3.5	Wrinkle problem. Too stiff for suit, difficulty for overfeed operation
S2 > 1.5[a]	Dimensional instability problem during tailoring
S4 > 1.0[b]	Dimensional change of fabric during wearing of suit
Q^c < 30 and/or Q > 100	Difficulty in predicting fabric dimensional change after steam pressing

[a]S2 is the equilibrium offset shrinkage after the steam press in %. S2 is estimated by the shrinkage and the weight 2 h after the pressing.
[b]S4 is the fabric shrinkage offset (%), in the equilibrium state after drying in room conditions. S4 is estimated by the measurement after 24 h drying in room conditions (= relaxation shrinkage).
[c]Q is the rate of press-shrinkage recovery (%). This is estimated by the dehydration recovery, 2 h after the pressing.

Sundaresan, 1993) of fabric properties for tailoring are given in Table 3.2. Inspection looks for the following:

1. Optimum indications for the operation in case fabric extensibility is abnormal (too high or low).
2. Special techniques for preventing puckering during the sewing process.
3. Prevention of fabric distortion during spreading of the fabric in the cutting or grading process and in steam pressing.
4. Pressing pressure and amount of steam in steam pressing to minimize fabric distortion.
5. When tensile resilience and shear hysteresis are abnormal and the fabric is too elastic, control of shape is very difficult in both sewing and steam pressing; special notice is necessary.

After having the optimum indications for the operation, fabrics are sent to the suit-manufacturing factories from textile factories. Pieces of the same fabrics are sent for objective measurement and to the retail showroom. The customer selects the fabric and style of suit at the shop. His selection and statistics are sent to the factory and test house. The objective measurement data are also sent to the factory; then tailoring starts. A feature of this system is that the number of on-line control points is reduced by strict buying control of the fabric from textile mills based on mechanical properties. The cutting size of the fabric is adjusted by the fabric extensibility. Finally, properties of fabrics from textile mills are measured. The fabrics are separated into two groups: where online control is necessary and where no control is required. If all the fabric mechanical properties and shrinkage parameters fall into the central "non-control zone", no special control of the tailoring process is required. If one or more parameters falls out of the non-control zone, then control in tailoring process is required for the concerned parameter. The snake zone (the zig-zag zone in Figure 3.2) is the high-quality fabric zone in which parameters of high-quality fabric fall. These fabrics are comfortable to wear, and usually have high extensibility in the weft direction. High extensibility usually causes difficulty in tailoring.

Higher fabric extensibility in the initial region causes difficulty in the handling of fabrics during the cutting and sewing processes, for example, fabrics having high extensibility cause dimensional distortion. Thus the fabric tensile, longitudinal compression, and shear properties are the main mechanical properties relevant to tailoring performance.

3.6 Fabric tailorability, buckling and formability

3.6.1 Tailorability

Tailorability is the ability and easiness with which fabric components can qualitatively and quantitatively be sewn together to form a garment (Behera, 1999; Harlock, 1989; Lindberg et al., 1960). It mainly depends on two fabric characteristics, such as formability and sewability. Formability is a measure of the ability of a fabric to absorb compression in its own plane without buckling. Sewability refers to ease of formation of shell structure and styles, the absence of fabric distortion and seam damage, and achieving acceptable seam slippage and seam strength. Researchers use the terms "sewability" and "tailorability" interchangeably. However, tailorability covers a broader issue of garment appearance.

3.6.2 Buckling and formability

Buckling has considerable importance in a typical situation that occurs when, for example, the shoulder of a suit must be tailored by partially overlapping, curving, and sewing together two surfaces of a fabric that are initially flat. The fabric that is overfed to follow the new configuration is subjected to compression in the longitudinal direction. Fabric can be compared to a column, and buckling can quite easily occur. Sudden yielding of fabric during buckling results in the appearance of wrinkles, which affect the appearance of a tailored garment. The critical load, P_{cr} required for buckling is obtained by using Euler's formula:

$$P_{cr} = \pi E I/L^2$$

where, E is Young's modulus, I is the moment of inertia of the section, and L is the length of the column.

$$P_{cr} = KEI = KB$$

where, K is the constant and $B = E*I$

Compressibility, C of the fabric in the longitudinal direction is given by:

$$C = \varepsilon/P$$

where, ε is deformation and P is applied load. For critical load, P_{cr}

$$\varepsilon = P_{cr}C = KBC$$
$$= KF$$

where, F is formability.

Lindberg (Lindberg et al., 1960), the first to apply the theory of buckling to textile fabrics in garment technology, defined fabric formability as the degree of longitudinal compression sustainable by a fabric in a certain direction before the fabric buckles. Formability signifies the conformance of a particular shape during tailoring, and this is achieved by forcing a two-dimensional fabric to take on a simple or complex three-dimensional shape. In practice this type of compression is imposed upon the fabric by a combination of thread size, needle size, thread tension, and stitch rate; a fabric that buckles easily under these forces will form puckered seams. Formability is a direct indicator of the likelihood of seam pucker occurring either during or after sewing. It can be calculated by using values from Fabric Assurance by Simple Testing (FAST) instrument, which include FAST 2 and FAST 3 (FAST Manuals) modules. For lightweight fabrics, if $F < 0.25$ mm^2, the fabric is likely to pucker. Normally F varies between 0.25 and 0.75 mm^2. Some derived properties produced by FAST are not measured directly but are calculated using values from different FAST instruments, such as bending rigidity and shear rigidity.

3.7 Sewability

Sewability is the ability and ease with which the fabric components can be qualitatively and quantitatively seamed together to convert a two-dimensional fabric into a three-dimensional garment. The quality and serviceability of a garment depend not only on the quality of the fabric but also on the quality of its sewing. The quality of sewing is affected by the machine parameters, the sewing thread parameters, and the fabric to be sewn (Behera, 1999; Behera and Chand, 1997; Mehta, 1992; Stylios and Sotomi, 1993; Sundaresan, 1996; Chopra, 1997).

3.7.1 Quality parameters for sewability assessment

The quality of the seam depends on its strength, elasticity, durability, stability, and appearance. These characteristics can be measured by seam parameters such as seam strength, seam pucker, seam slippage, seam appearance, and seam damage. Each of these parameters is influenced by various material and machine variables, and can be quantitatively measured.

3.7.1.1 Seam strength

The strength of the seam should be equal to that of the material it joins in order to have a balanced construction that will withstand the forces encountered during use (Nayak et al., 2010). The transverse strength of a seam is determined by a number of factors, such as stitch type, stitch density, thread strength, thread tension, needle size, and needle type. It will be observed that the seam failure in a garment can occur because of:

1. Failure of the sewing threads, leaving the fabric intact.
2. Fabric ruptures, leaving the seam intact.
3. Both fabric and seam breaking at the same time.

It is desirable that the sewing thread, rather than the fabric yarns, should break, because the seam can be restitched. The durability of a seam depends largely on its strength and the elasticity of the material, and it is expressed in terms of seam efficiency:

$$\text{seam efficiency} (\%) = \frac{\text{seam tensile strength}}{\text{fabric tensile strength}} \times 100$$

It generally ranges between 85 and 90%. Seam efficiency depends on strength, count, and ply of sewing thread; it correlates well with the toughness index of sewing thread. The sewing thread loop strength correlates well with the measured strength of the seam in which thread breakage occurs. The reduction in the strength of sewing thread depends on (Chopra, 1997):

1. Shape and type of needle.
2. Fabric thread density, hardness of yarns, and number of plies to be sewn.
3. Number of passages of the thread through the needle eye and the fabric before it is incorporated into the seam.

The loss in strength due to abrasion in the sewing machine is 25% for the upper and 15% for the lower threads in single- and double-thread chain-stitch machines. During sewing, the thread is subjected to mechanical and thermal stresses, resulting in severe thread deformations. These deformations cause structural and mechanical damage to the thread and the fibers.

3.7.1.2 Seam pucker

Puckering is a disruption in the original surface area of a sewn fabric that gives a swollen and wrinkled effect along the line of the seam in an otherwise smooth fabric. Seam pucker is a differential contraction along the line of a seam and in most cases is caused by the tension from the thread of the seam or the yarns of the fabric (Chopra, 1997). The sewing thread displaces the yarns of the fabric into a new position and prevents them from recovering by the stitch bulk. A resulting crowding and compressive situation then exists around each stitch. The severity of this condition depends on the yarn and fabric density and the closeness of the weave (Chopra, 1997). Its severity is also directly related to the thickness of the thread and the number of stitches per inch. Another important cause of pucker is the high thread tension imposed during sewing. When a seam is made, the thread is stitched into the fabric under a certain amount of tension in order to form a good stitch. Pucker can actually happen either during the stitch formation or after stitching due to relaxation of the sewing thread.

During stitching, the needle along with the thread enters the space between two consecutive picks (yarns perpendicular to the seam). The extent of deformation of the weft yarns will take place depending on the spacing between these two picks and the diameter of the sewing thread.

When the needle enters the fabric, due to its impulsive action, deformation of the weft will be maximum; but when it comes out, this deformation may get reduced due to the readjustments in the fabric structure. The fabric structure plays a crucial role because the actual pucker in a fabric depends on its mechanical properties; that is, how and to what extent the deformation is caused and absorbed by the fabric structure.

Therefore, for a better understanding of the seam pucker behavior of fabrics, an understanding of the role of physical and mechanical properties of the fabrics in causing and preventing pucker is needed. The sewing thread forms the seam, so its contribution towards pucker is also required. The study of seam pucker formation by sewing threads of different tensile properties would not only help in better understanding the pucker mechanism but also guide in the selection of suitable sewing thread for fabrics having specific physical and mechanical properties. Analysis of the mechanism of seam pucker relates the cause of seam pucker to the compressive forces on the fabric generated during sewing by thread and interaction between the thread, fabric, and feed mechanism. Seam pucker depends on sewing-thread properties, stitch length and type, thread tension, sewing speed, presser foot pressure, needle size, and frictional properties of the fabric and its constituent yarns. The seam pucker can be quantified by thickness strain (Chopra, 1997; Nayak et al., 2013), as given below:

$$\text{thickness strain } (\%) = \frac{t_s - 2t}{2t} \times 100$$

where t_s is seam thickness and t is fabric thickness.

Seam pucker model

In order to visualize the development of pucker, a model (Figure 3.3(a) and (b)) is shown to understand the deformation of fabric structure due to the insertion of a seam in it. This stitching line or a seam has to find a position on and in the plane of the fabric. In order to accommodate the bulk of stitch, the fabric yarn in both directions gets displaced around this juncture. Depending on the physical and mechanical properties of the fabric, the sewing thread may have no resistance from the fabric and would manifest in different types of yarn movements (distortions) in the fabric structure; otherwise due to resistance from the fabric, only compression of the thread and fabric yarn may take place. The former causes severe pucker and the latter causes little or no pucker. It is either the yarns parallel to the seam, or the ones perpendicular to the seam, that absorb the longitudinal and lateral

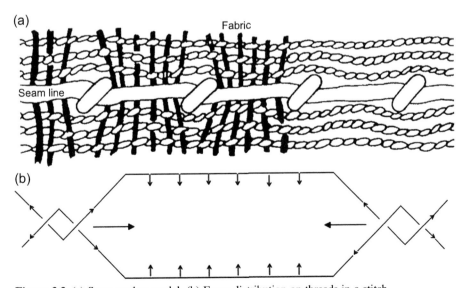

Figure 3.3 (a) Seam pucker model. (b) Force distribution on threads in a stitch.

compressive forces supposedly leading to a ripple effect in the body of the fabric. If the fabric bulk and properties are such that these forces get absorbed or diluted by the movement of the yarn around the stitch, then low pucker will result; but if the fabric properties allow the transfer of this deformation, maximum dislocation around the stitch would appear as severe pucker. The generation of forces also depends upon the tensile properties of the sewing thread. The different threads may give rise to different extents of pucker in the same fabric due to different tensions developed in them.

The physical movement of the fabric yarn may offer differential resistance depending upon the fabric properties such as cover factor, tensile extensibility, bending and shear rigidity, frictional hysteresis, thickness, and lateral compressibility. However, the extent of pucker would depend upon the amount of resistance to the yarn offered by the fabric properties and the tension developed in the sewing thread.

Process of pucker formation

The introduction of the sewing thread in the structure of the fabric triggers deformations, which may be dissipated into the bulk of the fabric or result in the development of pucker. The factors that would come into play are the physical and mechanical properties of the fabric, tensile properties of the sewing thread, and some sewing machine parameters.

Initial pucker The initial pucker is dependent upon the size of the sewing thread, the needle, and the physical and mechanical properties of the fabric. The diameter of the sewing thread and needle during stitch formation initiates dislocation of fabric yarn. If the adjacent warp and filling yarns have enough free space for the physical presence of a stitch, then no change in the fabric plane occurs and no pucker is formed; otherwise due to the impulsive action, force is generated to push these aside to make space for the needle and the sewing threads. Yarn mobility can be unidirectional or multidirectional depending upon the inter-yarn space available; the latter dilutes the impact, so low pucker is initiated; but the former—because of clustering, bunching, or overlapping— gives higher pucker. The extent of yarn mobility is determined by the properties of the fabric and the sewing thread. The yarns around the insertion may slide, rotate, or deform depending on the fabric mechanical properties.

Subsequent pucker When they enter inside a fabric, the sewing threads are under tension and a movement is caused in the yarns of the fabric. The type and extent of movement (pucker) is dependent upon the amount of tension in the sewing thread (high-modulus, low-extension thread would have higher tension compared to low-modulus, high-extension thread). The interaction between the horizontal component of the thread tension and the nature of the longitudinal fabric compression decides the extent of deformation. High thread tension and unfavorable fabric properties give very little yarn mobility, so the seam remains under tension until time-dependent stress relaxation. Therefore the initial pucker and the resultant pucker will be the same. The thread tension will be in equilibrium with the reaction from the yarns of the fabric and will decay as a function of time. High thread tension and favorable fabric properties increase initial pucker in thin fabrics due to low fabric cover. Decrease in initial pucker due to dissipation in thick fabrics (into the bulk of the fabric) occurs by yarn compression and does not cause physical yarn movement. Low thread tension and unfavorable fabric properties do not affect yarn mobility, so the initial pucker will not change. The possible yarn movements in the fabric deformations are as follows:

Yarn sliding would happen when bending rigidity is high but its frictional hysteresis is low; shear rigidity and its frictional hysteresis are high. The extent of yarn movement under these conditions will depend upon the cover factor, especially that of the filling:

1. Very low cover → no yarn bunching → very low pucker
2. Low cover → yarn bunching → high pucker
3. Moderate cover → some yarn bunching → medium pucker
4. High cover → sewing thread is embedded → low pucker

Yarn rotation takes place when bending rigidity and its frictional hysteresis is high but shear rigidity and its frictional hysteresis is low. The final pucker in this case would depend upon the extent of the cover factor, especially the filling:

1. Low cover → high pucker
2. Medium cover → medium pucker
3. High cover → low pucker

Yarn bending takes place when the bending rigidity and its frictional hysteresis are low or moderate; and shear rigidity and its frictional component are moderate. Under these conditions the actual pucker will depend on fabric cover factor; filling in particular:

1. Low cover → low pucker
2. High cover → high pucker

Sewing-thread properties and pucker

The relevant sewing thread properties affect the appearance and performance of a seam based on their interaction with the fabric properties. Thus the important thread properties from that point of view are:

Initial tensile modulus During sewing, the needle and bobbin thread, under certain tension, deform the fabric for stitch formation. High-modulus, low-extension threads develop more tension than the low-modulus, high-extension sewing threads when stitched at the same static tension. Figure 3.3(b) shows that the sewing thread exerts force perpendicular and longitudinal to the fabric for lateral and longitudinal fabric compression, respectively. Thus a high-modulus, low-extension sewing thread would have more lateral force to compress, thereby helping to reduce the extent of pucker. However, the fabric response to longitudinal compression induced by the horizontal force exerted by the sewing thread depends upon the fabric properties such as bending and shear rigidities (along with their frictional hysteresis), extensibility, and cover factor. If these properties are favorable, then their reaction to the horizontal force will determine the extent of pucker.

Diameter When the sewing thread and the needle enter the fabric for stitch formation, they cause localized deformation of the warp and filling yarns around it because actual physical space is required for two diameters of the sewing thread even after the needle pulls out. A larger thread diameter needs more space for its in-situ deposition, resulting in greater deformation and a higher level of pucker depending on the fabric cover.

Yarn diameter compressibility If the fabric properties are such that very little movement of the yarn takes place (high fabric cover) due to the insertion of the sewing thread, then the sewing thread and the adjoining yarns of the fabric are compressed. Therefore, a sewing thread that has greater compressibility will tend to cause less fabric yarn deformation and subsequently less pucker.

Fiber density Sewing threads spun from low-density fibers will have a larger diameter; polyester fiber has lower density than cotton.

In conclusion, among all the sewing-thread properties discussed above, the one that is most significant from the seam pucker point of view is initial tensile modulus and tensile extension.

Compressional behavior of fabrics and pucker

The sewing thread develops tension during stitching on the sewing machine, after which it tends to relax by shortening the thread path in a stitch. This can be achieved in two ways:

1. Lateral compression

When force is applied perpendicular to the plane of the fabric, a certain amount of fabric compression takes place, decreasing its thickness. These fabric properties are compressional energy and actual thickness compression. The extent of compression depends upon factors like the original fabric thickness, fabric construction (weave or sett), crimp, and fiber content. The sewing thread under tension compresses the fabric laterally in order to reduce the thread stretch and hence relaxes the tension developed during sewing.

If the fabric has a higher capacity to compress and it is stitched with a sewing thread that develops high tension, it would increase the compression of the fabric and smooth the pucker. Thus fabrics with a high degree of compression (usually thicker fabrics) are less vulnerable to pucker, especially when stitched with high-modulus sewing thread.

2. Longitudinal compression

The sewing thread under tension also compresses the fabric in its plane. At each stitch juncture, the longitudinal force causes in-plane compression of the yarns perpendicular to the seam, in a stitch, resulting in its deformation (bending). The resistance to movement and deformation (bending and shear) is due to frictional resistance at the cross-over points due to inter-yarn force. Therefore, fabrics respond differently depending on the fabric sett, yarn count, and the degree of setting, which depends on the finishing treatment given to the fabric. The deformation is facilitated by low values of shear and bending and higher values of extensibility. Due to crimp interchange, thickness within and around the stitch may increase. Thus the extent and the nature of longitudinal compression will depend on the fabric structure and the compressive force applied. If the fabric properties favor longitudinal compression, then during the insertion of the needle and the sewing thread, the fabric yarn (warp and filling) will be pushed aside easily and may lead to bunching or clustering of yarn, leading to an increase in pucker. This is more likely in thin, multifilament fabrics. Further, the horizontal component of the tension in the sewing thread would tend to compress them more, to further increase pucker; or its inability to compress longitudinally may either lead to an increase in pucker due to buckling, as very thin fabrics have low bending rigidity; or in thick fabrics may help in dissipating the pucker into the bulk of the fabric.

Correlation coefficients

To get a better understanding of the pucker mechanism, the correlation coefficient between fabric properties and the thickness strain values are required. By taking one fabric property at a time with thickness strain values, it is possible to find out the contribution of each fabric parameter. These fabric properties are fabric mass, thickness compression, bending rigidity, shear stiffness, extensibility, cover factor, and formability in the warp and filling direction.

It was found by Behera and Hari (2009) that most of the fabric properties gave insignificant correlation except fabric thickness. It gave fairly a high correlation coefficient (-0.751 to -0.810) for cotton and polyester sewing threads. Fabric thickness alone gave a good correlation with the thickness strain values, indicating the maximum contribution of this property. Since the other fabric properties were having an insignificant effect, though they were important for the actual deformation of the yarns within and around a stitch, it is logical to expect that these fabric properties may affect indirectly by interacting with the fabric thickness.

Correlation coefficients were calculated by combining each of other fabric properties with fabric thickness to examine their interaction; interestingly, there was a considerable increase in the correlation coefficient values. This clearly indicates that pucker formation, a very complex phenomenon, is directly as well as indirectly dependent upon various fabric properties. The decrease in initial tensile modulus of the sewing thread gave higher correlation coefficients, indicating that the interaction of sewing thread varied with fabric properties. The correlation values improved significantly when the effect of the secondary fabric properties along with fabric thickness was considered ($0.943-0.965$). These secondary fabric properties contribute indirectly through the primary fabric properties to affect pucker. Cotton fabrics even with low thickness gave low to medium pucker. The manmade filament fabrics (polyester and viscose) gave high pucker, but manmade spun fabrics gave low pucker due to more thickness and higher level of surface friction. Blends as a group gave low to medium pucker. Cotton sewing threads are more suitable for cotton and blended fabrics; polyester sewing threads are better for manmade fiber fabrics.

3.7.1.3 Seam slippage

When the seam is under some transverse strain, then displacement of the stitch relative to one or more of the fabrics can occur. This is called seam slippage. The stitch displacement produces some displacement of one yarn system in the fabric against the other, which causes opening in the fabric. This phenomenon is an adverse feature of some woven fabrics, and it decreases the range of possible end-uses and causes problems in the garment.

The amount of seam slippage or the fabric resistance to seam slippage depends on:

1. Yarn-to-yarn friction
2. Contact angle between threads
3. Stitch density
4. Yarn flexural rigidity

An increase in the values of these factors increases the fabric resistance to seam slippage. Seam slippage depends on weave, fabric raw material, type of seam, stitch density, and sewing thread tension.

Understanding seam slippage mechanisms

When two fabrics are sewn together parallel to a yarn system and the seam is open by applying some tension transverse to the seam axis, this tension imposes strain on the seam. The tension in the sewing thread in the seam depends on thread tension during sewing, fabric tension, and stitch density. The tension applied to the fabric through the seam creates a pushing force on the threads transverse to the seam axis. This causes deformation of stitch geometry and displacement of threads along the seam.

Seam slippage is basically a consequence of tensile elongation. Seams in the garments are constantly being subjected to stresses in various directions, among which those perpendicular to the seams are most common. When two pieces of woven fabric are joined by a seam and an increasing force is applied to the assembly at right angles to the seam line, rupture ultimately occurs at or near the seam line and at a load usually less than that required to break the unsewn fabric (Chopra, 1997). The set of yarns being pulled slip out of the seam assembly, overcoming the frictional resistance due to inter-yarn forces. In addition, the extent of slippage is determined by fabric mechanical properties and the tensile behavior of the sewing thread. The severity of seam slippage is generally observed in tight-fitted women's garments made from either lightweight or medium-weight fabrics. By and large these smooth fabrics are made from manmade filaments. The stretch on the seams in the perpendicular direction leads to a change in the basic structure of the fabric due to the extensions of yarns in the direction of the pull. This extension in one direction would most certainly lead to changes in the fabric geometry on the opposite side. How much change takes place, and with what consequences, would depend upon the physical and mechanical properties of the fabric. The presence of a stitch line (seam) acts like a barrier in the load elongation behavior of a seamed fabric.

When fabrics are stretched or load is applied, some changes are expected in the structure of the fabric, which are dependent not only upon the tensile properties of the fabric but also upon other mechanical properties like bending, shear (along with frictional hysteresis), and the surface coefficient of friction and roughness. In addition, the physical properties of the fabrics are also important to facilitate the mechanical properties to interact effectively.

As a load is applied to stretch the fabric, at first the crimp is straightened, and then the yarn extends and overcomes the inter-yarn friction. The higher cover factor in the cross-direction gives a greater number of cross-over points. Therefore, yarn slippage of such a fabric will be more difficult. Another important consideration is that decrease in crimp in the direction of the pull increase the crimp in the opposite direction. Therefore, fabric elongation becomes more difficult due to an increase in the frictional resistance resulted by an increase in the contact area between yarns.

When force is applied perpendicular to a seam joining two pieces of fabric, some of the above movements would take place but with a difference. The stitching line acts like a clamp, which is flexible. Its flexibility would therefore determine the load elongation behavior of a seamed fabric.

Process of seam slippage

When a tensile load is applied to a fabric seam, it has to overcome two types of frictional forces. One is the inter-yarn frictional forces within a fabric, and the other is the frictional force of the stitch assembly. The former is dependent upon the crimp, yarn diameter, fiber content, and number of cross-over points. The latter, however, is dependent upon the fabric properties like fiber content, type of yarn (spun or filament), thickness, lateral compression, cover factor (threads per cm), bending, shear, tensile and surface roughness, and coefficient of friction. It is also dependent upon the properties of the sewing thread like fiber content, diameter, coefficient of friction, initial modulus, and extensibility. All of these properties, together with the machine variables like needle and bobbin thread tension and the stitch length, make up the frictional force of the stitch assembly. Thus different combinations of these would be expected to provide different frictional resistance and hence different loads at which seam slippage may take place.

A stitch assembly acts like a clamp for the fabric as long as the pulling force is less than the frictional resistance of the stitch assembly; the strain is then shared by the fabric and the stitch assembly. But as soon as the pulling force exceeds the frictional resistance of the stitch assembly, the fabric yarn slips and the load is transferred indirectly through the cross-yarn to the stitch assembly. Thereafter, depending upon the fabric characteristics, the yarns slip out; or if that is not possible, then the force gets shared by the fabric yarn and the sewing thread. A more extensible sewing thread may break at a higher load in comparison to a less extensible one. The correlation of the resistance to seam slippage per thread (RSST) value with each fabric property is calculated individually, so that the interaction or influence of each fabric property or variable (primary properties) on the seam slippage behavior of the fabric is known.

3.8 Conclusions

The quality of fabric influences not only the quality of the garment but also the ease with which a shell structure out of flat fabric can be produced, which is indicated by sewability. During the sewing of a fabric, good sewability implies that the constituent yarns in the fabric are separated by the needle going through them without producing fabric damage. When thread densities are high, yarns are either broken or separated by the needle as a result of the pressure against the neighboring yarns. The fabric specifications for different end-use requirements are different, and the selection of an appropriate fabric is one of the most difficult jobs for the clothing manufacturer. The specifications of fabrics for apparel manufacturing can be considered in terms of primary and secondary characteristics. The primary characteristics are static physical dimensions whereas the secondary characteristics are the reactions of the fabric to the applied forces. With the advent of high-speed automatic line production systems, the interrelationship between fabric mechanical properties and fabric processability in tailoring has become very important. The selection of manufacturing processes requires selection of machine and process variables based on the specific properties of the fabric being processed.

References

Behera, B.K., Hari, P.K., 2009. Woven Textile Structure — Theory and Applications. Published by Woodhead, UK.
Behera, B.K., 1999. Testing and Quality Management, vol. 1. IAFL Publication, pp. 386—402.
Behera, B.K., Hari, P.K., September 1990. Indian J. Fibre Text. Res. 19, 168.
Behera, B.K., Chand, S., 1997. Int. J. Clothing Sci. Technol. 9 (2/3), 128.
Chopra, K., 1997. Seam Behaviour of Woven Fabrics (Ph.D. thesis). IIT Delhi, India.
FAST Manuals, Published by CSIRO Research Centre, Australia.
Harlock, S.C., 1989. Fabric objective measurement: 4. Production control in apparel manufacture. Text. Asia 20 (7), 89.
Hari, P.K., Sundaresan, G., 1993. Garment quality: interrelationship with fabric properties. In: GARTEX-NIFT'93 Conference, New Delhi, October 16—17, pp. F1—F7.
Kawabata, S., Niwa, M., 1989. J. Text. Inst. 8 (1), 19.
Kawabata, S., Niwa, M., Ito, K., 1992. Tailoring process control. J. Text. Inst. 83 (3), 361—373.
Kawabata, S., Niwa, M., 1994. Int. J. Clothing Sci. Technol. 6 (5).
Lindberg, J., Waertraberg, L., Svenson, R., 1960. J. Text. Inst 51, T1475.
Mori, M., 1994. Int. J. Clothing Sci. Technol. 16 (2/3), 7.
Mehta, P.V., 1992. An Introduction to Quality for Apparel Industry. ASQC Quality Press, Wisconsin.
Nayak, R., Padhye, R., Dhamija, S., Kumar, V., 2013. Sewability of air-jet textured sewing threads in denim. J. Text. Apparel Technol. Manage. 8 (1), 1—11.
Nayak, R., Padhye, R., Gon, D.P., 2010. Sewing performance of stretch denim. J. Text. Apparel Technol. Manage. 6 (3), 1—11.
Nayak, R.K., Punj, S.K., Chatterjee, K.N., Behera, B.K., 2009. Comfort properties of suiting fabrics. Indian J. Fibre Text. Res. 34, 122—128.
Postle, R., 1990. Fabric objective measurement technology. Int. J. Clothing Sci. Technol. 2 (7), 7—17. Nos. 3—4.
Potluri, P., Porat, I., Atkinson, J., 1996. Int. J. Clothing Sci. Technol. 8 (12), 12.
Stylios, G., Liods, D.W., 1990. Int. J. Clothing Sci. Technol. 6 (4), 6.
Shishoo, R.L., February 1989. Text. Asia 20 (2), 64.
Stylios, G., Sotomi, J.O., 1993. J. Text. Inst. 84 (4), 601.
Sundaresan, G., 1996. Studies on the Performance of Sewing Threads during High Speed Sewing in an Industrial: Ockstitch Machine (Ph.D. thesis). IIT Delhi, India.

Production planning in the apparel industry

S. Das[1], A. Patnaik[1,2]
[1]Faculty of Science, Nelson Mandela Metropolitan University, Port Elizabeth, South Africa;
[2]CSIR Materials Science and Manufacturing, Polymers and Composites Competence Area, Nonwovens and Composite Group, Port Elizabeth, South Africa

4.1 Introduction

Clothing is the quintessential global industry in which the world's largest retailers, branded marketers, and manufacturers without factories are the dominant players. These lead firms set up networks by shifting their production to contractors and subcontractors in a variety of exporting countries, which allow them to concentrate on the design for the next generation of clothing, and more importantly on branding, marketing, and advertising.

The apparel and textile industry sector is always under constant pressure and where competition is fierce, there is a chance of rival firms waiting to challenge them. During the heydays of garment production, manufacturers would name the price of a product including their cost price in making the product plus profit. They would offer a range of garments to the retailers. After the selection was made, the retailer would place an order for a particular product and ask for a bulk delivery on a fixed date, months later.

But the present scenario has completely changed. Now the retailers rule the market and drive the garment supply chain. Because they know exactly what they want in terms of the product, its design, color, pattern, etc., they now have a say on the price of the product, keeping in mind the consumer's needs and expectations. From these issues it was found that most manufacturers and suppliers could not meet the demands of the retailer, which gave rise to dealing with production issues such as lead times, responsiveness, costing factor, improper planning, etc. that could be affecting the timely delivery of the product to the ever-changing market place.

In the new world of ever-developing technology and techniques, organizations must consider how to deal with the issues of increased competition, rising customer expectations, and increase in product variety. This chapter will discuss about various production systems, production planning in the apparel industry, supply chain and inventory management and various tools to improve the productivity in the apparel industry.

4.2 Production planning

Production planning and control (PPC) are important aspects of the garment manufacturing industry. Precision in planning equates to on-time shipments, the best use of labor, and assurances that appropriate supplies and equipment are available for each order (Ray, 2014). Production planning involves everything from scheduling each task in the process to execution and delivery of products. In most cases the production of garments is very time sensitive in order to ship goods to stores and boutiques for the upcoming season. In the context of apparel manufacturing, the primary roles of the PPC department are listed below. Each function is explained briefly, just as an overview of the task (Sarkar, 2011).

Job or task scheduling: This requires preparation of time and an action calendar for each order from order receiving to shipment. The job schedule contains a list of tasks to be processed for the styles. Each task planner notes when to start a task and the deadline for that task. The name of the responsible person (department) for the job is listed, for example, scheduling planned cut date and line loading date.

Material resource planning (inventory): Preparation of a materials requirement sheet according to sample product and buyer specification sheet is necessary. Consumption of material (e.g. fabric, thread, button, and twill tape) is calculated and cost of each material is estimated.

Patterns and markers: Pattern making, grading, and markers are a crucial part in planning for production. Once markers of each style are finalized, one can easily calculate the yield of fabric needed for production and in turn can order the fabric. When planning for the production schedule one must speak to the fabric supplier about the turnaround of the fabric and any additional time needed for dyeing or washing the fabric (www.ehow.com).

Facility location: Where a company has multiple factories (facilities) for production and factories are set for a specific product, the planner must identify which facility will be most suitable for new orders. Sometimes there may be a capacity shortage in a factory; in that case the planner must decide which facility will be selected for those orders.

Estimating quantity and costs of production: The planner estimates daily production (units) according to the styles' work content. With the estimated production figure, production runs, and manpower involved, the planner also estimates production cost per piece.

Capacity planning: The PPC departments play a major role during order booking. They decide (suggest) how much order they should accept according to their production capacity, that is, allocating of total capacity or deciding how much capacity to be used for an order out of the total factory capacity.

Line planning: A detailed line planning with daily production targets for the production line is prepared. In most cases line planning is made after discussions with the production team and industrial engineers.

Cost control: Garment manufacturers cannot afford to lose time or materials in the production process. Raw material prices rise consistently, and poor planning can lead to missed opportunities and higher costs. With styles changing rapidly and vendors making increasingly smaller orders to keep up with changing trends, the planning phase of each production piece must be as accurate as possible.

Reduce loss: About 60–70% of the cost of a garment is in the fabric. As such, it is vital that one orders appropriately and tracks the cutting room processes to keep errors to a minimum. Effective production planning relies on the ability to order the appropriate amount of fabric for a run and to realize no more than a 2–3% rate of error in cutting. That means that the number of garment components produced should equal 97–98% of the garment components cut (Ray, 2014).

Deliver timely shipments: Early delivery can be as harmful to the company's future as late delivery, because buyers then must accommodate early deliveries with additional storage capacity. Ideally, the planning should allow for exact delivery when the customer demands. At the same time, there must be sufficient labor and raw material delivery in the exact proportions to meet the deadlines without having to pay extra for overtime.

Follow-up daily: Once the plans are set for a garment run and delivery deadlines, one must institute strict follow-up procedures to ensure that the plan is followed correctly. If the cutting room, for example, falls behind in its production schedule, then the sewing and finishing lines must wait, leading to backlogs and missed delivery deadlines for shipment. To avoid this, one must have a daily oversight of each step and keep the rest of the line updated with any delays so that plans can be adjusted to pick up the slack.

4.3 Production systems

Production system is the framework within which the production activities take place. A production system comprises attributes with the function to transform inputs into desired and predicted outputs (Jacobs and Chase, 2013). The attributes can be human labor, machines, or tools. For the apparel industry, the production system is defined as "an integration of material handling, production processes, personnel and equipment" (Vijayalakshmi, 2009; Kincade et al., 2013). In the apparel industry, the most basic apparel production system is the whole garment system (Kincade et al., 2013). This system involves one operator who sews all cut pieces into the final apparel product (Babu, 2006; Solinger, 1988; Kincade et al., 2013). Commonly this is now followed by traditional tailors and haute couture seamstresses. This system is labor intensive, low in productivity, and rarely seen in modern mass production facilities. Designing production system ensures the coordination of various production activities. There is no particular production system that is universally accepted, yet there are different types of production systems followed by different organizations as discussed in the following section.

4.3.1 Types of production systems

The different types of production systems are distinct and require different conditions for working. However, they should meet the two basic objectives, that is, to meet the specification of the final product and to be cost-effective in nature. The main aim of any production system is to achieve a minimum possible total production time. This automatically reduces in-process inventory and its cost. The subassembly system reduces temporary storage time to zero by combining temporary storage time with transportation time.

Any production system has four primary factors that make up the system. Processing time + Transportation time + Temporary storage time + Inspection time = Total Production Time (Babu, 2006). Processing time is the sum total of working time of all operations involved in manufacturing a garment. Transportation time involves the time taken to transport semifinished or finished garments from one department to another or from one operation/machine to another. Temporary storage time is the time during which the garment/bundle is idle as it waits for the next operation or for completion of certain parts. Inspection time is time taken for inspecting semifinished garments for any defects during manufacturing or inspecting fully finished garments before packing.

In the apparel industry, four types of production systems are commonly used: bundle system, progressive bundle system (PBS), unit production system (UPS), and modular system (Oliver et al., 1994; Vijayalakshmi, 2009; Kincade et al., 2013; Bailey, 1993; Kanakadurga, 1994). The bundle system and PBS are categorized into mass production and UPS and modular systems are categorized into flexible specialization (Bailey, 1993). Kanakadurga (1994) used five attributes of the production systems to classify them in the apparel industry into three categories (i.e., bundle system, PBS, and modular system). Those five attributes are: workflow, method of retrieval between workstations, work in progress (WIP) inventory, number of tasks per operator, and interaction between workers. The study found that one production system could be distinguished from another according to these attributes.

4.3.1.1 Bundle system and PBS

Most apparel manufacturers believe that the best way to be efficient and achieve economy of scale is by mass production. The bundle system enables manufacturers to gain economies of scale. The bundle system, often referred to as the traditional garment production system, has been widely used by garment manufacturers for several decades. The bundle system is a dedicated system comprising cut parts, tied into bundles, to complete one or more sections of an apparel product (Oliver et al., 1994; Vijayalakshmi, 2009; Kincade et al., 2013). The American Apparel Manufacturers Associations (AAMA) Technical Advisory Committee (AAMA, 1993) reported that 80% of the apparel manufacturers use the bundle system of garment production. The bundle system of apparel production consists of garment parts needed to complete a specific operation or garment component. Bundles of garments are assembled in the cutting room where cut parts are matched up with corresponding parts and bundle tickets. The bundles are moved to the sewing room and given to the garment operator who is scheduled to complete the operation.

PBS is a variation to the bundle system. The main characteristics of the systems are one worker with a single skill at a single operation, no interaction between operations, piece rate compensation, maximization of productivity of individual operators, need for extra spaces for material storage, straight lined equipment layout, and manual material movement in large batches (Bailey, 1993; Glock and Kunz, 1995). The system helps operators to make more units of garments at faster speeds, with high productivity but with more quality problems. To stabilize the workflow, higher levels of WIP add to production costs, not value, and lead to longer lead times. In these systems, the difference between actual run time and total manufacturing throughput time is significant (Oliver et al., 1994). The success of PBS depends on how the production system is set up and used in a plant.

Advantages of the PBS of garment production (Babu, 2006)
- This production system may allow better utilization of specialized garment production machines, as output from one special purpose automated garment machine may be able to supply several garment machine operators for the next operation.
- Small garment bundles will allow faster throughput until and unless there are any bottlenecks or long lead times within operations.

Disadvantages of PBS of garment production (Lee, 2000; Babu, 2006)
- The PBS is cost-efficient for individual garment operations. As the garment operators perform the operation on a daily basis it allows them to increase their speed and productivity, and as a result they are not willing to learn a new garment operation because it reduces their efficiency and earnings.
- Operators working in a PBS of garment production are independent of other operators and the final product.
- Absenteeism, machine failure, and slow processing are also some major bottlenecks in this system.
- WIP is seen in large quantities in this production system, which leads to longer lead times.
- Poor quality of garments are seen in bundles, which were hidden because of large inventories in the work process.
- Leads to extra handling and difficulty in controlling inventory in the garment industry.
- Line balancing becomes difficult most of the time.
- Proper planning is requires for each batch and for each style, which takes a lot of time.
- Improper planning leads to poor labor turnover, poor quality and less production.
- Variety of styles of lesser quantity are not effective in this system.

4.3.1.2 Unit production system

The UPS is a response to competitive pressure because of consumer demands and increasing global competition. This system requires an automated overhead transporter system to move individual units from one operation to another instead of human handling of materials in the bundle system. The resulting WIP and manufacturing throughput time of the UPS are greatly reduced compared to those of the bundle system (Bailey, 1993; Glock and Kunz, 1995). Although it is seen that UPS can solve problems relating to inventory, material handling, and quality, it is basically adopting the characteristic of a PBS with assembly line, individual piece rate compensation, and increased supervisor monitoring capabilities. Most manufacturers also argue that UPS is an efficient form of a mass production system and not suitable for a product line that needs style variations. Some authors discuss this as a mechanical system or a method of work retrieval (Babu, 2006; Kincade and Gibson, 2010; Kincade et al., 2013), while other sources propose this system as a fourth production system (Textile/Clothing Technology Corp (TC^2), 2005; Vijayalakshmi, 2009; Kincade et al., 2013). UPS requires substantial investments, which are not always justified by conventional payback calculations. Most importantly, they provide a clothing factory with the capability to respond quickly to any changes that might occur (www.garmentmerchandising.com). In the fast moving fashion and apparel industry this is highly essential.

Advantages of the UPS of garment production (Babu, 2006)
- Bundle handling system is completely eliminated.
- Time involved in the pick up and drop is reduced to a minimum.
- Output is automatically recorded, eliminates the operator to register the work.
- Computerized systems in this UPS automatically balance the work between two work stations.
- Up to 40 styles can be produced simultaneously on one system (www.garmentmerchandising.com).

Disadvantages of the UPS of garment production (Babu, 2006)
- UPS requires high investments where the payback period is long.
- Planning needs to be effective every time.
- Sometimes the orderly and controlled flow of work via computerized control of the whole production process becomes difficult to measure.

Comparison between the PBS and the UPS
In garment manufacturing companies various types of sewing systems are installed. The owner of an apparel manufacturing unit buys these systems depending on the production volume, product variety, and the cost involved in buying and maintaining these machines. Among them the PBS is the mostly installed sewing system to date (Sudarshan and Rao, 2013). In this system, the bundles of cut pieces are moved manually on the feeding line by the helpers. The operators drag the bundles on their own, finish their required operation, and move forward to the next operator. With advancement in technology, mechanical material transport systems are bought in. An overhead transport system, UPS, is the one that transports cut pieces hanged in hangers (one hanger for one piece) by an automated system. It reduces manual transport and handling and has many other benefits over the PBS.

4.3.1.3 Modular system

The modular system of production is basically obtained from the Japanese auto manufacturing and is regarded as one way to meet the flexibility in demands required in the apparel industry. The modular system, although dates back to the 1980s, is one of the newest production systems in the apparel industry. This system is also called a team or cellular system (Kincade et al., 2013). Pressure from increased market segmentation and the need for shortened production development cycles and greater operating flexibility, plus intense competition in standardized product lines from low-wage countries, have forced apparel manufacturers to experiment with this new production system. Modules use a cross-training technique, which requires multiskilled workers and a small set of machines to produce a finished garment. The key principle here is operator/worker involvement and team work, which requires important changes in the industry's human resource (HR) department. In this system the workers are responsible for quality. The main features of the modular system are groups of workers with multiple skills in one module, group piece rate or hourly rate compensation, U-shaped module, and single piece hand offs; this system is called Group Technology or cellular manufacturing (Nahmias, 1997). Less material handling drastically reduces inventory levels and throughput time, which saves the cost of inventory and material handling.

Berg et al. (1996) studied the differences in performance between two widely used production systems in the apparel industry (i.e., bundle and modular systems). They interviewed managers from different apparel companies and surveyed workers to understand their reaction towards the use of production systems. The result revealed that the modular system is the most beneficial to manufacturers in every aspect.

Advantages of the modular system of garment production (Sudarshan and Rao, 2013; Bailey, 1993)

- Modular system leads to cost savings.
- There is reduction in throughput time.
- It helps in improving a company's ability to provide exactly what retailers want in colors, sizes, and styles in a timely manner.

Disadvantages of the modular system of garment production (Sudarshan and Rao, 2013)

- Costs are involved in training workers to deal with several tasks at one station.
- Number of machines is higher than that of mass production.
- This system requires a higher degree of communication and cooperation among operators, mechanics, and supervisors.
- High turnover rate typical in the apparel industry also hinders the cooperative teamwork approach of the modular system (Sudarshan and Rao, 2013; Bailey, 1993).

Among these systems, PBS is reported to be the most commonly used system in the apparel industry (Berg et al., 1996; Glock and Kunz, 1995; Kanakadurga, 1994). Various studies carried out in different apparel industries on product line characteristics and sewing systems involved to calculate productivity have shown that most of them use PBS. Bundle system has been suggested for a company whose strategic focus is on meeting consumer needs. These studies even concluded that the volume and frequency of the product line are directly related to the type of sewing system used. Studies have also revealed that only few manufacturers use computer technology for determining the quantity of fabric, fabric description, purchase order preparation, and monitoring the purchase order while doing business with their vendors (Berg et al., 1996; Glock and Kunz, 1995; Kanakadurga, 1994). Ko and Kincade (1998), in their study on the relationship between product line characteristics and quick response (QR) implementation with more than 100 manufacturers, found that apparel manufacturers of fashion products adopted high levels of QR technologies. From these previous studies, product line characteristics are known to be related closely to the managerial decisions in the apparel industry, although the exact relationship is not clear. Therefore, product line characteristics must be considered as an important characteristic of the apparel industry that influences the effect of an innovation such as supply chain management.

4.4 Production planning and control in the apparel industry

To cope with the short lead time and small but frequent orders, apparel manufacturers strive to improve their production processes in order to deliver finished products within the expected time frame at the lowest production cost. Production planning is therefore gaining importance in contemporary apparel manufacturing.

Production planning can be defined as the technique of foreseeing every step in a long series of separate operations, each step to be taken at the right time and in the right place and each operation to be performed with maximum efficiency (Kumar, 2008). It in a way helps the organization to work out the quantity of material, manpower, machine, and money required for producing a predetermined level of output in a specified period of time. An outline of an apparel production cycle is shown in Figure 4.1 (Mok et al., 2013).

Managing production in an organization mainly involves planning, organizing, directing, and controlling production activities. It deals with converting raw materials into finished goods along with proper decision making regarding the quality, quantity, cost, etc. involved in it.

The basic planning process in clothing manufacture includes:

- Receiving the order.
- Proper planning to check if there is sufficient plant capacity is available to achieve the delivery date specified.
- Checking availability of cut parts and panels in the nonsewing areas (cut embroidery if any, print, wash, and pack).
- Checking if there is ample time to order and receive fabric, trims, approve sample, lab testing, etc.
- Confirmation of the delivery date to the customer.
- Proper communication between departments for smooth flow of the process.
- Monitoring progress against plan.
- Replan if required.

However, skills and efficiency vary for different teams and absenteeism is also always taken into account. For instance, if a team normally makes woven garments, a change to knitted T-shirts means that they are less skilled at handling that item. Automatically there is a fall in the rate of production, which impacts planning.

The clothing industry is still very much dependent on human labor, despite increasing usage of automatic machines and processes. Now, with the ever-changing

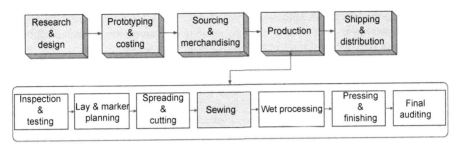

Figure 4.1 Outline of apparel production cycle.
Source: Mok et al. (2013). With kind permission from Springer Science and Business Media.

fashion market, which demands constant style changes, it becomes a challenge in achieving efficiencies and optimizing operator skills. So the trick here is to have the same team of operators working on a similar type of product for as long as possible, hence minimizing production loss. The planning must also consider special areas where loading might impact resources. Also, there should be planning for postsewing operations, otherwise it leads to WIP buildup in other areas. Following critical path analysis is a vital tool in this process.

Most business systems offer some capacity planning, while many of them are not graphical, complex to use, and not user-friendly. Using spreadsheets has its own drawbacks: lack of transparency, not visual, cannot be shared on a network, which limits coordination, and is cumbersome and difficult to manage with large orders.

To overcome these hindrances companies can change (Collins and Glendinning, 2004):

- From flow line production to production in cells, where there is more team work and more skilled labor, which saves time.
- Just-in-time (JIT) approach to deal with the inventory.
- Strategically having partners to gain short lead times.
- Service orientation rather than product orientation.
- Emphasis on accuracy of output than volume of output.
- Long-term capacity planning.
- Short-term detailed planning, that is, planning of cutting room and sewing room activities.
- Inventory control—raw material purchasing, finished goods, etc.
- Critical path control.
- The management must allocate HRs in the form of Planning Executive (PE), Merchandising Executive, Factory Planners, and Planners so that planning and control activities are well carried out. The PE will liaise with customers and merchandising executives to establish requirements for this season and next; agree to delivery schedules as planned; respond to the changes that must be made on the basis of consumer demand; allocating garments to appropriate factories to achieve customer requirements and control of critical path (Collins and Glendinning, 2004).
- The factory planner takes information from the PE and works on them in the factory, such as scheduling and sequencing work lines with delivery dates; working closely with the production department at the factory; liaising with fabric suppliers for procurement of fabric.

The planner does the detailed planning of work for each line as per the customer information supplied; purchasing of fabric and trimmings; loading the cutting room in time; production control, that is, to ensure that quantities ordered match with the cut quantities.
- Before a garment can go onto the production line, there is a huge amount of preproduction activity that must take place. Sometimes it is done as specified by the customer or on a general basis. Activities are allocated to individuals who can cross-examine them before sending them to the sewing floor, for example, label information, wearer trials, quality seals, and fabric approval. By doing this it becomes easy for the planning department to track where the product is on the planning board and where it needs to be pushed forward or backward, and hence maintain a critical path to attain the target. If by any case the production is delayed by some weeks or days, the problem is highlighted on the planning board. So the PE must postpone the start date and try to expedite the preproduction process (Collins and Glendinning, 2004).
- Therefore, only the fittest and leanest organizations survive in the race for manufacturing quality apparel in the desired time by working efficiently, cost-effectively, responsively, and flexibly.

4.4.1 Production control

Production means total number of garment pieces produced by operators in a line or batch in a given time (for example, 8 h day). Production is also termed as daily output. To estimate production the following information is necessary (Sarkar, 2011).

1. Standard allowed minutes (SAM) of the garment. It means how much time is required to make one complete garment including allowances.
2. How many operators are working in each line?
3. How many hours each line will work in a day?
4. Average line efficiency level.
5. Total break time.

Formula used for production estimation:

Daily production = Total man minutes available in a day/SAM × average line efficiency.

Total available man minutes = Total No. of operators × working hours in a day × 60.

Suppose, SAM of a particular garment is 20 min, with 30 operators line, which works 8 h shift in a day. Line works at an average efficiency of 50%. Operators are given a total 45 min for lunch and tea break.

So, total available man minutes = 30 × (8 × 60 − 45) = 13,050 min.

Daily estimated production = 13,050/20 × 50% = 326 pieces.

Above output from that line is achievable if everything goes well. It can be seen that the production of a line is directly proportional to the line efficiency, number of operators, and working hours. Production is inversely proportional to the garment SAM. If efficiency of a line increases higher production can be achieved. Similarly, if the SAM of a style is reduced, then higher output can also be achieved.

Any one of the following factors can reduce production of the assembly line. For an estimated output, the following points should be considered.

1. Machine breakdown
2. Imbalance line (WIP control)
3. Continuous feeding to the line
4. Quality problems
5. Individual operator performance level
6. Operator absenteeism

SAM is used to measure the task or work content of a garment. This term is widely used by industrial engineers and production people in the garment manufacturing industry. For the estimation of cost of making a garment, SAM value plays a very important role. In the past, scientists and apparel technicians did research on how much time to allow to do a job when one follows a standard method when doing a job. According to research, standard minute value has been defined for each movement needed to accomplish a job. Synthetic data are available for each movement, while performing specific steps in a garment manufacturing.

For example, general sewing data (GSD) have a defined set of codes for motion data for SAM calculation. There are also other methods through which one can calculate SAM of a garment without using synthetic data or GSD. Both methods are explained in the following section.

Method 1: Calculation of SAM Using Synthetic Data

In this method "Predetermined Time Standard" (PTS) code are used to establish "Standard Time" of a garment or other sewing products.

Step 1: Select one operation for which one wants to calculate SAM.
Step 2: Study the motions of that operation. Standing by the side of an operator (experienced one) one can see the operator performing the operation and hence noting down all the movements used by the operator in doing one complete cycle of work.
Step 3: List all motions sequentially. Refer to the synthetic data for time measuring unit (TMU) values. For synthetic data one can refer to GSD (without license use of GSD code prohibited but for personal use and study one can refer to GSD code and TMU values) or sewing performance data (SPD) table. Now one can obtain a TMU value for one operation (for example, say it is 400 TMU). Convert total TMU into minutes (1 TMU = 0.0006 min). This is called as basic time in minutes. In this example it is 0.24 min (i.e. 400×0.0006 min).
Step 4: SAM = (basic minute + bundle allowances + machine and personal allowances). Add bundle allowances (10%) and machine and personal allowances (20%) to basic time. Now one can obtain standard minute value (SMV) or SAM. SAM = $(0.24 + 0.024 + 0.048) = 0.312$ min.

Method 2: Calculation of SAM through time study

Step 1: Select one operation for which one can calculate SAM.
Step 2: Taking one stop watch stand by side of the operator. Capture cycle time for that operation (cycle time—total time taken to do all work needed to complete one operation, that is,

time from pick up part of first piece to next pick up of the next piece). Time study can be done for five consecutive cycles. Discard if abnormal time is found in any cycle. Calculate average of the five cycles. Time obtained from time study is called cycle time. To convert this cycle time into basic time one might need to multiply cycle time with operator performance rating (Basic time = cycle time × performance rating).

Step 3: Performance rating—Now one can rate the operator at what performance level he was doing the job seeing his movement and work speed. Suppose that operator performance rating is 80%. Suppose cycle time is 0.60 min. Basic time = (0.60 × 80%) = 0.48 min.

Step 4: SAM = (basic minute + bundle allowances + machine and personal allowances). Add bundle allowances (10%) and machine and personal allowances (20%) to basic time. This gives the SMV or SAM. SAM = (0.48 + 0.048 + 0.096) = 0.624 min.

Standard allowed hours

The concept is most commonly used in manufacturing operations where a large number of units are to be produced, and attaining a profit requires close attention to the number of hours of production (www.accountingtools.com).

The concept of standard hours allowed is usually based on a reasonable estimate of hours required to produce a product (sometimes called an *attainable standard*). However, some organizations prefer to use theoretical standards, which are only attainable under perfect conditions where there is no scrap, no setup inefficiency, no breaks, and no rework. If a company is using theoretical standards, the calculated amount of standard hours allowed will be reduced, which means that there is likely to be an unfavorable variance between that number and the actual number of hours allowed.

As an example of standard allowed hours: ABC International produces 500 green widgets during April. The labor routing states that each unit should require 1.5 h of labor to produce. Therefore, the standard hours allowed is 750 h, which is calculated as 500 units multiplied by 1.5 h per unit.

Production control is the process of planning production in advance of operations, establishing the exact route of each individual item part or assembly, setting, starting, and finishing for each important item, assembly or the finishing production, and releasing the necessary orders as well as initiating the necessary follow-up to have smooth functioning of the enterprise (Kumar, 2008).

4.4.1.1 Dispatching (Kumar, 2008)

Dispatching involves issue of production orders for starting the operations. Conformation is given for:

- Movement of materials to different workstations.
- Movement of tools and fixtures necessary for each operation.
- Beginning of work on each operation.
- Recording of time and cost involved in each operation.
- Movement of work from one position to another in accordance with the route sheet.
- Inspection or supervision of work.

Dispatching is an important step as it translates production plans into production.

4.4.1.2 Follow-up

Follow-up is done to check if there are any bottlenecks in the flow of work and to ensure that the production operations are occurring in accordance with plans. It spots delays or deviations from the production plans. It also helps to reveal defects in routing and scheduling, underloading or overloading of work, etc. All remedial measures are taken to ensure that the work is completed by the planned date.

4.4.1.3 Inspection

This is mainly done to ensure the quality of the product, which is discussed in Chapter 16.

4.4.1.4 Corrective measures

Corrective measures may involve any of those activities of adjusting the route, rescheduling of work, repairs and maintenance of machinery or equipment, and control over inventories. Alternative methods are suggested to handle peak loads.

4.4.1.5 Production strategies in the garment industry

The identified production strategies in the garment industry are (Sudarshan and Rao, 2013):

- Flexible manufacturing strategy
- Value-added manufacturing strategy
- Mass customization.

4.4.1.6 Flexible manufacturing strategy

This strategy is responsive to consumer demand for small orders and short lead times. Flexible manufacturing strategy means the capability of quickly and efficiently producing a variety of styles in small production runs with no defects. The manufacturing firm adopting this strategy will be able to operate with the flexibility needed to meet the demands of its consumers and the ability to adapt to immediate changes in the apparel market (Sudarshan and Rao, 2013).

4.4.1.7 Value-added manufacturing strategy

This strategy is a quick response strategy that focuses on the handling of unnecessary operations that do not add value to the product but rather leads to delay in production. The rationale behind this strategy is that every operation performed on a style should add value. Operations such as inspection, handling, sorting, warehousing, etc. require extra time, handling, and personnel but these activities do not add any value to the product (Sudarshan and Rao, 2013).

4.4.1.8 Mass customization

The strategy here is to produce products that can be made to order rather than made to plan. Product life cycle is short and the strategy requires processing single orders with immediate turnaround. Considering the complexity of many apparel products and the number of processes that a style may require, the equipment, skills, information, and processes must be highly integrated. This may involve single ply cutting, single piece continuous floor manufacturing, and integral information technology. This has the opportunity of having garments fully customized including style, fit, fabric, and trim with delivery direct to the home in a few days at a price similar to the mass produced garments. Body scanning technology will be the basis of custom fit. Mass customization reduces the risk associated in trying to anticipate consumer demand months ahead of the point of sale to the ultimate consumer.

4.5 Supply chain management in the apparel industry

The apparel industry stands out as one of the most globalized industries in the world and it is a supply driven commodity chain led by a combination of retailers, contractors, subcontractors, merchandisers, buyers, and suppliers; each plays an important role in a network of supply chains that spans from fibers to yarn, to fabrics, to accessories, to garments, to trading, and to marketing (Ramesh and Bahinipati, 2011). Moreover in today's competitive environment, markets are becoming more global, dynamic, and customer driven, where customers are demanding more variety, better quality, and service, including reliability and faster delivery. Therefore, to ensure growth, it has become mandatory for the apparel industry to be more participative and adaptive.

Traditional, supply chains are viewed as a flow line, where input enters at one end and transforms to output at the other end (Wang and Chan, 2010). This is quite static and is applicable for products that are changing less frequently. However, advances in information technology have changed the supply chain configuration from 2004 to 2014 (Boyle et al., 2008). One of the phenomenal changes is globalization and the other one is the concept of mass customization (Wang and Chan, 2010).

4.5.1 Responsiveness of supply chain

In a rapidly changing competitive world, there is a need to develop organizations and supply chains that are significantly more responsive than the existing ones (Ramesh and Bahinipati, 2011). The responsiveness of the supply chain system is defined by the speed with which the system can adjust its output in response to an external stimulus, for example, a customer order (Reichhart and Holweg, 2007). Apparel markets are synonymous with rapid change and as a result, commercial success or failure is largely determined by the organization's responsiveness.

Across industry sectors, such as fashion products, personal computers, consumer electronics, and automobiles, companies are contemplating strategies to increase their responsiveness to meet consumer needs by offering high product variety with short lead times. A survey by Hitachi Consulting and AMR[1] Research reported that

---

[1] AMR Research, Inc. was an independent research and industry analysis firm founded by Tony Friscia in 1986 and sold to Gartner Research in 2009.

developing collaborative processes, both within their company and with partners and customers, will improve the supply chain responsiveness (Ramesh and Bahinipati, 2011).

In order to enhance the responsiveness of the whole supply chain, time management and the use of technology become crucial topics in the industry (Choi et al., 2013). One of the strategies to improve responsiveness in the apparel industry was to adapt QR strategies and it was implemented in the mid-1980s in the American apparel industry.

The impact of QR is reported to be especially substantial in supply chains with products of short shelf-life and highly volatile demands (Ramesh and Bahinipati, 2011). A QR environment for a supply chain enables speed to market of products by moving them rapidly through the production and delivery cycle, from raw materials suppliers, to manufacturers, to retailers, and finally to consumers. Many fashion retailers, such as Zara, New Look, and George clothing at ASDA[2], have developed a variety of QR strategies that increase their responsiveness to the volatile market (Choi et al., 2013). This trend has also spread to China and many brands in China, such as Vancl, are also adopting quick response in their fast fashion concept. Responsiveness requires information sharing among all members across the supply chain and thus how to facilitate coordination is a major problem to address (Choi et al., 2013).

The enterprise resource planning (ERP) system has been employed to deal with this issue by reengineering the supply chain within and beyond an organization (Koch and Mitlohner, 2010; Li, 2012; Choi et al., 2013). ERP is a method of effective planning of all the resources in an organization. ERP covers the techniques and concepts employed for the integrated management of businesses as a whole, from the viewpoint of the effective use of management resources, to improve the efficiency of an enterprise. ERP software is designed to model and automate many of the basic processes of a company, from finance to the production floor, with the goal of integration of information across the company and eliminating complex, expensive links between computer systems that were never meant to talk to each other. The ERP platform is specifically necessary in the apparel industry. This is because the apparel industry and its supply chains face a demand-driven market and it becomes of upmost important to obtain the latest market information and share the information among the channel members.

4.5.2 Root cause analysis of three echelons of the apparel supply chain (Ahmad et al., 2012)

4.5.2.1 Cutting

- Manual laying
- Manual pattern making
- Finished item arrangement without visual display
- Manual material movement

---

[2]ASDA is a British-based, American-owned supermarket chain which retails food, clothing, general merchandise, toys and financial services.

- Manual ticketing
- Excess manpower
- Excess time consumption
- Excess cost

4.5.2.2 Sewing

- Unavailability of storing system
- Unnecessary overtime
- Unnecessary manpower
- Unnecessary waste (spare parts, fabric, thread, etc.)
- Rework
- Overproduction
- Inappropriate processing

4.5.2.3 Finishing

- Manual goods handling
- Ancient machine utilization
- Excess WIP
- Unavailability of finishing materials
- Ordinary quality of garments
- Irregular machine repairing system
- Stacking goods without proper arranging

Figure 4.2 shows the real state of the supply chain network in the apparel manufacturing industry and Figure 4.3 shows the proposed supply chain for the manufacturers (Ahmad et al., 2012).

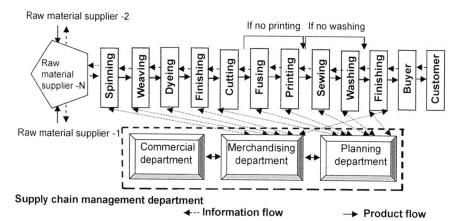

Figure 4.2 Supply chain of a typical apparel manufacturer.
Source: Ahmad et al. (2012). With kind permission from Leena and Luna International.

Production planning in the apparel industry

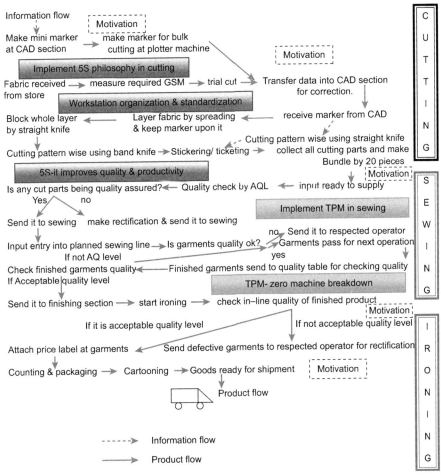

Figure 4.3 Proposed apparel supply chain network for manufacturer.
Note: 5S: sort, set in order, shine, standardize, and sustain.
Source: Ahmad et al. (2012). With kind permission from Leena and Luna International.

4.6 Inventory management

Inventory can be described as stock maintained between any two processes for uninterrupted operation. It can also be defined as assets that are intended for sale, or are in the process of being produced for sale, or are to be used in producing goods (Stitchworld, 2012). Inventory is primarily maintained due to two reasons—to optimize the sourcing cost (where requirement can be predicted) and to minimize the risk of stock out (where requirement cannot be predicted). The higher the inventory, the higher the capital blockage, and/or the higher the space requirement; on the other hand, lower inventory may lead to disruption in production or unsatisfied customers. Therefore, managing inventory is very important for every organization.

Knowing what you have, knowing what you need, knowing what you do not have—that is inventory control. The overseeing and controlling of ordering, storage, usage, and disposal of raw material, as well as finished goods' inventory, are the key aspects of inventory management. There are three inventory-costing methods that are widely used by both public and private companies (Stitchworld, 2012):

1. FIFO (first in, first out): The first unit bought in the inventory is the first to be sold or issued.
2. LIFO (last in, first out): The last unit bought in the inventory is the first to be sold or issued.
3. Average cost: It takes the weighted average of all units available for sale during the accounting period and then uses that average cost to determine the value of goods sold and ending inventory.

Most garment factories that buy fabric on an order basis may use the average cost method as the prices do not vary much in a small span of time. Since inventories represent a sizeable investment of company funds, and larger inventories mean higher costs (space, insurance, taxes, capital costs, etc.), a common inventory management goal focuses on improving inventory turnover. Low inventory turnover means that the company is carrying too much inventory, thereby unnecessarily restricting the access to cash that could be used in other profit-making activities.

COGS (cost of goods sold): It is the cost of the amount of inventory sold and can be deduced from the records.

Average inventory: The average inventory is calculated on a monthly basis. Companies can prefer to have it weekly or daily.

Dead stock: Material that has not been used for a significant amount of time and cannot be sold anymore is termed as dead stock. This material has gone out of fashion or more was bought than that was required or is damaged and hence cannot be used. Mostly it is sold at a reduced price. Regularly checking/monitoring the store room avoids storing the dead stock.

4.6.1 Inventory accuracy

Inventory accuracy is a measure of checking how closely official inventory records match the physical inventory. Operations and material management people have an interest in the accuracy of each item, as shortage may result in major production breaks or emergency buying and an excess can result in obsolete inventory. Inaccuracy in inventory can be because of many reasons such as improper data entry, incorrect unit used in calculation, poorly trained employees, stealing, or supplier errors.

4.6.2 Average days of inventory in hand

The time for which the organizations hold the inventory before selling or producing a product is called average days of inventory in hand. If these days are less, the organization requires less working capital to invest in inventory.

For the garment industry the average days of inventory in hand is the day fabric is procured until the planned cutting date. Manufacturers who buy fabric must keep an eye on the number of stock, so as to maintain a healthy cash flow.

4.6.3 Inventory carrying cost

Inventory carrying cost is the cost a company incurs over a certain period of time, to hold and store its inventory. The best way to avoid or reduce these costs is to sell them as soon as they become obsolete. Inventory carrying cost is expressed in percentage on the average inventory throughout a full year.

Cost of carrying inventory = total annual costs/average inventory value

Total annual costs include warehouse space, taxes, insurance, material handling, cost of money invested, etc.

4.6.4 Incidence of out of stock

For many businesses having a shortage of stock midway in production is a major factor leading to damage in business. Deciding on inventory levels is a delicate balance between not having too much and avoiding stock out situations. One simple stock out can lead to devastating effects on costs and may be losing a customer and therefore these should be tracked from time to time to control inventory levels.

As competition among manufacturers and distributors continues to rise, companies are constantly looking for ways to maximize their efficiency and gain better control over their inventory management. Some of the consulting firms (www.retailsmanagement.geowaresolutions.com/2013) have gathered predictions from leading industry domains and researchers on what future challenges they expect in the retail industry in terms of inventory management. Some of these challenges include inefficient processes, higher productivity requirements, fewer resources, and increasing competition. Say, for example, in the apparel industry, apparel retailers must manage continuously with the changing fashion, style and design. As the number of designs of products increases, the inventory management becomes more critical and complex.

4.7 Manufacturing performance improvement through lean production

As stated by Mannan and Ferdousi (2007) "the key to competing in the international marketplace is to simultaneously improve both quality and productivity on continual basis." In today's competitive and changing business world, lean production philosophy has brought changes in management practices to improve customer satisfaction as well as organizational effectiveness and efficiency (Karim, 2009; Ferdousi and Ahmed, 2009). Customers are now demanding a wide variety of products at a lower cost but with fast delivery (Ferdousi and Ahmed, 2009). They also expect more innovative products at competitive price as customers have more opportunities to choose from a variety of products.

Lean manufacturing can be defined as "a systematic approach to identifying and eliminating waste through continuous improvement by flowing the product at the demand of the customer (Silva, 2012)." The major purposes of the use of lean production are to increase productivity, improve product quality and manufacturing cycle time, reduce inventory, reduce lead time, and eliminate manufacturing waste (Ferdousi and Ahmed, 2009). To achieve these, the lean production philosophy uses several concepts such as one-piece flow, kaizen, cellular manufacturing, synchronous manufacturing, inventory management, workplace organization, and scrap reduction to reduce manufacturing waste (Russell and Taylor, 1999). As stated by Kilpatrick (2003), lean manufacturing makes an organization more responsive to market trends, deliver products and services faster, and produce products and services less expensively than a nonlean organization. There are studies that have been undertaken bearing on the relationship between lean practices and manufacturing performance of the firms (Ferdousi and Ahmed, 2009) and also have shown the improvements in manufacturing through lean practice. The garment industry has numerous opportunities for improvement using lean principles. Many countries have started to practice lean manufacturing tools in the garment industry and observed tremendous improvement (Ferdousi and Ahmed, 2009). Wastes in lean manufacturing are categorized into seven types such as overproduction, waiting, transportation, over or incorrect processing, defects, and motion, which were identified as part of the Toyota Production System as discussed below:

1. Overproduction—producing more than the customer demands. There are two types of overproduction (Silva, 2012):

- Quantitative—making more products than needed
- Early—making products before needed

Overproduction is highly costly to a manufacturing plant because it obstructs the smooth flow of materials and degrades the quality and productivity. Overproduction manufacturing is referred to as "just-in-case" whereas lean manufacturing is referred to as "just-in-time" (McBride, 2003).

Causes of overproduction

- Misuse of automation
- Just-in-case logic
- Long process setup
- Unleveled scheduling
- Unbalanced work load
- Redundant inspections

2. Waiting—typically more than 99% of a product's life cycle time in traditional mass production is spent in waiting (Silva, 2012). This includes waiting for material, labor, information, equipment, etc.

Causes of waiting

- Misuses of automation
- Unbalanced work load
- Unplanned maintenance
- Long process setup times
- Upstream quality problems
- Unleveled scheduling
- Poor communication

3. Transportation or conveyance—excessive movements and handling can cause damages and can lead to quality deterioration. Materials should be delivered to its point of use, only when needed.

Causes of transportation waste

- Poor plant layout
- Poor understanding of the process flow for production
- Large batch sizes, long lead times, and large storage areas

4. Over processing or incorrect processing—taking unneeded steps to process the parts, such as reworking, inspecting, and rechecking. Indistinct and unclear customer requirements cause the manufacturer to add unnecessary processes, which add cost to the product. Extra processing waste can be minimized by asking questions such as "why a specific processing step is needed?" and "why a specific product is produced?".

Causes of overprocessing waste

- Customer's true requirements not properly defined
- Product changes without process changes
- Overprocessing to accommodate expected downtime
- Lack of communication or extra copies/excessive information
- Redundant approvals

5. Excess inventory—any type of inventory (raw material or in process or finished goods) does not add value to the product and it should be eliminated or reduced. Excess inventory uses floor space and hides problems related to process incapabilities. Excess inventory results in longer lead times, obsolescence, damaged goods, transportation and storage costs, and delay.

Causes of excess inventory

- Compensating for inefficiencies and unexpected problems
- Product complexity
- Unleveled scheduling
- Poor market forecast
- Unbalanced workload
- Unreliable shipments by suppliers

6. Defects—can be either production defects or service errors. Having a defect results in tremendous cost to an organization. In most of the organizations the total cost of defects is often a significant percentage of total manufacturing cost.

Causes of defects

- Little or no process control
- Poor or inconsistent quality standards
- Inadequate education/training/work instructions

7. Motion—any motion that employee has to perform that does not add value to the product is considered unnecessary.

Causes of motion waste

- Poor people/machine efficiency
- Inconsistent work methods
- Poor facility or cell layout
- Poor workplace organization and housekeeping

The prime objective of waste elimination from the system is achieved with lean manufacturing techniques and tools. Based on this requirements, JIT, total quality management (TQM), flowcharts, and workplace redesigning techniques are used to reduce the waste. Some of the commonly used tools are:

> **JIT**—it is a management idea that attempts to eliminate sources of manufacturing waste by producing the right part in the right place at the right time.
> **Kanban Systems**—it is an information system that is used to ascertain that production occurs when the demand is created downstream (www.fibre2fashion.com). Kanban is a system to control the logistic chain from a production point of view, and is not an inventory control system. Kanban as a tool in lean manufacturing to achieve JIT.
> **Production Smoothing**—it is a concept adopted from the Toyota production system, where in order to decrease production cost, it was necessary to build no more cars than the number that can be sold. To accomplish this, the production schedule should be smooth so as to effectively produce the right quantity of parts and efficiently utilize manpower.
> **Kaizen**—it is a systematic approach to gradual, orderly, and continuous movement. Improvement in manufacturing can be in any form such as reduction of inventory and defective parts. One of the most important tools of Kaizen is the 5S, that is, Sort, Set in order/ Straighten, Shine, Standardize/Systemize, and Sustain.
> **Fishbone diagram**—one of the main tools in TQM is to analyze the root causes for a problem. This is the simple graphical method to represent the problems and root cause for these problems. The diagram has the ability to show many levels of causes simultaneously so it can be used to gain a full idea about the problem and to analyze it fully.

4.7.1 Six Sigma

Along with the extensive use of production systems, techniques, and tools to bring in huge benefits to the company, the application of management techniques in the business will bring long-term stability in the market and a better company reputation. Six Sigma is such a management tool that is viewed as a systematic, scientific, statistical, and smarter approach to create quality innovation and total customer satisfaction (Kumar and Sundaresan, 2010). It is a proven method for improving profits by pursuing perfection. It also allows us to draw comparisons among all processes, and tells how good a process is. It provides efficient manpower cultivation and utilization.

The 3Cs in Six Sigma are:

- Change—changing society
- Customer—power is shifted to customer
- Competition—competition in quality and productivity

Some of the Six Sigma models are DMAIC (Define, Measure, Analyze, Improve, Control), which is used to improve the existing process, and DMADV (Define, Measure, Analyze, Design, Verify), which is used to employ the new products.

4.7.1.1 Possible areas in textile industry for Six Sigma application (Kumar and Sundaresan, 2010)

- Reducing rejections in shipments
- Improving first sample approval percentages while working with buyer
- Improving supplier evaluation processes
- Improving acceptable quality level (AQL) performance in shipments
- Improving merchandiser performance
- Improving processes at the source (including fabric purchase and inspection, stitching, embroidery, packing, and shipping) to reduce rejections at later stages
- Eliminating manufacturing errors/defects

4.7.1.2 Seven basic tools used in Six Sigma process

1. Cause and effect diagram
2. Check sheet
3. Control chart
4. Histogram
5. Pareto chart
6. Scatter diagram
7. Stratification (Flow charts, Run charts, etc.)

4.8 Waste management

Ecological consciousness and increased environmental awareness are growing steadily worldwide among manufacturers, retailers, and consumers, who are all being encouraged to recycle and buy products made from recycled materials (Domina and Koch, 1998; Dowdle, 2005). According to Domina and Koch (1998), apparel is one of the largely unexploited consumer commodities with strong recycling potential. The most popular global approach to waste management seems to culminate in the

3Rs: reduce, reuse, and recycle (Fletcher, 2008). El-Haggar (2007), however, refers to 7Rs, adding the concepts of regulations, recovering, rethinking, and renovation. Globally, many of the bigger textile and apparel manufacturers are starting to take an environmental awareness position in the market (Larney and van Aardt, 2010). Incentives for recycling include savings on resources and their costs, reduction of waste, savings with respect to landfilling, and creation of jobs. The apparel industry has been identified as the fastest growing waste recycling concern across the country and processes to encourage public participation and improved education have been some of the industry's key issues. Recycling is a dynamic process and, as markets become saturated with specific products such as wiping cloths and mop heads, new products must be developed and introduced. Furthermore, special equipment is, for instance, necessary to shred fabrics, to separate waste, and to extract fibers. A third and very important aspect is the cost of products as a barrier to entering the marketplace. Many of the processes are labor-intensive, for example, the sorting of waste (Larney and van Aardt, 2010).

The use of landfill sites seems to be the most important practice used by the apparel industry to dispose the solid waste (Larney and van Aardt, 2010). Lack of technology and suitable equipment are considered to be the major factors hindering recycling. Consumers should be encouraged to use recyclable materials and appreciate the benefits it could generate to them as well as the generations to come.

4.9 Human resource management

There is no doubt that the "personnel" as the principal resource of a labor-intensive garment industry plays a crucial role in any productivity improvement and development. Further, this argument is supported by the facts that labor compromises about 20% of the cost price of a garment (Ghosh and Ghosh, 1995). Therefore, the way people are directed, motivated, or utilized will be decided on whether the organization will be prosperous and survive or fail. Hence, people are the key element for competitive advantage.

The uniqueness of the HR management lies in its emphasis on the people in work setting and its concerns for the well living and comfort of the HRs in the organization. For the textile and apparel companies it is compulsory that they treat the employees as an asset of the company. Many issues such as better productivity and production, social compliance, labor problems, and coordination are still grappling the industry and efforts at a small level are being made for betterment of the existing scenario. Thus HR plays the role and acts as a bridge between the management and the employees and provides a structure in such a way that it satisfies both.

HR must consider the following steps for a much needed organizational development in the apparel industry:

- HR should find ways to align its aims and objectives with the organizational mission and goals, for example, proactive participation in addressing the quality and productivity issues in various departments.

- HR functions should develop smart, user-friendly, and effective staffing, development and compensation procedures to support objectives of each functional area, for example, production, merchandising, and R&D (research and development).

4.10 New tools developed in production planning

- World on a hanger—provides easy to use Web-based product data, production, stock, and order management software priced on a monthly basis.
- Vietsoft ERP—apparel and footwear MIS (management information system) for windows, MRP (material requirements planning), purchasing, PPC, inventory, payroll, etc.
- Simparel—offers the quickest, most flexible, and affordable global supply chain management solution available today.
- AentezVogue—enables apparel firms to follow-up on the whole manufacturing processes from order to shipment.
- OnePlace PLM (product lifecycle management) and Workflow—a suite of business solutions for any organization involved in the production and development of retail/FMCG (fast-moving consumer goods).
- N41—apparel management solutions, sales, planning, production.
- Millennium—control production, efficiency, fabric, and raw materials.
- MerchanNet—garment software. Manage sampling, costing, orders, purchasing, production, warehouse, shipping, and accounts.
- GCS—sewing and production data.
- Apparel +—integrated apparel management, planning, and inventory management.
- ApparelMagic—sales, purchasing and production, accounting.
- AGMS—centralized control to production flow from sales order entry to shipping.
- Zedonk—affordable apparel and production management software.
- AIMS—order and production processing system for apparel manufacturers.

Acknowledgments

The authors acknowledged Springer and Leena and Luna International for granting permissions to reproduce some figures used in this book chapter from the following sources:
Mok, P.Y., Cheung, T.Y., Wong, W.K., Leung, S.Y.S, Fan, J.T., 2013. Intelligent production planning for complex garment manufacturing. J. Intel. Manuf. 24 (1), 133–145. With kind permission from Springer Science and Business Media.
Ahmad, S., Khalil, A.A.B., Rashed, C.A.A., 2012. Impact efficiency in apparel supply chain. Asian J. Nat. Appl. Sci. 1 (4), 36–45.

References

AAMA, 1993. Technical Advisory Committee Report. American Apparel Manufacturers Associations, Arlington, VA.
Ahmad, S., Khalil, A.A.B., Rashed, C.A.A., 2012. Impact efficiency in apparel supply chain. Asian J. Nat. Appl. Sci. 1 (4), 36–45.

Babu, V.R., 2006. Garment Production Systems: An Overview. Available from: http://www.indiantextilejournal.com/articles/FAdetails.asp?id=28 (accessed 12.03.14).

Bailey, T., 1993. Organizational innovation in the apparel industry. Ind. Relations 32 (1), 30−48.

Berg, P., Appelbaum, E., Bailey, T., Kalleberg, A.L., 1996. The performance effects of modular production in the apparel industry. Ind. Relations 35 (3), 356−373.

Boyle, E., Humphreys, P., McIvor, R., 2008. Reducing supply chain environmental uncertainty through e-intermediations: an organizational theory perspective. Int. J. Prod. Eco. 144 (1), 347−362.

Choi, T.M., Chow, P.S., Liu, S.C., 2013. Implementation of fashion ERP systems in China: case study of a fashion brand, review and future challenges. Int. J. Prod. Eco. 146, 70−81.

Collins, P., Glendinning, S., 2004. Production planning in the clothing industry: failing to plan is planning to fail. Control Mag. 1, 16−20.

Domina, T., Koch, K., 1998. Environmental profiles of female apparel shoppers in the Midwest, USA. J. Consum. Stud. Home Eco. 22, 147−161.

Dowdle, H., 2005. Recycling in styles. Nat. Health 35, 36.

El-Haggar, S., 2007. Sustainable Industrial Design and Waste Management: Cradle-to-Cradle Sustainable Development. Elsevier, Amsterdam.

Ferdousi, F., Ahmed, A., 2009. An investigation of manufacturing performance improvement through lean production: a study on Bangladeshi garment firms. Int. J. Bus. Mangmt. 4 (9), 106−116.

Fletcher, K., 2008. Sustainable Fashion and Textiles. Earthscan, London.

Ghosh, G.K., Ghosh, S., 1995. Indian Garments: Past and Present. A.P.H. Publishing House, New Delhi.

Glock, R.E., Kunz, G.I., 1995. Apparel Manufacturing: Sewn Product Analysis, second ed. Prentice Hall, Englewood Cliffs.

Goffin, K., Szwejczewski, M., New, C., 1997. Managing suppliers: when fewer can mean more. Int. J. Phy. Dist. Logst. Mangmt. 27 (7), 422−436.

Jacobs, F.R., Chase, R., 2013. Operations and Supply Chain Management. McGraw-Hill/Irwin, New York, USA.

Kanakadurga, K.S., 1994. A Comparative Study of the Flexibility of Three Types of Apparel Production Systems to Variations in Collar Designs (Master thesis). Virginia Polytechnic Institute & State University.

Karim, S., 2009. The Impact of Just-in-Time Production Practices on Organizational Performance in the Garments and Textiles Industries in Bangladesh (Doctoral thesis). Dhaka University.

Kilpatrick, J., 2003. Lean Principles. Utah Manufacturing Extension Partnership.

Kincade, D.H., Gibson, F.Y., 2010. Merchandising of Fashion Products. Pearson/Prentice-Hall, Upper Saddle River, NJ.

Kincade, D., Kim, J., Kanakadurga, K., 2013. An empirical investigation of apparel production systems and product line group through the use of collar designs. J. Text App. Tech. Mangmt. 8 (1), 1−15.

Ko, E., Kincade, D.H., 1998. Product line characteristics as determinants of quick response implementation for U.S. apparel manufacturers. Clothing Text Res. J. 16 (1), 11−18.

Koch, S., Mitlohner, J., 2010. Effort estimation for enterprise resource planning implementation projects using social choice—a comparative study. Ent. Info. Sys. 4 (3), 265−281.

Kumar, A., 2008. Production Planning and Control: Lesson 8. Course material. Delhi University. Available from: www.du.ac.in/fileadmin/DU/Academics/course_material/EP_08.pdf (accessed 14.03.14.).

Kumar, R.S., Sundaresan, S., 2010. Six Sigma in Textile Industry. Available from: http://www.indiantextilejournal.com/articles/FAdetails.asp?id=2896 (accessed 15.03.14).

Larney, M., van Aardt, A.M., 2010. Case study: apparel industry waste management: a focus on recycling in South Africa. Waste Mangmt. Res. 28, 36−43.

Lee, Y., 2000. Study of Relationships between Apparel Manufacturers' Supply Chain Management, Company Characteristics, and Inventory Performance (PhD thesis). Virginia Polytechnic Institute & State University.

Li, L., 2012. Effects of enterprise technology on supply chain collaborations: analysis of China-inked supply chain. Erp. Info. Sys. 6 (1), 55−77.

Mannan, M.A., Ferdousi, F., 2007. Essentials of Total Quality Management. The University Grants Commission of Bangladesh, Dhaka.

McBride, D., 2003. The 7 Manufacturing Wastes. Available from: http://www.emsstrategies.com/dm090203article2.htm (accessed 25.03.14).

Mok, P.Y., Cheung, T.Y., Wong, W.K., Leung, S.Y.S., Fan, J.T., 2013. Intelligent production planning for complex garment manufacturing. J. Intel. Manuf. 24 (1), 133−145.

Nahmias, S., 1997. Production and Operations Analysis, third ed. Irwin, Chicago.

Oliver, B.A., Kincade, D.H., Albrecht, D., 1994. Comparison of apparel production systems: a simulation. Clothing Text Res. J. 12 (4), 45−50.

Rajamanickam, R., Jayaraman, S., 1998. U.S. apparel distribution: where we stand and where we're going. Bobbin 51−53.

Ramesh, A., Bahinipati, B.K., June 28−30, 2011. The Indian apparel industry: a critical review of supply chains. In: International Conference on Operations and Quantitative Management (ICOQM). Nashik, India, pp. 1101−1111.

Ray, L., 2014. Production Planning for Garment Manufacturing. Available from: http://smallbusiness.chron.com/production-planning-garment-manufacturing-80975.html (accessed 05.08.14).

Reichhart, A., Holweg, M., 2007. Creating the customer-responsive supply chain: a reconciliation of concepts. Int. J. Opr. Prof. Mangmt. 27 (11), 1144−1172.

Russell, R.S., Taylor, B.W., 1999. Operations Management, second ed. Prentice Hall, Upper Saddle River, NJ.

Sarkar, P., 2011. Functions of Production Planning and Control (PPC) Department in Apparel Manufacturing. http://www.onlineclothingstudy.com/2011/12/functions-of-production-planning-and.html (accessed 05.08.14.).

Silva, S.K.P.N., 2012. Applicability of value stream mapping (VSM) in the apparel industry in Sri Lanka. Int. J. Lean Thinking 3 (1), 36−41.

Solinger, J., 1988. Apparel Manufacturing Handbook: Analysis, Principles and Practice. Bobbin Blenheim Media Corp, Columbia, SC.

Stitchworld, 2012. Performance Measurement Tools-5, Inventory Managements, pp. 34−37.

Sudarshan, B., Rao, N.D., 2013. Application of modular manufacturing system garment industries. Int. J. Sci. Eng. Res. 4 (2), 2083−2089.

Textile/Clothing Technology Corp (TC^2), 2005. Unit Production Systems. Bi-Weekly Technology Communicators. Available from: http://www.tc2.com/newsletter/arc/011905.html (accessed 12.03.14).

Vijayalakshmi, D., 2009. Production Strategies and Systems for Apparel Manufacturing. Available from: http://www.indiantextilejournal.com/articles/FAdetails.asp?id=1988 (accessed 10.03.14).

Wang, W.Y.C., Chan, H.K., 2010. Virtual organization for supply chain integration: two cases in the textile and fashion retailing industry. Int. J. Prod. Eco 127, 333—342.

http://www.accountingtools.com/questions-and-answers/what-is-standard-hours-allowed.html (accessed 05.08.14).

www.ehow.com (accessed 05.08.14).

www.fibre2fashion.com, Tiwari, A., Wanjari, S. Lean Manufacturing in Apparel Industry (accessed 28.03.14).

http://www.garmentmerchandising.com/ (accessed 20.03.14).

http://retailsmanagement.geowaresolutions.com/2013 (accessed 20.03.14).

Fabric sourcing and selection

A. Vijayan, A. Jadhav
RMIT University, Melbourne, VIC, Australia

5.1 Introduction

The impact of global trade in apparel manufacturing has evolved the concept of fast fashion in a scenario where apparel is produced quickly with a low cost and with the 'chic look' now dominating the retail world. Fast fashion is mass-produced, reasonable in price for most consumers and easy to obtain, making it simple for anyone to look stylish. However, to become dominant in this retail business sector, fabric sourcing and selection have become a pivotal stage in the manufacturing process. This chapter will encompass a critical review of the traditional and fast fashion approach with special focus on fabric sourcing, inspection, faults, grading and forthcoming trends.

Today, the clothing market changes rapidly based on the demographics, and easy and low cost sourcing. What is considered modern today, may be out of fashion tomorrow. This puts a lot of pressure on a clothing farm's ability to change the production stream quickly and thereby altering the flexibility of the supplier. Concerns about child labour, employees' work conditions, and environmental concerns have put addition pressure on clothing manufacturers and retailers.

Purchasing the right fabric can sometimes become a tough challenge encountered for apparel manufacturers. Of the total cost of a garment piece, the cost of fabric can contribute between 50 and 65%. Furthermore, even a minor oversight in selecting the right fabric and right manufacturer may spoil the entire apparel consignment. Interestingly, although the fabric constitutes the main part of a garment, many of the apparel merchandisers have a very limited knowledge about fabric. They consider fabric as another component of raw material such as buttons, care labels, hang tags or packing materials.

Inspection of the fabrics and grouping them according to shades are to be done with extra care. For example, if one particular colour fabric is used in large quantity, it is natural to have slight batch-wise variations. Mixing these batches on the cutting table can result in shade difference. To make the process easier, the fabric supplier should be instructed to supply the shade grouping charts with a roll number noted against each shade swatch. The garment buyers also should be informed clearly that for large volume orders that require fabric processing in more than one batch, there would be small batch to batch variation in properties. Fabric with special surface finishes, such as peach finish, microsanding, etc., needs special care to avoid the variations.

5.2 Fabric sourcing

5.2.1 Sourcing responsibilities

Depending on the size and type of the organisation, fabric sourcing responsibilities can be handled by the merchandiser or a combination of merchandisers, product specialists and marketing managers. Fabric sourcing is often a time-consuming process and requires controlled protocols to maintain schedules for the placement of orders; to seek delivery of orders from fabric manufacturers or agents that are contracted; and to follow through on shipment of orders. Broadly, fabric sourcing can involve several stages such as identification of potential sources, analysis of key factors that can influence the decision-making process, review of product specification, generation of samples, requisition of samples, contractual agreements, follow-up, quality inspection and procurement.

The key factors that can influence the decision-making process are product specification, production capacity, quality, cost and time. Based on the product specification, the buyer will determine which product will be purchased, will select the supplier and will negotiate on the delivery terms and price. The production capacity, sales forecasting, preseason style testing, sampling, merchandise line planning and development will ultimately govern the overall procurement process.

5.2.2 Sales forecasting

Sales forecasting is a common strategy practiced in the garment industry to plan production. Sales forecasting reflects the market potential rather than the production capacity. In a market pull system the production capacity is market driven whereas in a technology push system the sales are adjusted according to the production capacity.

Several forecasting methods have been adopted by the garment industry. The most commonly used methods are generic statistical time series models such as:

- Exponential Soothing Model (Brown, 1962)
- Holt–Winters Model (Winters, 1960)
- Box & Jenkins Model (Box and Jenkins, 1976)
- Regression Model or ARIMA (Papalexopoulos and Hesterberg, 1990)

Based on these models several forecasting systems are commercially available in the market such as Predicast of Aperia, SAP Demand and Forecasting, Forecast Management by Demand Solutions and Optimate of D3S (Thomassey, 2010).

More advanced systems based on soft computing techniques that include Fuzzy Logic, evolutionary algorithms and neural networks have been used for forecasting the textile–apparel market (Liu et al., 2013). These methods require complete historical data and are limited to sales forecast by item type. Data mining techniques, based on a clustering procedure of historical sales, that are capable of linking between prototypes of sales and descriptive criteria of historical items, and classifying sales from its descriptive criteria, have proven to be very effective to form homogeneous groups of items (Thomassey and Happiette, 2007). Other hybrid systems based on Fuzzy Logic (Vroman et al., 1998), neural networks (Wong and Guo, 2010), extreme learning

Fabric sourcing and selection

machine (ELM) (Yu et al., 2011) and seasonal auto-regressive integrated moving average (SARIMA) (Choi et al., 2011) have also been explored for predicting clothing, fashion and retail-based sales forecasting.

5.2.3 Merchandise planning

Merchandise planning is used to match the supply with the demand in the garment industry. This process looks after the maintenance of performance objectives for sales, inventory and other finance-related processes. The two fundamental steps involved in the planning are: (a) development of a sales forecasting based on the past demand, lead time in product supply, merchandising trends and market impact; and (b) determination of merchandise requirements based on budget and production plan. Planning decisions are based on historical trends and market foresight of the future demand changes. Merchandise planning process ensures that the product is available to the customer at the right time, price, and place.

There are several merchandising planning software solutions commercially available.

- FastReact are global providers of supply chain solutions such as Vision, Evolve and Align. Vision is a planning tool for sourcing and retailing. Evolve supports manufacturers in achieving effective capacity management and improving competitiveness. Align assists in raw material procurements quickly, accurately and effectively to apparel manufacturers.
- STAGE enterprise resource planning (ERP) production planning management developed by Eco Tech allows for streamlined business management and integration of information across departments and locations.
- APPS production planning and control module by Royal Datamatics allows for fixed capacity planning, determining machine utilisation and performing planned versus actual production.
- Software by Methods Apparel such as Pro-Merchant is used for effective vendor and buyer management to ensure on-time arrival of required materials and reduction in preproduction lead times; and Pro-Plan is used for planning systems that automatically predict order completion dates, comparing these dates to the requested delivery date.
- WFX provides software management products for business management and planning activities.
- Apprise software provides apparel and textiles ERP software that streamline and improve a wide range of enterprise activities.

5.2.4 Product specification

Product specifications are written descriptions or briefs that state the properties of materials, procedures, dimensions and performance attributes to provide control of the product during production. Product specifications may be based on a standard or a series of standards related to the product. Standards specify the performance requirements of a product and may include a set of characteristics or procedures that provides a basis for resource and production decision. Product specifications may be required on contractual agreements between buyer and supplier. In a manufacturing supply chain conscientious attention to specifications can lead to consistency in the quality of a product (Glock and Kunz, 2005).

The need for detailed and accurate specifications has increased with the advent of globalisation in the textiles and apparel industry. Specifications control the acquisition of materials and the production process to some extent, and will determine the performance characteristics of the final product. Specifications are used to describe a product, develop product consistency and negotiate sales contracts. Developing realistic specifications enhances the supplier−manufacturer relationship.

5.2.5 Source supplier identification

In the garment industry, suppliers are selected by the buyers, for the sourcing of fabric and trims. Manufacturing of fabric in-house along with garment manufacturing can be a convoluted and complex process to manage. It has become increasingly common in the garment industry to source the fabric from external sources rather than employing a large production capacity in-house. The main advantage of sourcing the fabric from external sources enables better control over the garment production (Keiser and Garner, 2008), and allows better choice of fabrics. However, other external sources can raise concerns about the quality of the source, due to lack of control over the production facilities and skills and training of suppliers' employees. Depending on the transparency of the supply chain, and the buyer's requirements and sourcing capacity, source identification can be classified into designated and nondesignated.

For both type of suppliers, the lead time remains the same irrespective of minimum order quantity (MOQ), total order quantity and quality parameters of the raw materials. Transit lead time will be dependent on the location of the supplier, that is, domestic and overseas suppliers.

5.2.5.1 Designated supplier

The concept of designated supplier is a common practice in the garment industry to achieve better control over the entire supply chain. The buyer designates a supplier mainly for products such as fabrics, accessories, and packaging material. This concept helps to control the supply chain, receive consistent quality and on-time deliveries. By designating a supplier the key advantage would be the standardisation of the product whereby the buyer obtains raw materials that have consistent quality with negligible defects or flaws. In addition, a designated supplier can deliver on time and can eliminate conflict between agents and garment manufacturers for delays in supply of trims, accessories or fabrics (Vasant Kothari, 2013).

Advantages of *designated* suppliers:

- Time savings for garment manufacturers as they only have to deal with one source to procure basic raw materials.
- Better quality assurance of the raw materials is consistent and it is controlled by the buyer.
- Designated suppliers can be involved in product development since they already have good experience in this field.

Disadvantages for *designated* suppliers:

- Designated suppliers may become overconfident about their product quality.
- There may be a monopoly of suppliers in the market.
- It may lead to unethical practices such as corruption by agents and traders.
- Loss of price competitiveness as suppliers can sometimes overcharge.

5.2.5.2 Nondesignated supplier

Nondesignated suppliers provide raw materials to the garment manufacturers directly. Sometimes buyers are employed to provide only the specification of the raw materials but the sourcing is still performed by the garment manufacturer independent of the buyer.

Advantages of *nondesignated* suppliers:

- Quality of the raw materials can be maintained due to competiveness among the suppliers and to sustain the business, nondesignated suppliers try their best to comply with the instructions given by buyers or garment manufacturers.
- The cost is competitive to market.
- Garment exporter may have an opportunity to explore the innovative products offered by the nondesignated supplier.

Disadvantages of *nondesignated* suppliers:

- There may be flaws in the quality of the raw materials due to lack of communication between the buyer and the supplier.
- The involvement of these suppliers in the product development process is a difficult task.
- It will be difficult for the buyers to control the social compliance and ethical business practice norms on these suppliers effectively.

5.2.6 Minimum order and limitations

MOQ plays an important role in sourcing of fabrics and other accessories as it determines the cost-effectiveness of the order. Every supplier requires a certain MOQ to produce the fabric economically, and if the order is less than the MOQ, the production cost is likely to be higher. Therefore, throughout the entire supply chain, the MOQ should be monitored in order to maintain the cost of production. The downside to minimum orders is the potential to end up with markdowns associated with placing orders larger than that can be supported by forecasts.

5.2.7 Materials testing

Various physical and performance attributes of the textile materials can be tested for required quality level. This may include the fibre, yarn and finished fabric testing (Hu, 2008). The details of some of the required tests for fabrics are noted in Table 5.1. The testing for other accessories are discribed in Chapter 16.

Table 5.1 Testing required for the approval of fabrics

Test type	Fabric type	
	Woven	Knitted
Fibre analysis	On request	On request
Yarn count	On request	On request
Weight	Essential	Essential
Dimensional stability to washing or dry cleaning	Essential	Essential
	All garments with fabric that can be washed	
	All garments with fabric that can be dry-cleaned	
Spirality	N/A*	Essential
Tensile strength	Microfibre fabrics, corduroy, satin, taffeta, flannel, flannelette, chemically treated fabrics	N/A
Tear strength	Essential	N/A
Colour fastness	Essential	Essential
Abrasion/pilling	Abrasion–martindale	Pilling–Pill box
Strength and recovery	Only for fabric with elastane	Essential
Water repellency	Rainwater and skiwear	N/A
Flammability	As per legal requirements of products	
Ultraviolet protection factor (UPF)	As per legal requirements of standard products	
Water absorbency	N/A	On request

Note: There may be additional tests as required by fabric buyer or this scheme may be modified.
* N/A, not applicable.
Hu, J. 2008. Fabric Testing. Elsevier.

5.2.8 Third party accreditation

Quality is very important in order to increase and maintain production efficiency and timely completion of orders. Poor quality can delay delivery of products and can cost money, time and reputation — it can cost future sales and possibly even lead to litigation. Although internal quality assessment is performed periodically by the manufacturers to assure product quality and reliability, checking the quality of the products from time to time with an independent third party is very essential (Plambeck and Denend, 2008). Fabric inspection performed by a third party (inspection/accreditation/organisation) will allow for unbiased product quality assessment and can help to reduce the risk of rejections.

Third party accreditation can be used to determine the defect level of the product during manufacturing. This practice can be adopted to prevent defective goods from entering the supply chain. Third party accreditation is based on product specification and compliance and can help to identify varying levels of defects in the product such as:

- Critical defect – noncompliant product, hazardous or unsafe product.
- Major defect – overall product failure, unstable product, damaged products.
- Minor defect – unlikely to reduce product usability.

The textile and apparel inspection procedure would ensure that fabric width, length, appearance and properties as mentioned in Table 5.1 comply with the relevant standards and regulations during the production process. The third party organisation should ensure all these parameters as per the specifications.

In addition to the product quality accreditation, certification programs such as worldwide responsible accredited production (WRAP) can be used to implement international workplace standards by taking into consideration local laws and workplace regulations; human resource management; health and safety and environmental practices; legal compliance including import and export; and custom compliance and security standards (WRAP, 2014).

5.3 Fabric inspection

Fabric inspection is an essential process for maintaining the quality of a fabric. Traditionally fabric inspection has been carried out manually through human visualisation performed by skilled inspectors. This process can be tedious and can add to the production cost. Besides many defects can be missed, and the inspection can be inconsistent whereby the output is dependent on the training and the performance and skill level of the human inspectors. In recent years automated machines have been developed based on adaptive and neural network systems. Automatic fabric systems are capable of providing consistent results that can be correlated with certain quality control standards.

Automated defect detection and identification systems can be classified as online and off-line systems. In the online monitoring system, the inspection is performed in the weaving machine while it is being produced. In the off-line monitoring system the produced fabric is taken from the weaving loom and mounted on an inspection frame where the quality of the fabric is analysed. Automated systems are capable of inspecting fabrics in full width either at the batcher for greige fabrics or at the exit end of the finishing machine, and in some cases are capable of being integrated on to weaving machine.

Automated fabric inspection systems are designed to find and catalog defects in a wide variety of fabrics including greige, apparel, upholstery, furnishing, dyed, finished, denim and industrial fabrics. Automated inspection systems such as the BMSVision Cyclops (Figure 5.1), Zellweger Uster Fabriscan (Figure 5.2), and Shelton WebSPECTOR (Figure 5.3) have proven to produce consistent and reliable results (Dockery, 2001). Some of the inspection systems are described in the following sections.

Figure 5.1 BMSVision Cyclops.
Source: www.visionbms.com.

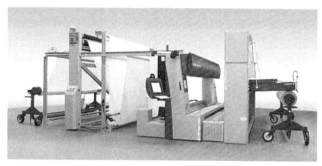

Figure 5.2 Zellweger Uster Fabriscan.
Source: www.techprt.co.uk.

Figure 5.3 Shelton WebSPECTOR.
Source: www.connectingindustry.com.

Fabric sourcing and selection

The BMSVision Cyclops (see Figure 5.1) has a traveling head that can scan at a speed of 18−54 m/min. The head includes a camera and illumination system, and can prevent production defects by stopping the weaving process. Defects such as stains, warp defects, and kinky and faulty weft can be detected. The Cyclops is designed to run with a BMS Sycotex weave room monitoring system where the fabric quality is stored in a database. At doffing, the on-loom system together with the monitoring system formulates fabric quality advice.

The Zellweger Uster Fabriscan (see Figure 5.2) online fabric inspection system, similar to the Cyclops provides automatic quality control directly to the weaving loom. The system can inspect fabrics at speeds up to 120 m/min and can handle widths up to 5 m. The Fabriscan has the capacity to inspect to the millimetre level of the fabric and check for defects. The scanner sensors are in direct contact with the fabric and are expected to deliver high-quality images of fabric defects. The Fabriscan classifies defects in a matrix called Uster Fabriclass, which is similar to the Uster Classimat systems used for yarns. This system differentiates between disturbing defects from nondisturbing defects and makes overdetection virtually nonexistent. The Fabriscan unlike other systems that are commercially available has a patented system for compensating the vibrations of the loom.

The WebSPECTOR surface inspection system is an adaptable fabric inspection system that can be operated online over a weaving loom or off-line as a standalone for batch to batch high-speed inspection. The system operates at speeds ranging from 20 to 200 m/min. When defects are detected, an image of each defect is stored along with all the identifying data that allow for classification of the defect by type in real time prior to physically marking the roll in line with the defect. An electronic defect file is created for the entire roll of fabric, which can be reviewed prior to the fabric being cut. This allows for an efficient cut plan for a parent roll to be split into small rolls for shipment and enduse.

Other fabric online and off-line defect inspection models have been proposed that can acquire digital fabric images and convert the image into binary images by restoration and a threefold technique (Banumathi and Nasira, 2012). The converted binary image is used as a propagation algorithm to generate classified defects as an output.

5.3.1 Fabric grading

It is well known fact that the quality of a garment has a direct correlation with the quality of fabric. It is the responsibility of the fabric manufacturers as well as the garment manufacturers to provide the final quality product to the consumer. The word 'inspection' in the garment industries relate to the visual examination or review of raw materials (such as fabric, accessories, trims, etc.). Fabric grading is different from the fabric inspection, which is essential for eliminating rejection due to the poor quality of fabric. It is also a precaution to remove unexpected defects on finished goods. The quality of a finished garment is dependent on the quality of fabric when sourced from the supplier. Fabric manufacturing involves the use of many types of yarns and weave patterns. The complexity associated with the raw materials and the formation of textile structures can lead to various faults/defects. It is therefore essential to identify these diverse types of visual defects that affect overall quality of the fabric and in turn the garment. There are many formal systems for evaluating and grading the quality of the fabric such as:

1 Graniteville "78" system.
2 Dallas system.
3 Four-point system.
4 Ten-point system.

In these systems, the operator calculates the numbers of major and minor defects as point values per square metre and then grades the fabric quality as 'first' or 'second' quality (Anagnostopoulos et al., 2001). Due to the specific nature of textiles, the defects encountered within textile production must be detected and corrected at early stages of the production process. Thus, visual defect detection is of utmost importance for the product's overall quality and cost.

5.3.1.1 Graniteville "78" system

In the Graniteville system, fabric defects are sorted into major and minor categories. If fabric has numerous major defects, then the fabric might be considered second-quality fabric and sold for less or encouraged to be reworked and sold as first quality. With this fabric-grading system, small and minor defects are often overlooked and not counted in the total penalty points (see Table 5.2). This system is recommended for garments and other textile products with large pattern pieces on wider widths of fabric (Rana, 2012).

5.3.1.2 Dallas system

Dallas system was developed in the 1970s, especially for knitted garments. According to this system, if any defect is found on a finished garment, the garment would then be termed as a second. In regard to fabric, this system defines second as more then one defect per ten linear yards, calculated to the nearest ten yards. For example, one piece of fabric, 60 yards long would be allowed to have six defects. This is not widely used.

Table 5.2 **Graniteville system**

Defect length	Penalty points
9 in.	1 point
9–18 in.	2 points
18–27 in.	3 points
27–36 in.	4 points

From Rana (2012).

5.3.1.3 Four-point system

Of the above methods, the four-point system is widely used in industry. In this system, fabric inspection is performed in accordance with ASTM D5430-93 (Standard Test Methods for Visually Inspecting and Grading Fabrics). Fabrics are usually graded using the 'four-point system' where all defects that are clearly noticeable from 3 feet are scored as a defect and demerit points are assigned according to the severity. Certain specialty fabrics, styles, colours and products from new suppliers undergo examination up to 100% of delivered goods. In this system, points are assigned for every possible defect in the fabric as shown in Table 5.3.

5.3.1.4 Ten-point system

The ten-point system separates defects by the warp and weft directions, which adds another layer of fabric grading. This system is challenging in everyday use. Penalty points are determined by the number of defects and length of each defect (see Table 5.4). If the penalty point total does not exceed the total yardage, the fabric is

Table 5.3 **Four-point system**

Defect characteristic	Penalty points
Width wise	
Up to 76 mm	1 point
76–152 mm	2 points
152–228 mm	3 points
>228 mm	4 points
Length wise	
Up to 76 mm	1 point
76–152 mm	2 points
152–228 mm	3 points
>228 mm	4 points
Area wise	
10 × 10 mm float	1 point
Up to 10 × 10 mm hole/stain	2 points
Over 10 × 10 mm to 20 × 20 mm hole/stain	3 points
Over 20 × 20 mm hole/stain	4 points
Yarn variation/Barré effect	
Minor	2 points
Major	4 points

Table 5.4 Ten-point system

Warp defects	Penalty points
10–36 in.	10 points
5–10 in.	5 points
1–5 in.	3 points
Up to 1 in.	1 point
Filling defects	**Penalty points**
Full width	10 points
5 in. to half the width of fabric	5 points
1–5 in.	3 points
Up to 1 in.	1 point

From Rana (2012).

considered first quality. If the penalty point total surpasses the total yardage, the fabric is considered second quality (Rana, 2012).

Defect calculation

Total defect points per 100 m^2 of cloth are calculated and therefore the acceptance criteria are usually not more than 40 penalty points. Fabric rolls containing over 40 points are considered 'seconds'. The formula to calculate penalty points per 100 m^2 is given by:

Penalty points per 100 m^2 = (Total points scored in the roll × 100,000)/(Fabric width in mm × Fabric length in m (inspected)).

Example: A fabric roll, 150 m long and 142 cm wide contains the defects as shown in Table 5.5. The calculation of points based on the severity of defect is also described in Table 5.5. See Table 5.6 for rejection classification of fabric defects, which describes various defects and the severity of a defect. The presence of some of the fabric defects can lead to the rejection of the whole roll of fabric.

Some of the main defects in the woven fabrics are knots, harness wrong draws, wrong ends, reed marks, missed picks, soiled spots, thin or thick places (slubs), warp or weft pull, mixed wefts, broken picks, loom bars, foreign fibres, warp and weft floats and defects from printing (Anagnostopoulos et al., 2001). Various woven fabric defects are described in Table 5.7 and are some of them shown in Figure 5.4. There may be some other fabric faults not described in this table depending on the type and nature of fabric.

As far as knitted fabrics are concerned, the main defects are dye spot, slub, stain ladder, spirality, ends outs, dragging ends, mixed yarns, missing yarns, holes, compactor creases, missing plush loops, skipped stitches, colour differences, slub, stain, pin marks (Anagnostopoulos et al., 2001), etc. Some of the knitted fabric defects are shown in Figure 5.5.

Table 5.5 **Penalty points calculations for defects**

No. of defects	Points per defects × number	Total points
Five defects up to 76 mm length	5 × 1	5 points
Four defects from 76 to 152 mm length	4 × 2	8 points
Two defects from 152 to 228 mm length	2 × 3	6 points
One defect over 228 mm length	1 × 4	4 points
Two holes over 20 × 20 mm	2 × 4	8 points
Total defect points		**31 points**
Therefore, points/100 square metre	= (31 × 100,000)/ (1420 mm × 150) = 13,000	14.55 points

Table 5.6 **Rejection classification of fabric defects**

Defects	Penalty points
Patta	Removable defect
Count or composition variation	Removable defect
Short end	Reject roll
Selvedge loose	Reject roll
Wrong drawing	Reject roll
Reed marks	Reject roll
Temple marks	Reject roll
Damaged selvedge	Reject roll
Wrong weave	Reject roll

5.4 Future trends

In recent years sourcing continues to be one of the most critical success factors for the garment industry. The expanding global trade, increased consumer demand, fast fashion, increased product accessibility, and competitive pricing have forced suppliers, designers, product manufacturers, and retailers to seek alternative sources of supply.

Table 5.7 Various fabric defects

Defect	Description
Slub, uneven yarn (thick and thin)	Thick uneven spot on the fabric caused by lint or small lengths of yarn sticking on fabric.
Barré	Textured or colour bars in the direction of warp or weft that can be caused by imperfection in the yarns.
Yarn contamination/fly	Foreign fibres or soil on woven (or knitted) fabric surface.
Dead cotton	Damaged cotton fibre by the weather conditions, over- or undermature cotton, which is difficult to process, leaving behind white or black spots.
Missing pick	Weft broken when weaving; the harness misdraw results in two ends weaving as one, caused by one end of yarn missing from the reed and machine continues to run.
Holes	Missing yarn, leaving behind a space, caused by broken yarn.
Reed mark	Running lines in the warp direction, caused by bent reed wire causing warp ends to be held apart.
Streaks	Dark or light uneven lines, caused by faulty processing.
Stop marks	Lines in the weft direction, caused when machine stopped, the yarn elongates the tension, results in making across the width.
Knots	Uneven raised knot, two yarn ends are tied together.
Miss weave pattern	Pattern that is different than the other area due to a malfunctioning of needle or jack in a jacquard machine.
Puckered selvedge	Uneven selvedge surface, caused by the stretching of the selvedge in finishing or by uneven wetting out in sanforization process.
Selvedge torn/broken	A single or group of yarns broken in the selvedge due to various reasons.
Pilling	Fibre ball formation in the fabric surface due to abrasion with other surface.
Shading (selvedge to selvedge)	A change in shade either abrupt or gradual, caused by poor processing.
Dye streaks	Uneven streaks occurred during dyeing or finishing process.
Colour smear	Uneven colour application as the result of colour being smeared during printing process.
Crease streak	Uneven marks showing light or dark lines as a result from creased fabric passing through squeeze rollers in dyeing process.

Table 5.7 **Continued**

Defect	Description
Slippage	Uneven blotch marks caused due to improper dyeing process.
Bowing	Woven weft yarns lie in an arc across fabric width (in knits the course lines lie in an arch across width of goods).
Skewing/bias	Condition where weft yarns are not square with warp yarns on woven fabrics (courses are not square with wales lines on knits).
Crease marks/wrinkles/ fold marks	Appear where creases are caused by fabric folds under pressure in the finishing process.
Pin holes	Holes along the selvedge caused by pins holding fabric while it processes through stenter frame. It is major, if pin holes extend into body of fabric far enough to be visible in the finished product.
Snagging	A break, tear or pull of yarn in the fabric.
Abrasion mark	Improper scratch on fabric during finishing processes.
Dirt/soil/stain	Dirty stains during fabric handling.
Oil spots/grease spots	Fabric surface with an oil spot.
Water spots/water marks	Light marks, usually caused by wet fabric being allowed to remain too long before drying; colour migrates leaving blotchy spots.
Yellowing of fabric	This is a phenomenon which causes light coloured fabric to yellow over time.
Printer machine stop mark	Dye smudged along width of fabric as a result of stopping of the printing machine.
Print out of position	Print is out of fit, caused by print rollers not being synchronised properly; results in various colours of the design not printed in proper position.
Miss print	Missing colour in the pattern caused by colour feeding stoppage or faulty printing.
Float	The warp or weft threads are not properly interlaced in the weave.
Wrong end	A wrong thread being pulled through the reed, which clearly shows up in the fabric.

Source: Matrix sourcing report.

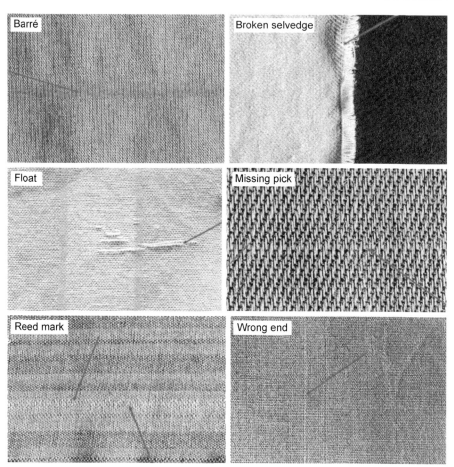

Figure 5.4 Woven fabric defects.
Singh (2012).

The early 2000s has witnessed the need for alternative countries such as Bangladesh as the next sourcing destination due to the increasing costs in China. The tragic series of incidents that happened in Bangladesh in 2010 has led the garment industry to revise sourcing strategies of large players in the apparel industry and to adopt sustainable and socially responsible practices. A survey conducted by McKinsey & Company indicated that sourcing costs are likely to increase with labour costs continuing to be the most important driving factor (Berg and Hedrich, 2014). The recognition and need for compliance was also emphasised.

The environmental policies for global apparel trade has changed significantly and is driven by changes in the regulatory system, and the changes in the strategies of global buyers and their sourcing policies. In the past, sourcing decisions in the garment industry were primarily dependent on labour costs. More recently, other

Fabric sourcing and selection

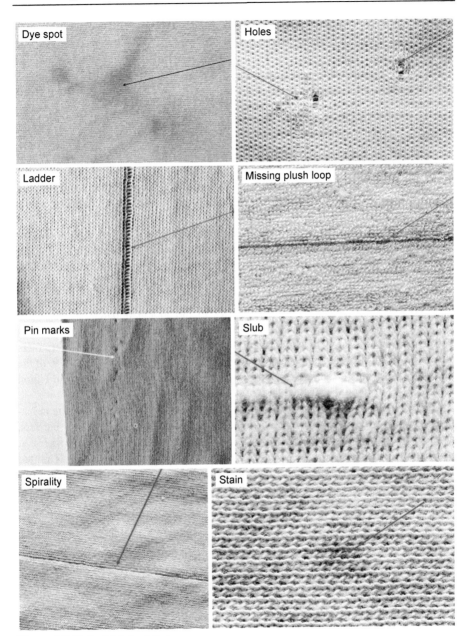

Figure 5.5 Knitted fabric defects.
Singh (2012).

factors such as cost, quality, reliability, lead times and compliance have influenced the sourcing decision-making process. Lead time in particular has influenced the buying process in the Western world where designers and retailers have shifted towards lean manufacturing and just-in-time delivery concept to accommodate fast fashion, and uncertain consumer markets by replenishing stock in very short cycles (Abernathy et al., 2006). Buyers have shown inclination towards sourcing, product development, inventory management, logistics, and finance, which have become more apparent (Staritz, 2012). Recently, effective supply chain management through consolidation of the supply base and streamlining of suppliers to obtain consistent quality, reliable delivery, flexible production, with competitive pricing, has started to evolve. The need for compliance with labour and environmental standards has increased importance in buyers sourcing decisions in the context of corporate social responsibility.

The 2000s has also witnessed significant turbulence in the global value chain for apparel. The changes in the supply and demand structures, asymmetric markets and power at the buyer's level have raised the need for supply chain rationalisation. The consolidation caused by the Multi Fibre Arrangement's (MFA) phase out has been intensified by the effects of the ongoing economic crisis. Competition between developing countries has led to overcapacity in the global apparel industry. The competitive advantage in products and activities relevant for low-income countries (LIC) exporters has shown by China and India has put constraints on Asian forerunners such as Japan and Korea (Kaplinsky and Morris, 2008). In 2012, China had 37% of the global market share for apparel exports. The rest of Asia had 20%, European Union (EU) shared 30% and the rest of the world 13% (Elms and Low, 2013).

The dynamics of the clothing industry is rapidly evolving. Traditionally, the industry was a platform for LIC to enter into the global market. With low entry barriers, simple technology, and a requirement for high number of unskilled workers, it provided the ideal entry point to diversify exports. The trend in recent years has shifted to technical sophistication, which may help some countries climb the value chain. However, cost is still a determining factor and is expected to impact on the global sourcing strategies.

There is a proposition by large retailers such as GAP and Marks & Spencer to expand their business in China in 2012–2016; however, it appears that China's 12th "five-year plan" has strategies for promoting domestic Chinese brands. Nonetheless, the future growth in the retail industry is predicted to be in China, India, Turkey, Brazil, Eastern Europe and Russia. This has led retailers from the US and EU to introduce sales outlets and stores in these countries to gain market share. High growth rates in countries such as China and India are slowing down and inflation is picking up. The future will require a better understanding of the local sourcing policies and power structures within the clothing value chains of buyers in order to stay competitive. In the regional and domestic markets, there may also be increased opportunities for suppliers to upgrade their activities to new product development, design, marketing, branding and to some degree retailing, to rise up the value chain.

5.5 Conclusions

The advances in sourcing strategies globally have impacted the process of new product development and manufacturing and the way the garment industry conducts its business in the supply chain. These changes have improved the capacity in business to meet the customer needs by controlling sourcing and providing on-time delivery and at a competitive cost. Factors such as cost, quality, reliability, lead times and compliance have influenced the sourcing decision-making process. Many of these strategies can be strongly affected by the international laws of tariffs and trade. Sourcing decisions are also made based on cost benefits, domestic market situation, international market situation, relation between the countries and the political stability of the country. Changes in the global scenario, including the emergence of powerful intermediaries and first-tier suppliers in the developing countries, shifting of end markets to large developing countries and the increasing importance of developing country buyers could revolutionise sourcing in favour of low income countries.

References

Abernathy, F.H., Volpe, A., Weil, D., 2006. The future of the apparel and textile industries: prospects and choices for public and private actors. Environ. Plann. A 38, 2207–2232.

Anagnostopoulos, C., Vergados, D., Kayafas, E., Loumos, V., Stassinopoulos, G., 2001. A computer vision approach for textile quality control. J. Visualization Comput. Anim. 12, 31–44.

Banumathi, P.N., Nasira, G.M., 2012. Fabric inspection system using artificial neural networks. Int. J. Comput. Eng. Sci. 2, 20–27.

Berg, A., Hedrich, S., 2014. What's Next in Apparel Sourcing? Mckinsey and Company, Chennai.

Box, G.E.P., Jenkins, G.M., 1976. Time Series Analysis: Forecasting and Control. Holden Day, San Francisco.

Brown, R.G., 1962. Smoothing, Forecasting and Prediction of Discrete Time Series. Prentice-Hall, New York.

Choi, T.M., Yu, Y., Au, K.F., 2011. A hybrid SARIMA wavelet transform method for sales forecasting. Decis. Support Syst. 51, 130–140.

Elms, D.K., Low, P., 2013. Global Value Chains in a Changing World. World Trade Organization, Geneva.

Dockery, A., 2001. Automated Fabric Inspection: Assessing the Current State of the Art. Available from: http://techexchange.com/thelibrary/FabricScanning.html.

Hu, J., 2008. Fabric Testing. Elsevier.

Kaplinsky, R., Morris, M., 2008. Do the Asian drivers undermine export-oriented industrialization in SSA? World Dev. 36, 254–273.

Keiser, S.J., Garner, M.B., 2008. Beyond Design: The Synergy of Apparel Product Development. Fairchild Books.

Glock, R.E., Kunz, G.I., 2005. Apparel Manufacturing: Sewn Product Analysis. Prentice-Hall, Upper Saddle River, NJ.

Liu, N., Ren, S., Choi, T.M., Hui, C.L., Ng, S.F., 2013. Sales forecasting for fashion retailing service industry: a review. Math. Probl. Eng. 2013.

Matrix, 2012. Material Quality Manual. Matrix Sourcing.

Papalexopoulos, A.D., Hesterberg, T.C., 1990. A regression-based approach to short-term system load forecasting. IEEE Trans. Power Syst. 5, 1535–1547.

Plambeck, E.L., Denend, L., 2008. Wal-Mart. Stanford Soc. Innovation Rev. 6, 53–59.

Rana, N., 2012. Fabric inspection systems for apparel industry. Indian Text. J. Available from: http://www.indiantextilejournal.com/articles/FAdetails.asp?id=4664 (15.07.14).

Singh, R., 2012. Common Fabric Defects (Online). Available from: http://textilelearner.blogspot.com.au/2013/07/common-fabric-defects-with-images.html.

Staritz, C., 2012. Apparel Exports—still a Path for Industrial Development? Dynamics in Apparel Global Value Chains and Implications for Low-income Countries.

Thomassey, S., 2010. Sales forecasts in clothing industry: the key success factor of the supply chain management. Int. J. Prod. Econ. 128, 470–483.

Thomassey, S., Happiette, M., 2007. A neural clustering and classification system for sales forecasting of new apparel items. Appl. Soft Comput. 7, 1177–1187.

Vasant Kothari, S.J., 2013. Fashion merchandising: sourcing (Online). Textile Today, October 2013. Available from: http://www.textiletoday.com.bd/magazine/727.

Vroman, P., Happiette, M., Rabenasolo, B., 1998. Fuzzy adaptation of the holt-winter model for textile sales-forecasting. J. Text. Inst. 89, 78–89.

Winters, P.R., 1960. Forecasting sales by exponentially weighted moving averages. Manage. Sci. 6, 324–342.

Wong, W.K., Guo, Z.X., 2010. A hybrid intelligent model for medium-term sales forecasting in fashion retail supply chains using extreme learning machine and harmony search algorithm. Int. J. Prod. Econ. 128, 614–624.

Wrap, 2014. Worldwide Responsible Apparel Production.

Yu, Y., Choi, T.M., Hui, C.L., 2011. An intelligent fast sales forecasting model for fashion products. Expert Syst. Appl. 38, 7373–7379.

Selecting garment accessories, trims, and closures

K.N. Chatterjee, Y. Jhanji, T. Grover, N. Bansal, S. Bhattacharyya
The Technological Institute of Textile and Sciences, Bhiwani, India

6.1 Part 1: introduction to garment accessories

With the fast-growing fashion consciousness, development of fashion accessories has gained momentum. Apparel is an ensemble of fabric and accessories. Without accessories, the garment remains incomplete. Various kinds of accessories are used on garments; some are part of the garments such as buttons, zippers, and interlining, while others are used for decorating and enhancing the product appearance, such as sequins and embroidery. These accessories are considered as garment accessories as they form an integral part of the garment. There are other type of accessories, which can be treated as fashion accessories, as they are not a part of the garment. However, they are used to complement the outfit. Hence, this chapter will deal with garment as well as fashion accessories.

The apparel and accessory industries work in close coordination, as today's fashion-conscious consumer regards the wardrobe as complete when both apparel and accessories complement each other. Accessories, although additional components worn along with garments, go a long way not just in enhancement and beautifying the overall look of a garment but also have functional aspects. Fashion accessories can be divided into different categories based on the chosen raw material, functionality, trims, and notions used for their construction and end-uses. Decorative or ornamental accessories like jewelry, bracelets, earrings, hair accessories, bands, and clips symbolize the status of the wearer in addition to completing the look of garment, but do not contribute much in providing protection to the wearer against external elements. Accessories like footwear, gloves, hats, and headwear are functional in nature, providing protection to the wearer in different environmental conditions. Likewise, accessories like handbags and wallet sets provide ease of storage of important belongings to the users. Eventually it can be concluded that both functional and decorative accessories are part and parcel of the garment industry by the way they contribute to the aesthetic appeal and functionality of the garment to the wearer.

6.2 Part 2: selecting garment accessories

Selection of accessories is generally carried out depending on types of accessories, types of products, and whether the accessories are to be used for men's wear, women's wear, or children's wear. Usually, for value-added products, garment accessories should be selected from approved sources after carrying out quality inspection as per international standards. Proper care should be maintained while selecting garment accessories for children's wear.

6.2.1 Introduction to garment accessories

A garment accessory is an item that is used to contribute, in a secondary manner, to the wearer's outfit. Accessories are often used to complete an outfit and are chosen to specifically complement the wearer's look.

6.2.2 Types of garment accessories

Generally, there are three types of garment accessories:

1. Basic accessories, as shown in Figure 6.1.
2. Decorative accessories, as shown in Figure 6.2.
3. Finishing accessories like the hang tag, price tag, poly bag, paper, carton, tape, etc.

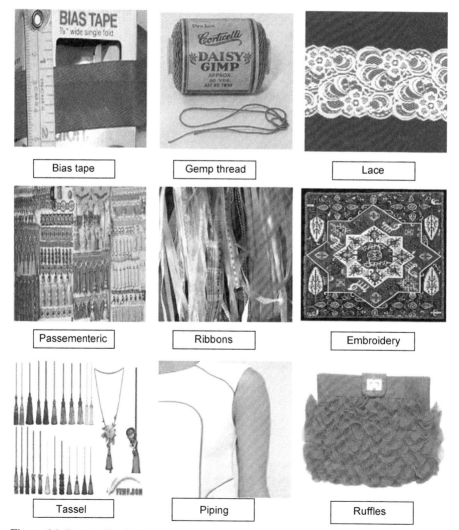

Figure 6.1 Types of basic accessories.

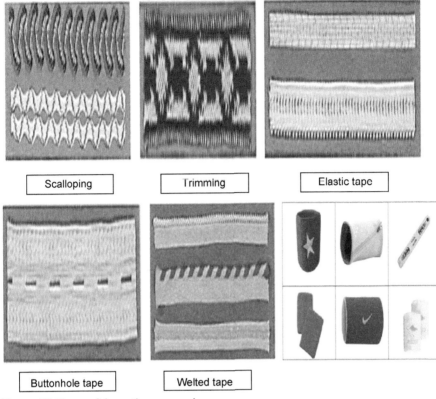

Figure 6.2 Types of decorative accessories.

6.2.2.1 Basic accessories

1. Buttons
2. Zippers
3. Linings
4. Interlinings
5. Ribbons
6. Toggles
7. Velcro
8. Elastic
9. Rivets
10. Labels
11. Motifs
12. Pocketing fabrics
13. Thread

Buttons

A button is a small disc, typically round, usually attached to an article of clothing in order to secure an opening, or for ornamentation. In modern clothing and fashion design, a button is a small fastener, most commonly made of plastic, but also frequently of seashell, which secures two pieces of fabric together.

Zippers

A zipper is a popular device for temporarily joining two edges of fabric. It is used in clothing, luggage, and other bags, sporting goods, camping gear, and other daily-use items.

Linings

Linings are one kind of trimming used on the underside of garments and next to the skin.

Interlinings

An interlining is a layer of flannel fabric sewn in between the face fabric and the standard lining. An interlining provides insulation and also adds a luxurious weight and softness, improves the drape of the fabric, and protects fragile fabrics.

Ribbons

A ribbon is a thin band of material, typically cloth but also plastic or sometimes metal, used primarily as decorative binding and tying. Cloth ribbons are made of natural materials such as silk, velvet, cotton, and jute; and of synthetic materials such as polyester, nylon, and polypropylene.

6.2.2.2 Decorative accessories

1. Bias tape
2. Elastic tape
3. Buttonhole tape
4. Seaming tape
5. Welted tape
6. Ribbed tape
7. Stamped tape
8. Cords
9. Braids
10. Soutache
11. Galloons
12. Embroidery
13. Gimp thread
14. Lace
15. Passementerie
16. Piping
17. Moiré ribbons
18. Velvet ribbons
19. Taffeta ribbons
20. Fringes
21. Rosettes
22. Pompons
23. Tassels
24. Ruffles
25. Rickracks

Bias tape
Bias tape is a narrow strip of fabric, cut on the bias. Many strips can be pieced together into a long "tape." The tape's width varies from about half inch to about three inches, depending on applications. Bias tape is used in making piping, binding seams, finishing raw edges, etc. It is often used on the edges of quilts, placemats, and bibs, around armholes and neckline edges instead of a facing, and as a simple strap or tie for casual bags or clothing.

Elastic tape
Elastic tape is an elastic cotton strip—a highly elastic, flat, braided band containing rubber or elastomeric fibers.

Buttonhole tape
Buttonhole tape is a broad elastic tape with buttonholes located in the center.

Seaming tape
Seaming tape is a cotton or viscose twill-woven tape used for stabilizing seams.

Welted tape
Welted tape is a cotton or viscose tape with a narrow welt at the edge.

Ribbed tape
Ribbed tape is a cotton, silk, or viscose tape with pronounced ribs for decoration or for waistbands.

Stamped tape
Stamped tape is an interlining tape with prestamped marks to show sewing width and seam allowance.

Cords
Cords are a trimming made by twisting or plying two or more strands of yarns together. Circular braided materials are of various thicknesses made from viscose, cotton, or synthetics. Cords are used in a number of textile arts including dressmaking, upholstery, macramé, and couching. Cords are also used as decoration for clothing, in household textiles, and in sporting goods.

Braids
A braid is a complex structure or pattern formed by intertwining three or more strands of flexible material such as textile fibers, wire, or hair.

Soutaches
A soutache is a moldable, flat braid with two ribs in silk or viscose used for formal clothes.

Galloons
A galloon is a particularly supple, plain, or patterned braided ribbon for piping or binding in outerwear.

Embroidery
Embroidery is the handicraft of decorating fabric or other materials with needle and thread or yarn. Embroidery may also incorporate other materials such as metal strips, pearls, beads, quills, and sequins.

Gimp Threads
Gimp is a narrow ornamental trim used in sewing or embroidery. It is made of silk, wool, or cotton and is often stiffened with metallic wire or coarse cord running through it. Gimp is used as trimming for dresses, curtains, furniture, etc.

Lace
Lace is an openwork fabric, patterned with open holes in the work, made by machine or by hand. A true lace is created when a thread is looped, twisted, or braided to other threads independently from a backing fabric.

Passementerie
Passementerie is the art of making elaborate trimmings of applied braid, gold or silver cord, embroidery, colored silk, or beads used for clothing or furnishings.

Piping
Piping is a type of trim consisting of a strip of folded fabric inserted into a seam to define the edges of a garment. Usually the fabric strip is cut on the bias.

Moiré Ribbons
Moiré ribbons are cotton, silk, or manmade fiber ribbons with a moiré pattern for hat bands and bows.

Velvet Ribbons
A velvet ribbon is made of cotton, silk, or viscose narrow-woven velvet. They are sensitive to handling.

Taffeta Ribbons
Taffeta ribbons are filament yarn ribbons, with a plain or check patterned, for ribbon bows.

Fringes
A fringe is a narrow edging of projecting yarns that is not woven into the fabric, made from viscose, wool, or silk.

Rosettes
Rosettes are decorative items used either alone or in combination with ornamental textiles.

Pompons
Pompons are bunches of wool, silk, or synthetics used as trimmings, hanging alone or in groups.

Tassels
A tassel is a finishing feature in fabric decoration. It is a universal ornament that is seen in varying versions in many cultures around the globe.

Ruffles
A ruffle is a strip of fabric, lace, or ribbon tightly gathered on one edge and applied to a garment as a form of trimming.

Rickracks
Rickracks are a flat, narrow braid woven in zigzag form, used as a trimming for clothing. Narrow bowed, zigzag, or scalloped ribbons are made using multicolored cotton or manmade material for edge trimming of traditional costumes and children's wear.

6.2.2.3 Finishing accessories

1. Hang tag
2. Price tag
3. Poly bag
4. Paper
5. Carton
6. Tape

Hang tag
A tag attached to a garment or other merchandise that includes information about the manufacturer the fabric or material used the model number, care instructions, and often the price of the garment.

Price tag
A price tag is a label or tag that shows the price of the garment, to which it is attached.

Poly bag
A polythene bag, especially used to store or protect the garment articles.

Paper
The items used to pack the garment which will hold the garment components in position. For example, the paper used at the collar and sleeves of a shirt.

Carton
A cardboard or plastic box used typically for storage or shipping of garments.

Tape
A long, narrow strip of linen, cotton or the like, used for tying garments, or garment components.

6.3 Part 3: selecting supporting materials

Supporting materials play an important role in deciding the quality of garments or products. Hence, proper care should be taken while selecting supporting materials for garments or home textile materials.

6.3.1 Linings and interlinings

6.3.1.1 Aims and objectives of linings

Linings are used to help hide the inner construction details of a garment, and also to help it slide off and on over other clothing with ease. Lining fabrics are usually slippery and silky, though other types may be used for effect. Linings are constructed separately from the garment and attached at facing or hem areas by hand or machine. Linings prevent the garment from stretching and reduce wrinkling.

6.3.1.2 Aims and objectives of interlinings

1. To make sewing easier and to increase production
2. Retaining shape and improving material appearance
3. Making a functional, lasting, easy-to-wear product

6.3.2 Definitions of lining and interlining

A lining is a separate, but attached, supportive or inner garment fabric or fabric construction that conceals or covers the inside garment construction. Garments may be fully or only partially lined; completely or partially attached to the fashion garment. Linings, as can be seen in Figure 6.3, are used in most tailored garments; however, they are not confined to use in these garments alone.

An interlining is a layer of fabric inserted between the face and the lining of a garment, drapery, or quilt. Interlining is similar to batting, a thick layer of fiber designed to provide insulation, loft, and body to quilts, pillow toppers, and heavy winter jackets.

Generally, interlinings are soft, thick, and flexible. Some interlinings are designed to be fused, while others are intended to be sewn to one or both layers of the textile. As an inner lining within textiles, interlining is used in a number of applications.

6.3.3 Applications of linings and interlinings

6.3.3.1 Applications of linings

It is important that the lining be the same weight or a lighter weight and softer than the fashion fabric so that it does not dominate the garment. Lining fabric should be preshrunk before using.

Figure 6.3 Linings.

6.3.3.2 Applications of interlinings

Interlinings can be used to protect fabrics, especially those used in drapes and consequently often exposed to direct light. Delicate fabrics like silk and velvet can suffer from sun damage if hung with a liner alone, and most drapers recommend the use of an interlining for the life of the fabric.

Interlinings are materials that are fused or sewn to specific areas on the inside of garments or garment components. They may provide shape, support, stabilization, reinforcement, hand, and improved performance for garments.

6.3.4 Wadding/batting

Batting is a layer of insulation used in quilting between a top layer of patchwork and a bottom layer of backing material. Batting is usually made of cotton, polyester, and/or wool.

6.3.5 Shoulder pads

Shoulder pads, as can be seen in Figure 6.4, are a type of fabric-covered padding used in men's and women's clothing to give the wearer the illusion of having broader and less sloping shoulders. In men's styles, shoulder pads are often used in suits, jackets, and overcoats, usually sewn at the top of the shoulder and fastened between the lining and the outer fabric layer. In women's clothing, their inclusion depends on the fashion taste of the day.

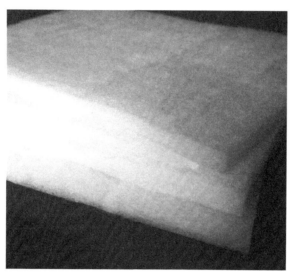

Figure 6.4 Wadding and shoulder pads.

6.4 Part 4: selecting closures

Closure selection should be carried out based on the types of products, that is, children's wear, women's wear, or men's wear. Quality inspection of the closures, and price, are the main criteria for selecting closures. Proper care has to be maintained while choosing types of closures for children's wear.

6.4.1 Zippers

A zipper, zip, fly, or zip fastener, formerly known as a clasp locker, is a commonly used device for binding the edges of an opening of fabric or other flexible material, as on a garment or a bag. It is used in clothing (e.g., jackets and jeans), luggage, and other bags, sporting goods, camping gear (e.g., tents and sleeping bags), and other items.

6.4.1.1 Components of zippers

The components of a zipper, as can be seen in Figure 6.5, are:

1. Top tape extension
2. Top stop
3. Slider
4. Pull tab
5. Tape width
6. Chain and chain width
7. Bottom stop
8. Bottom tape extension
9. Single tape width

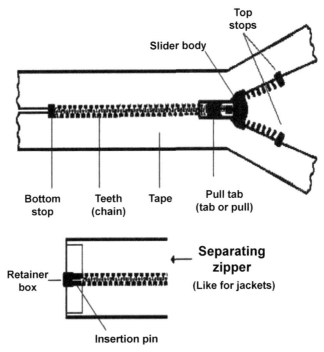

Figure 6.5 Components of zippers.

10. Insertion pin boll
11. Retainer box
12. Reinforcement film

6.4.1.2 Introduction to zippers

Tips for selecting the zipper

Different zippers suit different conditions, hence when buying zippers the following matters should be pointed out to the manufacturer:

1. Where the zipper will be used, or other special requirements
2. Requirements for the content of the zipper—whether it needs to contain AZO or nickel or a pass needle detector

Some more commonly used methods for judging the quality of a zipper are given next.

Tabs The tab of a zipper should be even without stain and scar. Touched by hand, it feels soft. The tape looks like a wave in the direction vertically and horizontally.

Teeth The surface of the teeth should be smooth; when used, it feels soft without noise.

Slider The brace should go smoothly and freely, fastened but will not come off.

Reinforcing tape The reinforcing tape fits the tape well without breaking and coming off.

Bolt and pin A square bolt can be adopted, freely fastening with the band.

Top stop and end stop

1. The top stop must come together with the first tooth on metal and nylon zippers. But the distance cannot be over 1 mm and should be firm and look fine.
2. The bottom stop must come together with the teeth or be fastened on the surface, and should be firm and look fine.

Allowable tolerance of zipper length

Due to the inertia caused by the speed of machine operation during the manufacturing of zippers as well as the completeness of the teeth, there will appear allowable tolerance. The longer the zipper is, the larger the allowable tolerance will be.

Zipper classifications

Zippers, as can be seen in Figure 6.6, can be classified into various categories, which are summarized in the next sections.

Coil zippers Coil zippers are made of polyester coil and are thus also termed polyester zippers. Nylon was formerly used, and though only polyester is used now, the type is still also termed a nylon zipper.

Invisible zipper

A coil zipper with slider

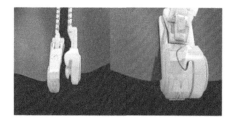
Metallic zipper

Figure 6.6 Types of zippers.

Invisible zippers Invisible zippers are usually coil zippers. They are also seeing increased use by the military and emergency services because the appearance of a button-down shirt can be maintained, while providing a quick and easy fastening system.

Reverse coil zippers Unlike an invisible zipper where the coil is also on the back, the reverse coil shows stitching on the front side and the slider will accommodate a variety of pulls (the invisible zipper requires a small, tear-drop pull due to the small slider attachment). Water-resistant zippers are generally configured as reverse coils so that the PVC coating can cover the stitching.

Metallic zippers Metal zippers are made in brass, aluminum, and nickel, according to the metal used for teeth making. All these zippers are basically made from flat wire. A special type of metal zipper is made from preformed wire, usually of brass but sometimes other metals.

Plastic-molded zippers Plastic zippers mostly use polyacetal resin, though other thermoplastic polymers are used as well, such as polyethylene.

Open-ended zippers Open-ended zippers use a box and pin mechanism to lock the two sides of the zipper into place, often in jackets. Open-ended zippers can be of any of the above-described types.

Closed-ended zippers
These are closed at both ends; they are often used in luggage.

Magnetic zippers These allow for one-handed closure and are used in sportswear.

6.4.2 Buttons

Buttons are small instruments used to fasten two parts of a garment. Buttons are an element that makes the difference and enhances a men's or women's garment, a leather item, a pair of jeans, or even nonapparel items. Elegant and classical buttons mean a unique style; the perfectly shaped and colored button is a design feature. Buttons are of different materials, colors, and shapes and they allow the wearer's personality to stand out, enhancing a garment, a leather item, jeans, or other secondary articles.

Buttons may be manufactured from an extremely wide range of materials, including natural materials such as antler, bone, horn, ivory, shell, vegetable ivory, and wood; or synthetics such as celluloid, glass, metal, bakelite, and plastic. Hard plastic is by far the most common material for newly manufactured buttons; the other materials tend to occur only in premium apparel.

6.4.2.1 Applications of buttons in apparel

- Buttons are used for opening and closing purposes in garments.
- Buttons are mostly used in all classes of men's, women's, and children's garments, as skirts, shirts, trousers, leather items, a pair of jeans, school bags, blouses, tops, T-shirts, etc.

6.4.2.2 Types of buttons

The selection of the buttons depends on the garment style, cost, and care of the garments. Various types of the buttons, as can be seen in Figure 6.7, are listed below:

1. Shank button
2. Two-hole button
3. Four-hole button
4. Snap button
5. Alloy button
6. Coconut button
7. Metal button

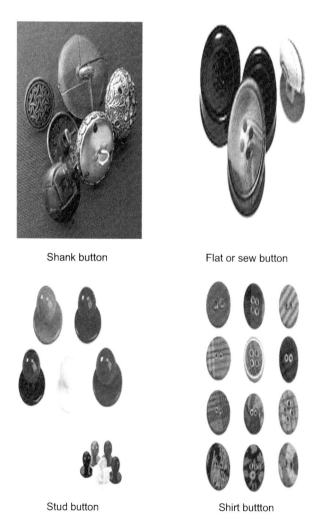

Figure 6.7 Types of buttons.

8. Packet button
9. Plastic button
10. Resin button
11. Wooden button

6.4.2.3 Importance of button attachment

The performance of the garment depends on the seams and stitches used for joining the different components of the garments. Along with this, the utility of the apparel is also affected by the trims and components used in them. Thus, the trims and buttons are important for study to determine the performance of the whole garment and the buttons as well.

6.4.2.4 Button-attaching mechanisms

Buttons can be attached by two main mechanisms: manual attachment and machine attachment.

Manual button attaching

In manual button attaching, buttons are attached by hand with the help of a hand needle and threads using a number of running stitches. In this mechanism, buttons are attached with the help of button-attachment machines.

Button-attaching machines

In these machines, the fabric is stationary during each button cycle, but the needle moves and performs every action. A single-thread chain stitch is made by the machine for button attachment.

6.4.2.5 Styles of attachment

1. **Shank** buttons have a hollow protrusion on the back through which thread is sewn to attach the button. Button shanks may be made from a separate piece of the same or a different substance as the button itself, and added to the back of the button, or can be carved or molded directly onto the back of the button, in which case the button is referred to by collectors as having a "self-shank."
2. **Flat or sew-through** buttons have holes through which thread is sewn to attach the button. Flat buttons may be attached by sewing machine rather than by hand, and may be used with heavy fabrics by working a thread shank to extend the height of the button above the fabric.
3. **Stud** buttons are metal round discs pinched through the fabric. They are often found on clothing, in particular on denim pieces such as pants and jackets. They are more securely fastened to the material. As they rely on a metal rivet attached securely to the fabric, stud buttons are difficult to remove without compromising the fabric's integrity. They are made of two couples: the male stud couple and the female stud couple. Each couple has one front (or top) and rear (or bottom) side (the fabric goes in the middle).

6.4.2.6 Types of fabric buttons

1. **Covered buttons** are fabric-covered forms with a separate back piece that secures the fabric over the knob.
2. **Mandarin buttons**, or frogs, are knobs made of intricately knotted strings. Pairs of mandarin buttons worn as cufflinks are called **silk knots**.
3. **Worked or cloth buttons** are created by embroidering or crocheting tight stitches (usually with linen thread) over a knob or ring called a **form**.

6.4.2.7 Button sizes

The size of the button depends on its use. Shirt buttons are generally small, and spaced close together, whereas coat buttons are larger and spaced further apart. Buttons are commonly measured in lignes, with 40 lignes equal to 1 inch. For example, some standard sizes of buttons are 16 lignes and 32 lignes (typical buttons on suit jackets).

6.4.3 Hooks, velcro and loop fasteners

A **velcro hook** refers to the material with tiny hooks along its surface that when pressed against loop material, creates a bond. Velcro hook tape is a roll or reel of hook material that can be cut to length and used in a variety of different applications.

Hook-and-loop fasteners consist of two components. The first component features tiny hooks; the second features even smaller and "hairier" loops. When the two components are pressed together, the hooks catch in the loops and the two pieces fasten or bind temporarily during the time when they are pressed together. When separated, by pulling or peeling the two surfaces apart, the Velcro strips make a distinctive "ripping" sound.

Different types of fasteners (hooks and Velcro) are described next.

Sew-on fasteners provide simple, easy-to-use fabric closures for clothing, accessories, and home decorating. These sew-on fabric and clothing fasteners can be used instead of buttons, snaps, or zippers and are washable by machine or dry cleaning. Simply hand- or machine-stitch around the edges of the tape and backstitch to secure.

Soft and flexible sew-on fasteners are softer and more pliable than our standard sew-on fasteners. The soft loop tapes reduce bulk and add comfort for lightweight fabrics and garments such as infant clothing. These apparel fasteners can be machine washed or dry cleaned.

Snag-free sew-on fasteners incorporate the hook and loop in one piece, with the soft loop buffering the hooks to reduce snagging and lint buildup. The one-piece design also means no leftover hook-only or loop-only pieces. These clothing fasteners can be machine washed or dry cleaned.

Durable closures/fasteners merely require positioning on the fabric, pressing firmly, and it is done. No sewing, ironing, or messy glues are required. The permanent adhesive bond withstands laundering. These are used for low-profile knit drapes with fabrics.

Fabric fusion tape bonds permanently to fabric with a steam iron. Perfect for fabrics that are thick or difficult to sew, Fabric Fusion works on most heat-tolerant fabrics, including cotton and blends, polyester, fleece, suede, canvas, denim, and nylon. It can be machine washed or dry cleaned.

6.5 Part 5: accessories for children's wear

6.5.1 Introduction

The value-added works in children garments mostly consist of:

- Embroidery
- Patchwork
- Attachment of accessories
- Printing and painting

Trim can make a garment special to a child. Decorative machine stitching, embroidery, smocking, ribbons, braid, appliqués, ruffles, lace, and bias binding are some trimming possibilities.

6.5.2 Different accessories used for children

- Metal fasteners
- Zipper fasteners
- Dungaree clips and sliders
- D-rings
- Functional and nonfunctional drawstrings, cords/ties
- Toggles
- Buttons
- Fringe
- Decorative trims and embellishments
- Beads
- Sequins

6.5.3 Design development for different accessories in children's garments

6.5.3.1 Mood boards for accessories in children's garments

Mood boards are basically collages of items such as photographs, sketches, clippings, fabric swatches, and color samples. A mood board can be actual or virtual, and can be an extremely effective presentation tool.

6.5.3.2 Storyboards for accessories in children's garments

A storyboard basically tells the story of the designer's idea. A storyboard shows not only the individual garment, but the sequence of making the garment as well.

The storyboard should include original illustrations and flats as well as additional materials. A storyboard enhances the creativity and originality in the designer's ideas and makes the visual communication of the design possibilities.

6.5.4 Safety issues for different accessories in children's garments

With a view to avoid potential hazards and to comply with safety standards, components used in children's apparel such as zippers, drawstrings, fasteners, and decorative attachments must not pose any harm to children during normal use.

6.5.4.1 Small parts: choking hazards

Small parts like snaps/rivets, buttons, appliqués, pompoms and fringes, bows and rosettes, dungaree clasps and sliders, zipper components, belt fasteners, toggles, decorative and functional loops, and decorative labels can cause choking hazards to a child.

Metal fasteners

All metal fasteners such as press fasteners, stud buttons, eyelets, and rivets must be from approved sources. They should be securely attached to the garments and free from contaminants, rust, or any toxic substances.

Zipper fasteners

Zippers on all children's products must not contain any toxic elements. If the zipper has a surface coating, it must comply with Consumer Product Safety Improvement Act (CPSIA) 2008. Zipper fasteners should not have rough or sharp edges and they must be free from rust, contamination, oxidation or all other types of degraded corrosion. Heavy zippers should be avoided, as they are uncomfortable for the child to lie on. Metal zippers get heated on exposure to the sun and should hence be avoided.

Dungaree clips and sliders

These must be selected from approved sources. They should be securely held and the coating must be capable of withstanding washing and dry cleaning in accordance with the garment care label. They must be free from rust, contamination, oxidation, or all other types of degraded corrosion. They must be metallic and nonferrous in nature.

D-rings

Selection of D-rings must be from approved sources. They should be securely held and must be free from rust, contamination, oxidation, and all other types of toxic elements. D-rings must be capable of withstanding washing and dry cleaning in accordance with the garment care label.

Functional and nonfunctional drawstrings, cords, and ties

Children's garments must not be designed to have functional and nonfunctional drawstrings, cords, or ties.

In the waist area, functional drawstrings must be secured to the garment with a bar tack to prevent the drawstrings from being pulled out of the garment. This also prevents one end of the drawstring from ending up longer than the other and thus becoming an entrapment hazard. Nonfunctional drawstrings must be bar tacked 1 cm to 0.5 in. from the exit point. No functional drawstrings, cords, or ties are allowed in any children's size range in the hood or neck area of garments.

It is recommended that children's upper outerwear have alternative closures, such as snaps, buttons, Velcro, and elastic.

Toggles
Wood, cork, leather, mother of pearl, glass, or other nondurable toggles must not be used on children's clothing. Toggles must only be used on functional and nonfunctional drawstrings, functional and non-functional ties or cords that have no free ends. Positioning of toggles must be seriously viewed and they must be securely attached.

Fixed bows
Fixed bows must be tested for color fastness to water. Bows must be securely attached with the bar tack at the center. However, there are maximum length restrictions to consider for the loop and tail lengths of bows when used on children's clothing.

Buttons
Wood, cork, mother of pearl, glass, or other nondurable buttons must not be used on children's clothing. Buttons should not have rough or sharp edges and must be free from rust, contamination, or any other toxic elements. Only core-spun polyester sewing thread should be used when attaching buttons.

Pom-poms and fringes
Traditional pom-poms and fringes made from hand knitting/sweater yarns and those constructed with metal components are not permitted for garments of children 3 years or under. Stuffed pom-poms made from fabrics are acceptable. They must be securely attached.

6.5.4.2 Decorative trims and embellishments
Appliqués should be edge stitched or stitched at the center. In the case of center stitching, they must be lockstitch attached. Embroidery is to be backed with interlining if the reverse is scratchy and comes in direct contact with the skin. In backing if lining is used, it must be permanently attached so that it cannot pose a potential choking hazard.

Beads
Individual beads can be stitched by hand in children's garments for 4 years and above, by using core-spun polyester thread. Those must be securely attached with double thread and the end of the thread is to be knotted. A maximum thread end of 1 cm and a minimum of 0.5 cm are acceptable, but floats over 1 cm are not acceptable.

Beads should not contain toxicity or any undesirable surface coating. There should not be any loss of color or loss of beads after washing or dry cleaning.

Sequins
Sequins must be securely attached with double thread and the end of the thread is to be knotted. Sequins must be lockstitch attached and secured. Sequins should not contain toxicity or any undesirable surface coating.

6.6 Part 6: evaluation of quality of trims and accessories

Trims and accessories are used to enhance the aesthetic appeal of the garment. But any defect in the accessories used in the finished product is bound to invite customer complaints about a garment that is otherwise considered to be of good quality.

In the garment industry, trims and accessories are used in a wide variety and in a huge quantity. This makes the 100% inspection of trims and accessories an almost impossible task. Moreover, inspection alone cannot assure quality. Hence, there is a need to establish an effective and efficient procurement system for trims as a preventive measure. This involves the following steps:

1. **Establishing the purchasing data.** Before procurement of trims and accessories, it is necessary to develop specifications for the items to be purchased. It is always better to have standard trim cards (duly approved by buyers) for each style in duplicate, a copy of which may be sent to the vendors along with the purchase order. The purchase order should clearly mention the specifications of trims like color, shade, and measurement (width and length) with tolerance limit, and where applicable the appropriate standard to which the product should conform.
2. **Evaluation of vendor and preparation of approved vendor list.** The organization should establish a suitable system for evaluation of vendors based on certain criteria with suitable weightage, which may include, but are not limited to, the following:
    a. Quality of supplied item (expressed in terms of percentage rejection)
    b. Price
    c. Delivery schedule
    d. Availability of item both in terms of quantity and variety
    e. Responsiveness, etc.

    Wherever necessary, the capability of the vendor may be judged on the basis of on-site assessment and review of the quality-management system.

3. **Updating the approved vendor list.** Based on the above-mentioned criteria, the organization may prepare an approved vendor list item-wise and may update the list at regular intervals. The vendors performing poorly may be deleted and a few new vendors may be added in the list.
4. **Verification of the purchased product.** Like garments, trims are not checked by 100% inspection. However it is required to confirm that bulk trims are sourced as per the required specifications. Only trims or accessories that are made manually are required to be checked. All the trims that are made by machine, for example, printed text on labels, lace, tapes, etc.,

are considered to be of the same quality. Wherever possible, the trims may be purchased along with certificates stating that the products conform to the required specifications. For trims, visual inspection of the trims is not sufficient but sometimes it is necessary to check the physical properties. To ensure that physical properties are as per standards, trims are inspected for in-house tests involving store personnel, testing investigators (if a test lab is available), and the production supervisor.

Prior to inspection, approved trim cards or reference samples as sent along with the purchase order should be made available to the quality checkers. Wherever possible, samples must be approved by the buyer or representative of the buyers or factory merchant before bulk purchase.

6.6.1 Inspection and testing of trims

6.6.1.1 Inspection of embroidery yarn

1. Shade matching: Check visually the color of the yarn against the standard in a color matching cabinet under D-65 and TL-84 light.
2. Visual assessment: Select at least five yarn packages from a lot and carry out the process. Inspect visually the appearance. The appearance should be consistent, uniform, and free of visual defects. Also during a trial run check for the presence of knots, which may create trouble during operations and impair the appearance of the product.
3. Snarling: Hold the yarn loosely at the two ends at least 50 cm apart. There should not be any snarling tendency.

6.6.1.2 Inspection and testing of zippers

1. Dimensions: Check for the correct width of tape and length of the zipper. The maximum tolerance in width of the tape should be 0.8%.
2. Color matching: Check if the color of the tape attached with the zipper matches the reference sample.
3. Check that the pull tab is affixed firmly on the slider body.
4. Check that the slider rides freely, and at the same time it should not be loose on the chain.
5. Check that the slider locks tightly.
6. Move the slider rapidly and repeatedly and observe that there should not be popping up of the zipper at any point.
7. Check the color fastness to washing of the zipper tape. The change in color of zipper tape and degree of staining of the multifiber fabric are to be checked using the AATCC gray scale.

6.6.1.3 Inspection and testing of buttons

1. Check that the sew holes of the buttons are clear and free from flash and burrs.
2. Verify that the color of the buttons is as per the approved samples.
3. Check the buttons visually and ensure that there is no crack-melting surface.
4. Attach at least 100 buttons to 10 fabrics (at least 8 buttons per fabric). Now perform dry cleaning, laundering, and pressing. Check the appearance of the buttons after these operations. There should not be any change in color or development of cracks.

6.6.1.4 Inspection and testing of buckles

1. Check the size of the buckles.
2. Check visually if there are any defects like sharp, burred edges.

6.6.1.5 Inspection and testing of snap fasteners

1. Check visually the appearance of the snap fasteners.
2. Check that the socket is clean.
3. Check that the bottom of the post is flat.
4. Mount at least 50 snap fasteners onto the fabrics. Check manually how easily they can be fastened or unfastened and their holding power.

6.6.1.6 Inspection and testing of labels

1. Check the text of the labels and font size against the approved samples.
2. Check the size of the labels to determine if they follow the standard.

6.6.1.7 Inspection and testing of fusible interlinings

Fusible interlinings, before going into a mass-scale production process, should be assessed by a trial run. At least three pieces of the fabric are cut and subsequently fused following the procedure recommended by the manufacturers. The appearance and feel of the specimens are then checked. The appearance of the fabric will be impaired (bubbled appearance) in the case of excessive differential shrinkage.

6.6.1.8 Inspection and testing of lace or tape

1. Dimension: Check the dimension of the tape/lace against the reference standard. If any variation is observed, it is accepted if it is within acceptance level after reporting it to buyer or his representative.
2. Shrinkage: If the shrinkage percentage of the trims varies from the fabric's shrinkage percentage, it will cause defective garments. Trims such as tapes and laces must be tested for their shrinkage percentage by dipping a predefined length of material in water for 12 h and measuring the same after drying The shrinkage found in testing is compared against the tolerance label to determine whether the trims will be accepted or not.
3. Color fastness: Dyed trims are checked for color fastness. In this test, lace or tape samples are washed in a wash wheel along with multifiber fabrics for the number of cycles mentioned in the testing methods (AATCC or ASTM). If the white fabric becomes tinted with the trim's color, then the trims should not be used in production.

6.7 Part 7: fashion accessories

Jewelry is the most popular fashion accessory. A popular piece of jewelry consists of large pendant necklaces, many of which exhibit a cross or other popular or meaningful symbol. Trendy pieces of fashion jewelry may include earrings, rings, necklaces, bracelets, pins, and so on.

Another kind of fashion accessory is a purse or even a handbag. Young girls most commonly prefer their own purses and handbags. A purse is a container used for carrying money and small personal items or accessories often used as a bag that is smaller or compact in size, and handbags are often a little bigger. Handbags and purses come in different varieties with elegant styles to describe the status of the user; thus, among women and teenagers, it is common to possess more than one purse or handbag. In conjunction with handbags and purses, travel bags are sometimes considered a fashion accessory. Travel bags are similar to purses and handbags, but they are larger and are often designed similarly for both women and men.

Shoes and boots are also considered a fashion add-on. Like handbags and purses, many women attempt to coordinate their shoes, especially for work, with the rest of their ensemble.

Another important type of fashion accessory is belts. Females' belts are available in a number of different sizes, shapes, and styles. Belts, handbags, purses, travel bags, jewelry, and boots and shoes are just a few of the countless fashion accessories in popular use today. Hence, fashion accessories are an effective way to spice up any wardrobe, especially one that could use an updating.

6.7.1 Footwear

Footwear, including different types of socks, stockings, boots, and shoes, form the largest category of accessories. These are the garments that are worn on the feet for several purposes, such as adornment and protecting the feet against the environment.

6.7.1.1 Socks

Socks, as shown in Figure 6.8, are manufactured in a variety of lengths. Bare or ankle socks extend to the ankle or lower and are often worn casually or for athletic use. Bare socks are designed to create the look of bare feet when worn with shoes. Knee-high socks are sometimes associated with formal dress or as part of a uniform, such as in sports or as part of a school's dress code.

6.7.1.2 Stockings

A stocking is a close-fitting, elastic garment covering the foot and lower part of the leg. Stockings, as can be seen in Figure 6.9, vary in color, design, and transparency.

Shoes and boots
Design and product development/manufacturing Shoe designers need to study current fashion trends so that the shoes coordinate well with the apparel. Style, size, and fit characteristics that will make up the finished shoe are important.

Pattern-making A shoe consists of a number of different parts, each of which must be made into a perfectly accurate pattern to ensure a problem-free final product. The patterns may be created by using different methods.

Die manufacturing Cutting dies are prepared to cut several layers of material to make the shoe. Sharp steel strips are bent around the various paper patterns to form individual dies for each part of the shoe.

Figure 6.8 Socks.

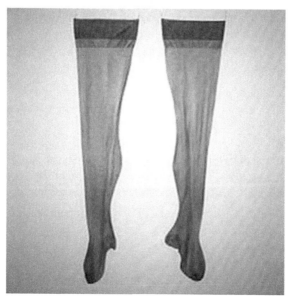

Figure 6.9 Stockings.

Cutting After manufacturing the dies, they are used to cut the materials. The dies are first placed over the materials and then pressed to cut through a number of layers.

Fitting or stitching A number of individual operations are performed in the fitting room where all individual parts are sewn together.

Lasting The fitted upper is pulled over the last in a series of operations that make the upper conform to shape of the last. Different lasting techniques can be adopted depending upon quality, appearance, and function of shoe required.

Assembling the remaining components After the upper is assembled and lasted, various parts/components such as counters, stiffening materials, sock linings, outsoles, shanks, and heels are assembled.

Bottoming In the bottoming room, the lasted upper receives shanks and fillers. It is prepared for the outsole to be permanently attached to the upper of the shoe. After this, heels are nailed through the insole for strength, which completes the shoe.

Finishing Various finishing operations are performed on a shoe before it is sent to the packing room to be shipped to retailers.

After the product is considered to be ready for the final delivery to the customer, it is boxed and sent either to a warehouse for storage or directly to the purchaser.

Components and styles of footwear

Footwear components Footwear consists of a number of different components (as shown in Figure 6.10) and described in the next sections. Some terms refer to parts that all shoes have, such as the sole; while other terms may only apply to certain types or style of shoes.

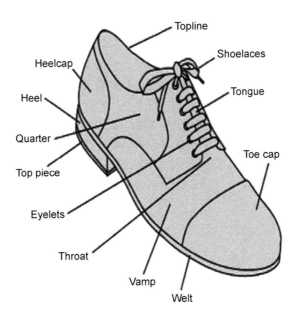

Figure 6.10 Footwear components.

Insoles An insole is a layer of material that is placed inside the shoe, which acts as a layer between the sole and the wearer's foot. Along with performing the function of hiding the joint with the upper, an insole adds comfort to the wearer.

Heels The heel is that part of the sole that raises the rear part of the shoe in relation to the front. The heel seat is typically shaped to match the form of the upper. The part of heel that comes in contact with the ground is known as the top piece.

Outsoles Outsoles are the exposed parts of the sole that come in contact with the ground. Outsoles are produced using a variety of materials. Various properties required from the outsole are grip, durability, and water resistance.

Breasts The breast is the forward-facing part of the heel that lies under the arch of the sole.

Counters Between the lining and upper of the shoe, the counter, a stiff piece of material, is placed at the heel of a shoe to help in maintaining the shape of the shoe.

Feathers The feather is that part of the shoe where the upper's edge meets the sole.

Linings Most shoes have linings on the inside of the shoe around the vamp and quarter for the purpose of improving comfort and increasing the lifespan of the shoe.

Puffs A puff is a reinforcement inside the upper, which provides the toe with its shape and support.

Quarters The quarter is the rear and side part of the upper that covers the heel.

Seats Seats are that part of the shoe where the heel of the fit sits in the shoe.

Shanks A shank is a piece of metal inserted between the sole and insole of the shoe lying against the arch of the foot.

Soles The sole is the entire part of the shoe that sits below the wearer's foot.

Uppers The upper is the entire part of the shoe that covers the foot.

Top piece A top piece is that part of the heel that comes in contact with the ground.

Vamps Vamps are the section of the upper that cover the front of the foot and join to the quarter.

Welts A welt is the strip of material that joins the upper to the sole.

Throats The front of the vamp next to the toe cap is called the throat.

Toe caps Shoes may have a toe cap in the front upper of the shoe.

Footwear styles Figure 6.11 shows different types of shoes.

Men's shoes

1. Oxfords
2. Loafers
3. Cap-toes
4. Monk straps

Men's Boots

1. Chukka boots
2. Cowboy boots
3. Platform boots

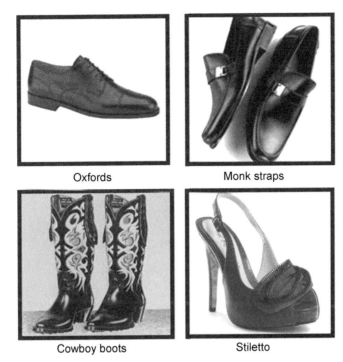

Figure 6.11 Types of shoes.

4. Motorcycle boots
5. Dress boots

Thigh-high boots, women's shoes

1. High-heeled footwear

   Types of heels found on high-heeled footwear include:

   1. Cone
   2. Kitten
   3. Prism
   4. Puppy
   5. Spool or Louis
   6. Stiletto
   7. Wedge
   8. Mule
   9. Slingback
   10. Ballet flats or dolly shoes
   11. Court shoe or pump

Women's boots

1. Booties
2. Ankle boots
3. Flat boots
4. Knee-high boots
5. Rain boots

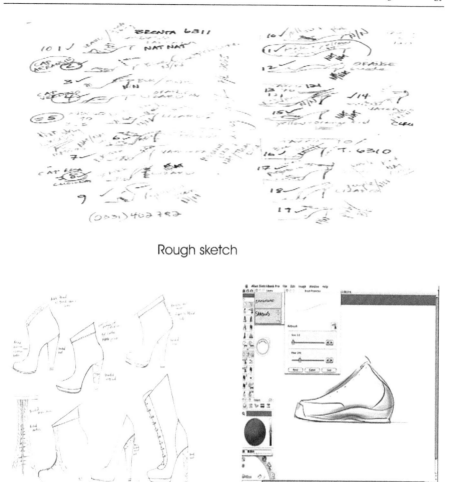

Figure 6.12 Sketches of shoes.

Sketching techniques for footwear Examples of hand sketches and computer sketches are shown in Figure 6.12.

Tips for choosing the right footwear There are two key factors to consider when buying a pair of shoes: (1) choose a shoe with a solid construction that provides all the support that customers need and (2) make sure the shoe fits well.

6.7.2 Handbags and small leather goods

Handbags must be both decorative and functional; they must hold necessities conveniently as well as fit into the fashion picture. Large bags such as totes, satchels, portfolios, or backpacks tend to be functional; smaller bags such as clutches or envelopes are usually decorative. Handbag styles, as can be seen in Figure 6.13, range from classic,

Figure 6.13 Styles of handbags.

constructed types to soft shapes. Leather, including suede and reptile, still represents approximately half of handbag material; vinyl, fabric, and straw make up the other half.

6.7.2.1 Variants of handbags

Nowadays a variety of handbag styles, as seen in Figure 6.14, can be described as having some of the following features.

Cross-body bag
A cross-body bag is a handbag of any size, but usually small, with a long strap, designed to be carried over one shoulder across the chest.

Drawstring bag
Any bag that can be closed or cinched with a drawstring on top can be called a drawstring bag.

Purse
The term "purse" started out as a small container holding coins, and now may contain a wallet, keys, cell phones, makeup, and anything else one might be able to squeeze into it.

Shoulder bag
A shoulder bag is any handbag with a long shoulder strap.

North–south bag
"North–south bag" is a designer term for any bag that is taller rather than wider.

Figure 6.14 Variants of handbags.

East—west bag

"East—west bag" is a designer term for any bag that is wider rather than taller.

Styles of handbags

The common styles of handbags, which can be seen in Figure 6.15, are described next.

Figure 6.15 Styles of handbags.

Clutch A clutch is a woman's small purse that can be carried in the hand and usually has no handle or strap.

Doctor's bag A doctor's bag is a large handbag shaped like the traditional doctor's bags, usually made from leather, with two handles.

Fold-over bag A fold-over bag is a zippered-top bag, usually flat, that may be folded over and carried under the arm or in a hand.

Hobo Hobos are single-strap women's bags, with a curved crescent shape and slouchy look, usually with a zipper closure on top.

Pouch A pouch is a relatively small bag of soft fabric or leather gathered with a drawstring as a means of closure, which usually extends as handles.

Satchel A satchel is a rigid-bottom bag of varying sizes with one or two handles. It is usually carried on the arm rather than the shoulder.

Tote bag Tote bags are large handheld bags with two handles, open top, and simple structure. Often made from canvas, they are usually used to carry groceries, beachwear, or books.

Method of construction for recycled tote bag
Designing First, a sketch is drawn on the ivory sheet with the help of pencil, and after that some fashion details are included in it.

Pattern-making Different pattern pieces are prepared according to the components of the bag.

Cutting The fabric pieces are cut according to the patterns with the help of scissors.

Sewing Two equal pieces of denim fabrics are taken for both the upper and lower sides of the bag.

The base of the bag is prepared by using cardboard for stiffness with denim fabric and attaching a piping all around the base of the bag in order to maintain the shape of the base of the bag. Two rectangular pieces of another type of denim fabric are attached, which has a slub effect. Slub denim is used for the sides of the bag. A welt pocket is made on both sides of bag by using a brass chain.

Trims and notion attachment Decorative wool threads are attached diagonally on both sides of bag. Two side buttons are attached on each side of bag, which enhances the aesthetic look of the bag.

Briefcase A briefcase is usually a flat, rectangular case, designed to carry documents, books, files, and personal items like a wallet and electronics. It is most often made from leather and has a single carrying handle.

Other types of bags Other types of bags include laptop bags, full-moon bags, and belt bags.

Travel-related bags
Travel-related bags range in size from large luggage bags, enough to hold a wardrobe and other necessities for a long trip, to smaller bags for transporting items on the daily route to and from home to work or school.

Duffel bag A duffel bag is a large, cylindrical bag with a zipper opening on top, usually made out of canvas or leather for carrying personal belongings.

Fanny pack A fanny pack is a small zippered pouch suspended from a belt around the waist.

Messenger bag A messenger bag is a rectangular bag with a single flap closure and an adjustable long shoulder strap.

Suitcase Suitcases are rectangular large bags with solid structure and reinforced corners, used to carry luggage on trips.

Travel bag A travel bag is generally a rectangular bag with multiple compartments and pockets, used as carry-on luggage on airplanes or short trips.

Backpack bag Backpacks are bags of varying shapes and sizes carried on one's back and secured with two straps that go over the shoulders.

Small leather goods

Small leather goods, as can be seen in Figure 6.16, encompasses a variety of containers that women typically carry in their handbags and men may carry in a pocket or briefcase to hold small items.

Checkbook case A checkbook case is a single-purpose case to hold and protect a checkbook. It has the signature rectangular shape and might have a few more slots or pockets for credit cards, ID, or a pen attachment, as well as an external pocket for change and bills.

Coin purse Coin purses are small, one-compartment purses used to carry coins, bills, and cards.

Cosmetic bag A cosmetic bag is a small bag with a zipper closure usually lined in plastic, which holds cosmetics.

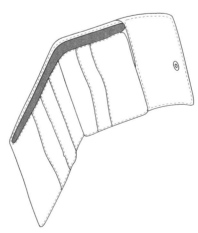

Figure 6.16 Small leather goods.

Credit card holder A flat, smaller version of a wallet, a credit card holder is designed to carry only a small number of credit cards and an ID card.

Eyeglass case An eyeglass case holds sunglasses and reading glasses and protects them while carrying them in a bag or pocket. Shapes and construction vary widely, but it always resembles and closely fits the eyewear.

Key holder There are quite a few variations in key holders. They can look like a wallet that opens up to expose a metal frame with hanging rings where one can attach keys.

Cell phone case With the advancements in technology and the fact that almost everyone carries a cell phone, there are a variety of leather cases that suit them.

Wallet A wallet is a small, flat case, usually with multiple compartments and a zipped coin section, used to carry money, credit cards, coins, and other personal items like ID and checks.

Wristlet A wristlet is a small, usually flat bag with a very short strap that only fits around the wrist. The most common shape is a rectangle with a zipper closure.

Design and product development

The elements of fabrication, silhouette, and color, as well as current trends in ready-to wear and footwear, are the most important components of handbag design, which may be seen in Figure 6.17. From an initial sketch, a sample is made from muslin or imitation leather. Felt, foam, and fabric interlinings are layered to give the bag a nice hand and cushion.

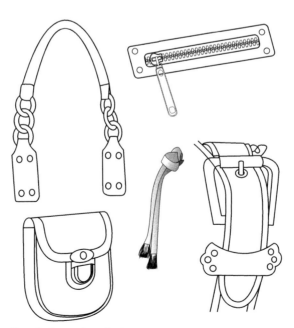

Figure 6.17 Handbag design and product.

Components of handbags

Although handbags vary in shapes and sizes, they can be distinguished by some characteristics and components, as may be seen in Figure 6.18. Most of these features are also found on travel-related bags, and some apply to small leather goods.

Handbag construction varies depending on the style of the bag. The minimum required components of a handbag is outer covering with frame, lining, handle, closure, gussets, trimmings, etc.

Outer covering The outer covering is comprised of fashion fabric, straw, leather, etc.

Frame The frame is a heavy-gauge steel or brass structure giving the handbag its shape.

Padding Padding may be used to cover the hard edges of the frame and protect the outer covering from the metal frame.

Lining Lining covers most of construction details like the seams and frame. It is most durable and found at higher price points.

Underlining Underlining refers to a heavy paper or cardboard placed between the outer covering and lining for additional support.

Handle A handle can be attached at the top or sideways on the bag to support the weight of the content.

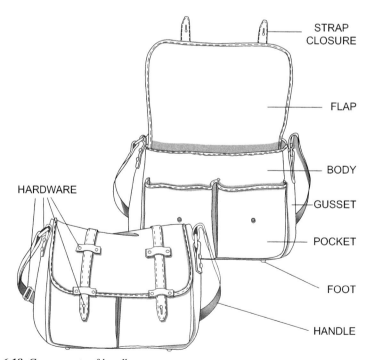

Figure 6.18 Components of handbags.

Fastener and closure The fastener can be defined as the hidden or decorative closing device, such as a flap, snap, drawstring, lock, or zipper.

Gusset Gussets are expanded gores or panels on the sides of the handbags to make it larger.

Trimmings Trimmings include all decorative features that are used to ornament the handbag.

Design process behind a handbag collection: conceptualizing, design, and development of mood board and handbags

Finding the inspiration A source of inspiration is found in the initial stage. A collection of photos is made on the mood board. For example, a theme for handbags might be conceptualized as "blossoms." For this particular theme, the designer has been mostly inspired by the season, nature, and flowers according to which photos are collected until the final decision upon the collection concept, which may be seen in Figure 6.19.

The mood board is essential for a designer. Not only does it charge him or her emotionally, but it helps define the course of the collection, the color palette, and to find the collection's story.

Figure 6.19 Findings.

Figure 6.20 Sketching of ideas.

Sketching of ideas The creative part comes naturally. Different ideas are mixed and collaborated, to be expressed through sketches, as may be seen in Figure 6.20. If the designer sees something inspiring, he or she will try to recreate it in bags, through drawings.

Deciding on colors and fabrics The concept comes to life when the right fabrics are finally found. Various fabric stores are explored, and different kinds of materials are tested. It is believed that design, including color and fabric, go hand in hand, as can be seen in Figure 6.21, and they depend on each other. When a product is created from leather, then various shapes and volumes are experimented with, until the desired form is obtained. This is something a mere drawing cannot give.

Documentation Documentation is one of the most important aspects of design, which helps the designer to set up the entire concept and story. A brief research is done on the inspirational source regarding the forms and proportions of flowers.

Choosing the final styles for the collection Various styles of handbags, as seen in Figure 6.22, are studied and observed, and are then finally included in the collection.

Creating patterns and prototypes Final styles are selected for the collection from the model drawing ideas, from which the patterns are created, as can be seen in Figure 6.23.

Production Finally the manufacturing process is completed, as can be seen in Figure 6.24. The process does not stop here. Next comes the photo session, distribution, and the evaluation.

Selecting garment accessories, trims, and closures 165

Figure 6.21 Colors and fabrics.

Figure 6.22 Selection of final style.

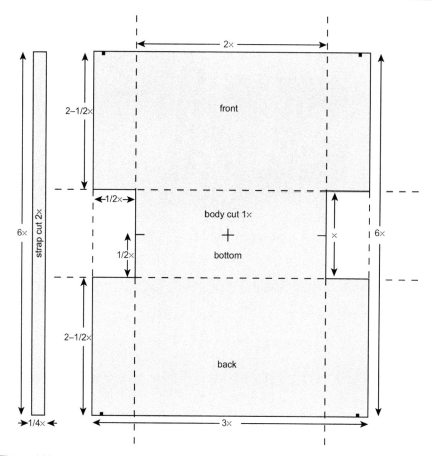

Figure 6.23 Prototype patterns.

Exploring fashions, designs, and embellishments for the handbag project

Shape Bags run from tall and slender to short and wide. The shape of a bag affects the handling and appearance.

Color Textiles and other materials are available in a wide variety of colors. Using coordinating and contrasting colored panels or trims, many designs can be created, as seen in Figure 6.25.

Pattern Patterns can be subtle, loud, abstract, geometric, floral, or simply the pattern created by the construction and trim on the bag.

Material The material affects the bag in many ways, including appearance, handling, weight, and feel.

Tips for choosing the right handbag
Most bags can be purchased with wheels, backstraps, and retractable handles. While handles and backstraps are consistently useful features, some travelers get frustrated by stiff luggage wheels.

Selecting garment accessories, trims, and closures 167

Figure 6.24 Production of handbags.

Figure 6.25 Exploring designs.

6.7.3 Other accessories

Headwear

Hats and caps protect the head from sunburn and possible sunstroke. They also provide shade for the eyes. Hats and caps are two of the best-selling fashion accessories for any age and either gender, because they are a relatively inexpensive wardrobe builder.

Hats A hat is a head covering. It can be worn for protection against the elements, for ceremonial or religious reasons, for safety, or as a fashion accessory.

Parts of hats The basic parts of a hat include the crown, visor, brim, hatband, and trim, as can be seen in Figure 6.26. Both the crown and brim are present in most hat styles and in innumerable combinations.

Crown The crown is the portion of a hat covering the top of the head.

Peak/visor/bill The peak, also known as the visor or bill, is a stiff projection at the front, to shade or shield the eyes from sun and rain.

Brim The brim is an optional projection of stiff material from the bottom of the hat's crown horizontally all around the circumference of the hat.

Puggaree/sweatband/hatband This is a ribbon or band that runs around the bottom of the torso of the hat. The sweatband may be adjustable with a cord or rope at the top and is on the inside of the hat touching the skin, while the hatband and puggaree are around the outside.

Decorative trim Decorative trim provides ornamentation, including ribbons, feathers, flowers, netting, and any other suitable materials.

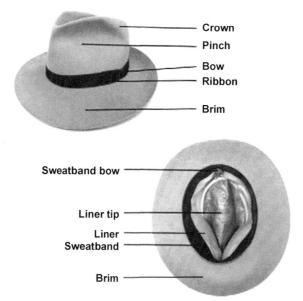

Figure 6.26 Parts of hats.

Designing and production of hats Designers create hats and hair accessories by drawing a detailed sketch, complete with trim, from which a sample is made. Designers consider the appeal, wearability, proportion, and price point when creating hat and hair accessories. Hats may be woven, felted, knitted, or crocheted. Buckram may form the base of a fashion hat that will ultimately be covered in fashion fabric.

Basic styles of hats Facial shape, head size, and hairstyle should determine which hat style a customer selects. Two general rules of thumb for wearing hats are (1) the hat style should contrast with the facial shape and (2) the hairstyle should be subordinate to the hat style. Some basic styles of hats are shown in Figure 6.27.

Caps Caps, as seen in Figure 6.28, have always been different than hats. Hats have a broad brim and can be made of material like straw. Caps fit closely to people's heads and are usually made of soft fabrics. Therefore, caps are generally soft and often have no brim or just a peak. For many centuries women wore a variety of head-coverings, as seen in Figure 6.29, which were called caps.

Styles of caps

Baby caps Babies and adults alike lose a large amount of heat through their heads. Babies have a hard time regulating their body temperature until about 6 months of age. Covering a baby's head with a baby hat or baby cap helps to reduce heat loss

Figure 6.27 Style of hats.

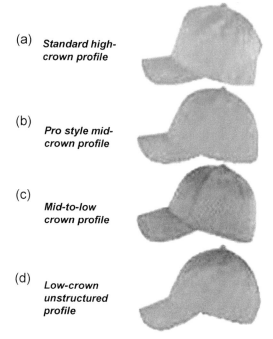

Figure 6.28 Caps.

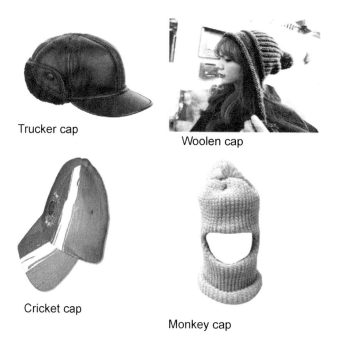

Trucker cap

Woolen cap

Cricket cap

Monkey cap

Styles of cap

Figure 6.29 Types of caps.

and keep them warm. For people living in a cold climate, it is important to cover your baby's head until babies are back in a warm environment.

Knit/winter caps A knit cap, originally of wool, though now often of synthetic fibers, is designed to provide warmth in cold weather.

Ski caps Ski caps are a popular staple in winter wardrobes everywhere there is cold weather. A ski cap protects the wearer with head-hugging material that insulates and keeps the heat from escaping the body.

Beanies Beanies are another tight-fitting winter cap. These caps fit much tighter than ski caps, acting like a glove for the head. They have no brim and come in several types of materials, including acrylic, wool, polyester, and fleece.

Trucker caps Trucker caps are similar to a baseball cap, but the construction is slightly different. A trucker cap is made of six mesh panels that meet with a button on the top. Trucker caps utilize foam. This foam makes the cap more breathable and comfortable to wear.

Russian ushanka caps These caps have earflaps that look great worn down over the ears, or tied up on top of the head. These earflaps come lined with the fur, faux fur, or sheepskin for added warmth and comfort.

Wool caps Wool caps offer the wearer one of the highest levels of warmth and protection from the windy conditions of winter. Caps made from wool come in several styles, including beanies, snowboard, berets, and church.

Cotton caps Cotton caps offer the most comfortable wear possible with an excellent warmth level. Cotton comes from a flowery shrub with soft, downy fibers. These fibers surround oily seeds, which makes for some of the softest pieces of clothing. Cotton caps come in several styles, including beanie, trooper, combat, and aviator.

Fleece caps Fleece caps are some of the softest caps due to the fabric itself. It comes from the wool of a sheep, or similar animal. It has deep, soft piles that make it feel luxurious against the skin. Fleece caps come in many styles, including trapper, snowboard, hood scarf, half-face mask, and hoods.

Polyester caps Polyester is a wrinkle-resistant fabric made from a number of synthetic fibers. Polyester caps are comfortable to wear and smooth to the touch.

Cricket caps A cricket cap is a type of soft cap, often made from felt, that is a traditional form of headwear for players of the game of cricket, regardless of age or gender. It is usually a tight-fitting skullcap made of six or eight sections, with a small crescent-shaped brim that points downwards over the brow to provide shade for the eyes.

Football caps In the early days of football, the concept of each team wearing a set of matching shirts had not been universally adopted, so each side would distinguish itself from the other by wearing a specific sort of cap.

Baseball caps Baseball caps are a type of soft light cotton cap with rounded crown and a stiff bill projecting in the front.

Golf caps A golf cap is a rounded men's or women's cap with a small stiff brim in front. Cloths used to make the cap include wool, tweed, and cotton. The inside of the cap is commonly lined for comfort and warmth.

Monkey caps The monkey cap is a form of cloth headgear designed to expose only part of the face. Depending on the style and how it is worn, only the eyes, mouth, and nose, or just the front of the face, are unprotected.

Turbans A turban is a kind of headwear based on cloth winding. Featuring many variations, it is worn as customary headwear, usually by men. Turbans worn in South Asia are known as Pagri.

Barretinas A traditional style, in red, a barretina is now used as a symbol of Catalan identity. It is worn with the top flopping down.

Balmoral bonnets The balmoral bonnet is a traditional Scottish bonnet or cap worn with Scottish highland dress.

Ayam caps An ayam cap is a Korean traditional winter cap mostly worn by women in the Joseon period, from 1892 to 1910.

Gloves

A glove is a garment covering the whole hand. Gloves have separate sheaths or openings for each finger and the thumb; if there is an opening but no covering sheath for each finger, they are called fingerless gloves.

Materials used for making gloves

Natural rubber: latex Durable natural rubber gloves are commonly used in industrial applications, for cleaning, and in food processing and handling. They offer a high level of protection against cuts, but are not recommended for use in oils. Latex gloves are offered in varied thicknesses for high sensitivity or extra protection.

Synthetic rubber: nitrile Gloves made from nitrile are available in four different thicknesses as ambidextrous examination type, and in hand-specific styles, both lined and unlined. Heavier nitrile gloves are routinely used in manufacturing, cleaning, and the food-processing industry.

Synthetic rubber: neoprene Neoprene offers superior protection against oils, acids, solvents, and caustic chemicals, but is somewhat less effective against abrasions and cuts than either natural rubber or nitrile.

Polyvinyl alcohol Polyvinyl alcohol gloves are very resistant to most chemicals and ketones, but are not recommended for use in water and alcohol. They protect well against snags, cuts, punctures, and abrasion.

Polyvinyl chloride Synthetic polyvinyl chloride (PVC) gloves offer economical protection for workers exposed to chemicals, fats, oils, abrasion, and punctures.

Polyethylene Polyethylene (PE) gloves are available in a variety of packaging options for the food-service industry.

Polyurethane coating Polyurethane (PU) coating on fine-gauge knit gloves offers superior dexterity and performs well in alcohols, grease, animal fats, acids, and bases, but is not recommended for use in organic solvents.

Polyester/cotton blends Varied combinations of yarn blends increase machine knit strength and durability, while still allowing for worker dexterity. These gloves are often coated with rubber, nitrile, or other substances to improve grip and longevity.

Kevlar One of the strongest of all fibers, this aramid material offers excellent cut protection and heat insulation, but offers little protection from abrasion.

Spectra This cut-resistant polyolefin fiber is offered for use in food-processing and industrial environments. These gloves can be repeatedly laundered and bleached.

Ultra-high-molecular-weight polyethylene Ultra-high-molecular-weight polyethylene (HMWPE) is a highly cut-resistant fiber yarn offering maximum cut protection and a high level of flexibility.

Teflon Teflon coating on quilted woven mittens offers protection with a thin fluorocarbon film that features microbial and stain resistance, excellent cleanability, and flame-retardant properties.

Pyroguard Pyroguard-woven mittens are flame retardant, repel moisture and stains, and also offer vapor protection.

Glove components and measurements

Components The component parts, as seen in Figure 6.30, that may be found in a glove are one pair of tranks, one pair of thumbs, four whole fourchettes, four half fourchettes, two gussets, and six quirks. Depending on the style of the glove there may also be roller pieces, straps, rollers, eyelets, studs, sockets, and domes. Finally, linings will themselves consist of tranks, thumbs, and fourchettes.

Measurements Using a measurement tape, measurements of hands are done all around, but not tightly. The number of inches shown on the tape measure is the correct glove size, as seen in Figure 6.31.

Glove styles
Commercial and industrial

- Aircrew gloves: fire resistant
- Barbed-wire-handler's gloves

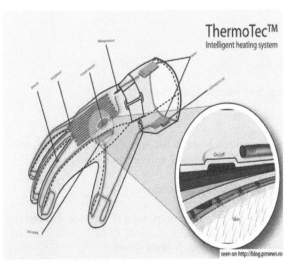

Figure 6.30 Glove components.

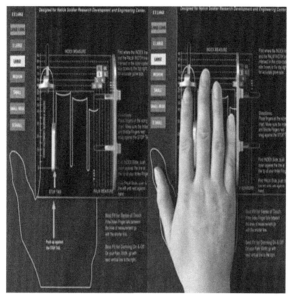

Figure 6.31 Measurement of gloves.

- Chainmail gloves: used by butchers, scuba divers, woodcutters, and police
- Chainsaw gloves
- Cut-resistant gloves
- Disposable gloves: can be used by anyone from doctors making an examination to caregivers changing diapers
- Firemen's gauntlets
- Food-service gloves
- Gardening gloves
- Impact-protection gloves
- Medical gloves
- Military gloves
- Rubber gloves
- Sandblasting gloves

Sport, recreational, and specific-use gloves

- American football various-position gloves
- Archer's gloves
- Baseball gloves or catcher's mitts: in baseball, the players in the field wear gloves to help them catch the ball and prevent injury to their hands
- Billiards glove
- Boxing gloves: a specialized padded mitten
- Cricket gloves
    - The batsmen wear gloves with heavy padding on the back, to protect the fingers in case of being struck with the ball
    - The wicket-keeper wears large, webbed gloves.

- Driving gloves, intended to improve the grip on the steering wheel. Driving gloves have external seams, open knuckles, open backs, ventilation holes, short cuffs, and wrist snaps. The most luxurious are made from Peccary gloving leather.
- LED gloves
- Motorcycle gloves
- Oven gloves, or oven mitts, used when cooking
- Paintball gloves
- Racing driver's gloves with long cuffs, intended for protection against heat and flame for drivers in automobile competitions
- Scuba-diving gloves
  - Cotton gloves: good abrasion but no thermal protection
  - Dry gloves: made of rubber with a latex wrist seal to prevent water entry
  - Wet gloves: made of neoprene and allowing water entry
- Touch screen gloves: made with conductive material to enable the wearer's natural electric capacitance to interact with capacitive touch screen devices without the need to remove one's gloves
- Ski gloves: padded and reinforced to protect from the cold but also from injury by skis
- Underwater hockey gloves: with protective padding, usually of silicone rubber or latex, across the back of the fingers and knuckles to protect from impact with the puck; usually only one, either the left or the right hand, is worn depending on which is the playing hand.
- Washing mitt or washing glove: a tool for washing the body
- Webbed gloves: a swim-training device or swimming aid
- Weightlifting gloves
- Wired gloves
- Power gloves: an alternate controller for use with the Nintendo Entertainment System
- Wheelchair gloves: for users of manual wheelchairs

Fingerless gloves Fingerless gloves used in motorcycling are commonly made of leather and padded in the palms, while the backs might be made of netted mesh. Riding gloves of this type allow a solid grip on the hand controls while absorbing some of the vibration of the road, protecting the palms, and making the ride more comfortable. Typically the backs are made with Velcro for easy placement and removal as seen in Figure 6.32.

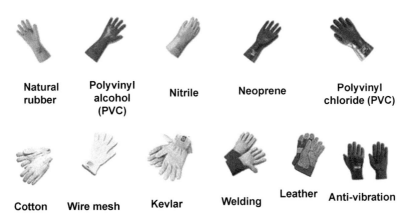

Figure 6.32 Types of gloves.

Leather gloves A leather glove is a fitted covering for the hand with a separate sheath for each finger and the thumb. This covering is composed of the tanned hide of an animal, though in recent years it is more common for the leather to be synthetic.

Driving gloves Driving gloves are designed for holding a steering wheel and transmitting the feeling of the road to the driver. They provide a good feel and protect the hands. They are designed to be worn tight and to not interfere with hand movements. The increased grip allows for more control and increased safety at speed.

Mittens Gloves that cover the entire hand but do not have separate finger openings or sheaths are called mittens. Generally, mittens still separate the thumb from the other four fingers. They have different colors and designs.

Belts

A belt is a flexible band or strap, typically made of leather or heavy cloth, and worn around the waist. A belt supports trousers or other articles of clothing and holds the garment in place without slipping down.

Belts, include:

- Casual belts
- Dress belts
- Dress casual belts
- Designer fashion belts
- Uniform belts
- Workwear belts

Types of belts

- A studded belt is typically made of leather or similar materials, and is decorated with metal studs.
- Instead of wearing a conventional belt, skateboarders often wear shoelaces in belt loops to hold their pants up.
- Belts are also used in judo, karate, and other martial arts, where different colors may indicate rank or skill.
- A breast belt is a belt worn by women that holds their breasts up, making them appear larger or more prominent underneath clothing.
- The leg belt is another fashionable article of clothing which, as the name suggests, one wears upon the leg.
- One specialized type of belt is the utility belt or police duty belt, which includes pockets for carrying items that the wearer needs for prompt use and loops to hang larger items.

Belt manufacturing design process

Process of manufacture After selection of a suitable material, the belts of different sizes are cut by a strap-cutting machine and skived from the edges. Similarly, the lining of the same sizes of required material is also cut. The skived edges are folded. Then the lining is attached by paste. After pasting, the belts are stitched with the help of stitching machine. The excess lining is trimmed. The other operations, like buckle attaching and edge-setting punching, are then carried out according to the design. The belts are then finally inspected and packed.

Selecting garment accessories, trims, and closures 177

Figure 6.33 Manufacturing process for belts.

The different steps involved in this process are shown in Figure 6.33 and summarized here.

Step 1: Layout
Step 2: Making the angle

Using the angle tool, scrub edge of leather that is going to be against body.

Step 3: Making belt border

It is done with the fork punch and a hammer. The border is punched all the way around the belt.

Step 4: Making belt holes

Holes are punched with a hole punch.

Step 5: Making designs on the belt

With a fork punch, a design is punched.

Step 6: Make holes for snaps

Punch holes for snaps.

Step 7: Punch snaps in place

Put snaps in hole and then punch them to lock them in.

Step 8: Finished belt

Snaps together and belt is completed.

Neckwear

Neckwear refers to various styles of clothing worn around the neck. Neckwear is worn for fashion, combat, or protection against the influences of weather. Common neckwear today includes bow ties, neckties, scarves, feather boas, and shawls. Historically, ruffs and bands were worn.

Neckties The tie is an accessory. It adds to the look. A tie can be the centerpiece. It can be a "piece of cloth" in the center or it can be a short tie, as seen in Figure 6.34.

Bows The bow tie is a type of necktie. It consists of a ribbon of fabric tied around the collar in a symmetrical manner such that the two opposite ends form loops. Bow ties may be made of any fabric material, but most are made from silk, polyester, cotton, or a mixture of fabrics, as can be seen in Figure 6.35.

Scarves A scarf is a piece of fabric worn around the neck, near the head, or around the waist for warmth, cleanliness, fashion, or for religious reasons. They can come in a variety of different colors, designs, and textures, as can be seen in Figure 6.36.

Figure 6.34 Neckties.

Selecting garment accessories, trims, and closures

Figure 6.35 Bows.

Figure 6.36 Scarves.

Figure 6.37 Shawls.

Shawls A shawl is a simple item of clothing, loosely worn over the shoulders, upper body, and arms, and sometimes also over the head. It is usually a rectangular or square piece of cloth, or can be triangular in shape. Other shapes, as can be seen in Figure 6.37, include oblong shawls.

The majority of the woolen fabrics of Kashmir, and particularly the best-quality shawls, are made of Pashmina.

Stoles A stole is a woman's shawl, especially a formal shawl of expensive fabric used around the shoulders over a party dress or ball gown.

A stole, as can be seen in Figure 6.38, is typically narrower than a shawl, and of simpler construction than a cape; being a length of a quality material, wrapped and carried about the shoulders or arms. Lighter materials such as silk and chiffon are simply

Figure 6.38 Stoles.

finished, that is, cropped, hemmed, and bound; heavier materials such as fur and brocade are often lined as well.

Mufflers A warm, cozy muffler wrapped around the neck on a cold winter day helps keep the chilly winds at bay. "Muffler" is another name for a heavy winter scarf, but produced in a narrower width. A wide variety of fabrics, colors, and patterns are used for mufflers. For use in very cold climates, fabrics such as fleece and wool can be good choices. Mufflers, as can be seen in Figure 6.39, are inexpensive and easy to make.

Handkerchiefs and pocket squares

A handkerchief, also called a hanky, is a form of a kerchief, typically a hemmed square of thin fabric that can be carried in the pocket or purse, and which is intended for personal hygiene purposes such as wiping one's hands or face, or blowing one's nose. A handkerchief, as can be seen in Figure 6.40, is also sometimes used as a purely decorative accessory in a suit pocket.

Figure 6.39 Mufflers.

Figure 6.40 Handkerchief.

6.8 Conclusions

Apparel and accessories are worn together, so one should not be considered independent of the other—as one changes, the other must also change to complement it. To understand the directions of fashion, accessories are important.

The relationship between accessories and apparel is usually complementary. The obvious connection is that they are worn together and should complement each other. When clothing is highly ornamented, accessories are more conservative; during a fashion season in which apparel is tailored or conservative, accessories are more decorative and embellished. Both are influenced by the same environmental factors, such as political, economic, and social climates. Apparel manufacturing companies also diversify into related product areas that include accessories, while accessories manufacturing companies get involved in the production of apparel.

While selecting accessories for children's wear, special attention should be given to the safety of the kids. Quality inspection and evaluation are required before purchasing any accessories by the garment industry.

Acknowledgments

The authors are thankful to the Director, TIT&S, Bhiwani for his encouragement in this work. The authors are indebted to Shri Sanjay Sharma, Senior Programmer, for his help in documentation work. They are also thankful to all PG students for their support.

Bibliography

Abling, Bina, 2011. Fashion Sketching Book. Fair Child Publication.
Anna, J., 2002. Handbags: The Power of the Purse. Workman publishing Co., New York.
Bhargav, R., 2005. Design Idea and Accessories, first ed. B.Jain Publishers(P) Ltd. pp. 130–165.
Brand, J., Teunissen, J., De Muijnck, C. (Eds.), 2007. Fashion & accessories. TERRA; ArtEZ Press.
Carr, H., Latham, B., 1994. The Technology of Clothing Manufacture, second ed. Blackwell Science, Oxford. 177.
Chenouna, F., 2004. Carried Away, All about Bags. Vendoina Press, New York.
Consumer Goods: Global Industry Guide, Datamonitor, March 2009. The Apparel, Accessories and Luxury Goods Market Consists of Mens, Womens and Infants Clothing, Jewelry, Watches and Leather Goods.
Designing and Development of Kids Wear from Union Woven Fabrics Produced from Yarns Using Cotton and Regenerated Cellulosic Fibers with the Help of Modern CAD Techniques, 2014. M.Tech thesis of Kavita.
Diamond, J., Diamond, E., 1993. Fashion Apparel and Accessories. Delmar Publishers Inc., New York, pp. 333–379.
Diamond, J., Diamond, E., 2008. Fashion Apparel and Accessories and Home Furnishing. Pearson Publications, Delhi, pp. 215–272.

Diamond, J., Diamond, A. Fashion Apparel Accessories and Home Furnishings.
Frings, Gini Stephens, 2007. Fashion from Concept to Consumer.
Genova, A., 2012. Accessory Design. Fairchild Publications, New York, p. 299.
Geoffroy, S., 2004. Berenice Bags. Editions Assouline, New York, Paris.
Gerval, O., 2009. Studies in Fashion: Fashion Accessories. Page one Publishing Pte Ltd., China pp.10−201.
Jarnow, J.A., 1987. Inside the Fashion Business. Macmillan Publication Company.
Kings well, T., 1998. Children wear trends for spring/summer. ATA Journal 9, 54−56.
Loalbos, 2009. Vintage Fashion Accessories. Krause Publications, p. 69.
Peacock, J., 1998. The complete 20th century sourcebook. Fashion Accessories. Thames and Hudson.
Rasband, Judith, 2001. Wardrobe Strategies for Women. Delmar Publication.
Sakthivel, S., Ramachandran, T., Chandhanu, R., Padmapriya, J., Vadivel, Vignesh, R., August 2011. The Indian Textile Journal.
Stall, C., 2004. Know Your Fashion Accessories. Fairchild Publications Inc., USA, 147−287.
Stanly, B., Sep. 1995. Textile Research Journal 21, 453−459.
Steele, V., 1999. Bags: A Texicon of Style. Scriptum Editions, London.
Stone, E., 2013. Fashion Merchandising, fourth ed., p. 208.
Stacy Lo Albo, Vintage Fashion Accessories.
Wilcox, C., 1997. A Century of Bags: Icons of Style in the 20th Century. Presparo, London.
Zippers Made Easy, Guide C-221. New Mexico State University.
http://en.wikipedia.org/wiki/Footwear.
http://en.wikipedia.org/wiki/Shoe.
http://www.clicktoconvert.com.
http://en.wikipedia.org/wiki/High-heeled_footwear.
http://shoes.about.com/od/boots/tp/womens_boots.htm.
http://shoes.about.com/od/boots/tp/womens_boots.htm.
http://www.besthealthmag.ca/get-healthy/health/choosing-the-right-shoes.
http://en.wikipedia.org/wiki/Handbag.
http://en.wikipedia.org/wiki/Hat#Hat_styles.
http://www.summitglove.com/definitions.html.
http://www.davidmorgan.com/glovesize.html.
http://www.glove.org/Modern/glovemeasure.php.
http://en.wikipedia.org/wiki/Necktie.
http://en.wikipedia.org/wiki/Scarf.
http://en.wikipedia.org/wiki/Shawl.
http://en.wikipedia.org/wiki/Handkerchief.
http://www.wisegeek.org/what-is-an-interlining.htm.
http://www.questoutfitters.com/zipper_tips.htm.
http://www.teonline.com/articles/2009/02/designing-party-dresses-with-b.html.
http://web.archive.org/web/20100617014018/http://www.ykkfastening.com/global/products/zs/structure.html.

Part Two

Garment design and production

Garment sizing and fit

R. Pandarum[1], W. Yu[2]
[1]University of South Africa, Florida, South Africa; [2]Hong Kong Polytechnic University, Hong Kong SAR, China

7.1 Introduction

The definition of garment sizing and fit has evolved over time. According to Burns and Bryant (1997), with the introduction of ready-to-wear garments categorised into sizes, fashion has changed from the era of tailor-fitted garments to mass-produced garments. Current garment sizing is based on the standard measurements of one single fit model of the ideal customer that is representative of the target market (Keiser and Garner, 2003) for a given style of garment. The garment patterns are adapted from basic block patterns that have been created to 'fit' a standard average-sized person or a person within the target market of the specific retailer.

A sample garment, produced from the adapted patterns, is then visually evaluated on a fitting model that also represents the average size of the different retailers' target markets for that garment style. Once the fit is approved, the patterns are produced in a range of different sizes according to a set of grade rules derived from a size chart. The range of garments within the different clothing sizes is then created by an apparel firm to fit a targeted range of customers (Ashdown et al., 2004). The size chart, comprised of key body measurements, is used to define a range of garment sizes within a fashion line, and this is crucial to ensure fit consistency across the size increments (White, 2011).

In an attempt to standardise garment sizing, and hence customer satisfaction with the fit of ready-to-wear garments, various garment-sizing systems have been developed worldwide. These systems define aspects such as the key body dimensions that should be used, how garment types are grouped together, how figure types are defined and how garment sizes are described. However, Stamper and associates (Stamper et al., 2005) have reasoned that fashion remains a major factor in how garment-fitting standards are developed in any given era. Factors such as the fabric, imperfections in body posture and body asymmetry, combined with customer errors in purchasing the correctly sized garment, contribute to either a well- or ill-fitting garment choice.

Since 1980s, advances in the clothing sector in terms of obtaining body measurement data quickly and nonintrusively, have heralded the use of 3D body scanners in the clothing sector. Worldwide, this technology has been used extensively for conducting national body sizing surveys, and has been instrumental in advances in body shape analysis; made-to-measure clothing; clothing style and size recommendations; and in virtual fashion within the clothing sector. Nonetheless, worldwide, the problem of ill-fitting garments still persists and this is a major concern for retailers and consumers (DesMarteau, 2000).

7.2 Geometry of the human form

7.2.1 The golden ratio – A generalised definition

For much of recorded history, humanity has been obsessed with seeking beauty and aesthetically pleasing proportions in surroundings such as buildings, in the shape of human faces (Laitala et al., 2011) and in the human form associated with clothing (Pheasant, 1986; Anon, 2013a, 2014a; James, 2005; Livio, 2002). Lehner (2013) comments that the first recorded use of a golden ratio, for defining visually pleasing proportions, was by the ancient Egyptians in their construction design, including that for the pyramids between 2630 BCE and 2611 BCE, using a system of measurement that was based on the human body. Furthermore, Kelly (2005) notes that 524 mm was considered to be the approximate length of an ancient Egyptian man's forearm, which was said to be a common unit of measure used in that period. However, the author elaborates by saying that the Egyptian artists drew in overviews and elevations, by dividing the human figure into 14 equal parts that corresponded to 14 body landmarks. Nevertheless, it was the Greek mathematician Euclid who started to define our current thinking about the golden mean, also known as the golden ratio or the golden number, in the third century BCE, by dividing a line into two sections as described by the formula, $\frac{AB}{BC} = \frac{AC}{AB}$, and represented by the Greek letter phi (Φ) (Kelly, 2005; Figure 7.1).

7.2.2 The golden ratio – associated with the 'ideal' human body

Pheasant (1986) reasons that it was only around 15 BCE that Vitruvius, a Roman architectural theorist, mentioned the body-part ratios that we are familiar with today. Vitruvius used antiquated units of measurement such as the cubit, which was said to be 'equal to six palms, or one and a half Roman feet'. Vitruvius used this in his desire to develop a system of body-part ratios, to convey perfection in the human body. Renaissance architecture is also said to be of harmonious form, and in mathematical proportions based on the ideal human body measurements (Anon, 2014a). Articles (Anon, 2013a) on the human body and phi, and the golden ratio, further rationalises that the human body, for example, has five

Figure 7.1 Golden ratio.

appendages to the torso, in the arms, legs and head; and there are five appendages to each of these, in the fingers and toes; with five senses in sight, sound, touch, taste and smell; and five openings in the face, in the nose, ears and the mouth that can be calculated in 'fives' as follows: 5^0.5*0.5 + 0.5*0.5 = phi. However, it was not until 1955 that Le Corbusier (1980) published 'Modular 11', incorporating the golden ratio within the human body by generating golden rectangles using three squares.

Pheasant (1986), in his description of the idealised male figure proportion, saw the proportions of the body in terms of regular relationships: interalia the palm is the width of the four fingers; the length of a man's outspread arms is equal to his height; a foot is the width of four palms; the distance from the hairline to the top of the breasts is one-seventh of the man's height; the maximum width of the shoulders is one-quarter of a man's height; the distance from the elbow to the armpit is one-eighth of a man's height; the distance of the nose to the chin is one-third the length of the face; and the length of the hand is one-tenth of the man's height. His explanation for dividing the body in this manner was that a man with these proportions would fit perfectly into a circle within a square. Nevertheless it was Eberle et al. (2002) who described the concept of the golden ratio in clothing design, as the division of body proportions into eight equal parts. Cooklin (1999) mentions that dividing the body in this manner makes sketching the ideal proportions of the body easier when placing the waistline at 5/8 of a woman's total height, with the crotch placed at half the height, as indicated in Figure 7.2.

7.3 The human figure divided into body proportions

In Western societies, endeavors towards describing an idealised figure type (Anon, 2013b; Roache and Eicher, 1973) related to body dimensions have been recorded throughout history and are said to have been developed over time, as seen in artefacts, and in paintings and sculptures displayed around the world today. This was also portrayed in body proportions. In ancient Egypt, in 1370 BCE, Nefertiti was said to have personified the female ideal with her 'large deep set eyes, long neck, and an oval head with straight nose' (Roache and Eicher, 1973). The author further mentions that Apollo, depicted as a beardless athletic youth and the 'full and the voluptuous' Venus de Milo, created between 130 and 100 BCE, whose bust/waist/hip ratio were 37/26/28, represented the Greek perception of the ideal male and female forms, respectively (Anon, 2014b).

In the Middle Ages, from the fifth to the fifteenth century, artwork also shows the female figure portrayed in a 'stretched oval' form (Roache and Eicher, 1973), where the distance between a women's navel to the crotch was half the measurement from that of the bust to the crotch. Nevertheless, the eighteenth century heralded huge changes in Western dress to favour form-fitting clothing for both men and women. Morton (1964), in her article on the art of costume and personal appearance, mentions that a few of the American female ideals of the period were said to be above five feet

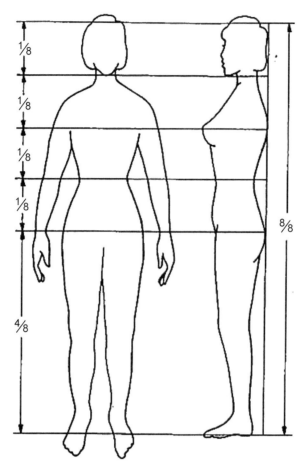

Figure 7.2 The body divided into eight equal parts.
From Cooklin (1999).

tall, oval-faced and slender-armed with a curved waistline and shoulders and hips of the same width, while the male ideals were illustrated as slender in build. However, such Western figures do not represent a universally accepted cultural ideal form, as this differs not only from one culture to another but also within the same culture.

The views held by the Americans, the British and the mainland Europeans as to what is considered attractive and beautiful today, may differ from those on the African continent (Kaiser, 1998). As an example, Elicher et al. (2000) described the Ghanaian female body, in 1993, as 'egg-shaped ovals' with a long-stretched neck, that is regarded as culturally beautiful; while in the Padaung tribe of Burma's Kayan ethnic group, the necks of the women were stretched with metal rings into what is called the 'giraffe neck', which was considered to be beautiful (Roache and Eicher, 1973). In South Africa, particularly among the Nguni, Sotho, Tsonga and Venda tribes, the ideal figure form is said to be wide hipped with prominent buttocks, as illustrated in Figure 7.3. It is considered to be both beautiful and a mark of the women's fertility

Garment sizing and fit 191

Figure 7.3 Figure with wide hips and prominent buttocks.

(Tyrrell, 1968). These body variances will ultimately influence body dimensions used in size charts and hence the ready-to-wear garment size choice.

The twentieth century also saw emphasis being placed on the concept of a perceived ideal figure form divided into body dimensions. Sometimes this was associated with health ideals linked to body mass indices, and not necessarily for clothing manufacturing, as garment manufacturers and retailers do not all use the same figure type classifications for their target markets. Fashion experts may, in their need to drive the fashion market, change their ideals more than once in an attempt to standardise garment sizes, and three or four different fashion ideals may exist during one's lifetime.

Dale (2013) explains this further saying that, for example, three of the twentieth century Western female figure images, recognised worldwide as fashion models and iconised as ideal figure forms, were Marilyn Monroe in the 1950s, who was said to have a voluptuous hourglass figure with bust/waist/hip measurements of 37/23/36; Twiggy in the 1960s, with 'boyish' bust/waist/hip measurements of 32/22/32; and Kate Moss in the 1990s, with bust/waist/hip measurements of 33/23/35. As indicated by these twentieth century models' body dimensions, the ideal fashion forms can vary significantly in bust, waist and hip measurements. While some females are said to be more athletic, hence having a more 'masculine' body profile, others are classified as either pear-shaped or hourglass-shaped, with no two individuals alike in stature, height or in girth measurements. In support of this, Thirty (2008) argues that any five women, at any given time, may have the same hip measurements but different waist to hip ratios.

The twentieth century also saw the emergence of the black African fashion models with very Westernised body proportions. Iman Abdulmajid has bust/waist/hip measurements (in inches) of 34/28/38, Alek Wek has 34/23/35, Liya Kebede has 33/23/34 and Oluchi Onweagba has 34/24/35. Hence the body proportions and African fashion ideal figure forms today on the African continent also reflect the Western-styled clothing worn in everyday life by the people of African or of Southern African descent that are driven largely by the adoption of Western norms. The lack of a uniquely South African or African garment sizing system; the lack of availability of cultural clothing in retail stores; and the global village concept of free trade with worldwide clothing imports are all contributing factors in this. Such trends pose a huge challenge for the local and the global garment manufacturing industries, as retailers and garment manufacturers still predominantly design for the hourglass figure type, using anthropometric data measured in the 1940s (Simmons et al., 2004). The same holds true in South Africa, as the garment/retail industry, and therefore the population, cannot lay claim to a uniquely South African garment sizing system.

Garment sizing systems currently in use by manufacturers and retailers have been adopted from the British, mainland European and American systems and adapted over time. This is further complicated by the fact that the South African garment manufacturers, at any given time, are not aware of the changing shape and sizes of the South African population (Papa, 2010). Nonetheless, the basic figure types (Anon) suggest that the underlying 'selective principals' of how figure types are classified in the Western world, as idealised figure forms, are based on the shoulder/bust/waist/hip ratios. This is said to be a common practice worldwide and that 'in garment measurement, body proportions are often used to express two or more measurements based on the body' (Anon, 2013c).

Although variations are said to exist in the number of figure types classified worldwide, for example, as either normal, hip heavy or top heavy, or as in the seven general categories found by a North Carolina University study of 6300 American women, as being of a rectangular shape that is characterised by not having a clearly defined waistline with the figure balanced from top to bottom. Rasband's (1998) study also identified additional figure types such as the spoon shape, also known sometimes as the triangle shape, or pear shape. This shape appears unbalanced from top to bottom, with more weight carried below the waist. The inverted triangle body shape has the appearance of being heavy or wider above the waist, and smaller or narrower below the waist. The hourglass shape appears to be larger at the bust area and is proportionally very small at the waist but is generally balanced from top to bottom (McCormack,

2005). The perfect or ideal hourglass figure shape is said to have a bust larger than the hips with a well-defined waist (Ashdown et al., 2004). In spite of these varied body classifications, research suggests that manufacturers and retailers still manufacture for the hourglass figure type, as previously stated (Knight, 2012).

However, Cooklin (1999), in his book on pattern grading for women's clothes, mentions that for centuries 'the eight head principle' was used by artists when sketching the female body. The head was 1/8 of the total height of the woman, and the bust, hips and waist were likewise also in 1/8 proportions, equally spaced, down the length of the body (see Figure 7.2) and were considered to be the ideal measurements. However, it is unclear how these ratios reflect the true body proportions of the majority of individuals living in any given era, as these idealised body dimensions may only be useful in drawing figures for pattern illustrations.

7.4 Garment size charts

White (2011) defines a size chart as a 'table of data showing measurements either of the body or of the clothing attached with a size label'. Similarly, Knight (2012) states that a 'size chart lists the average measurements of body sizes for a range of garment sizes'. However, this aside, the general components that constitute a size chart, for example, garment sizes 10–14, are as explained in Table 7.1.

Table 7.1 General components of a garment size chart for women's trousers

(a) The header indicates the type of garment to which the chart relates, in this example for women's trousers.
(b) Size labels represent a range of body measurement defined by the girth and height measurements. In this example, the waist and inseam measurements to fit garment sizes labelled 10, 12 and 14.
(c) The individual point of measurement, such as the waist and inseam (taken in centimetres for this chart).

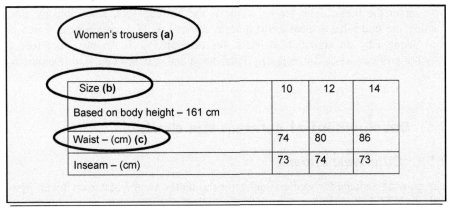

Women's trousers (a)			
Size (b) Based on body height – 161 cm	10	12	14
Waist – (cm) (c)	74	80	86
Inseam – (cm)	73	74	73

Table 7.2 **General components of a bra size chart for under-band sizes 34, 36 and 38**

Under-band measurement (cm)				
		34	36	38
		74–78	79–83	84–88
Over-bust measurement (cm)	90	A		
	92	B	A	
	94	C	B	A
	96	D	C	B
	99	DD	D	C
	102		DD	D
	104			DD
	107			E

Adapted from 3D intimate apparel data (Pandarum, 2009).

On the other hand, as shown in Table 7.2, intimate apparel such as the bra, is notated differently from outerwear. Bra size charts use two dimensions of fit, such as the under-band and the over-bust (bra cup size) measurements. As note in Table 7.2, the bra band size is the rib cage measurement that is notated numerically for example 34, 36 etc. and the bra cup size is measures alphabetically for example A,B,C etc. which refers to the size of the breast in relation to the rib cage. Every inch increase in the difference in over-bust and under-band measurement equates to an increased alphabetically notated bra cup size.

In the apparel industry, bra size can be described by two values. The first value is the under-band, which is a number based on the circumference of the ribcage under the breast. The second value is the bra cup size, calculated by subtracting the under-band measurement from the over-bust measurement, which is then denoted by an alphabetical letter, for example, A, B, C, etc. Therefore a 34B bra size denotes a 34 inch-size under-band and a B bra cup size (Pandarum, 2009).

7.5 Development of garment size charts

7.5.1 Outerwear apparel

The growing demand for soldiers' uniforms during the United States civil war, from 1861 to 1865, initiated the concept of standardised clothing sizes, as an observation made by the measurers at that time was that certain sets of garment measurements kept recurring. This, then, established the first commercial sizing scales for men after the war (Woolman, 2003). However, commercial sizing scales for women's clothing sizes only occurred after the 1920s, driven by factors such as industrialisation, the growth of the advertising industry, chain stores and mail-order catalogues that advertised affordable and trendy ready-to-wear garments. This, however, created considerable confusion in the clothing market as the sizing data had no clear scientific foundation because women with different body dimensions were categorised as having the same size label in the ready-to-wear garment sizes (Adu-Boakye et al., 2012). Kidwell (1979) notes that the first grading system of garment sizes used a single body measurement, for example, bust girth only, and the other sizes for grading pattern pieces were graded proportionally based on the change in this single measurement. Furthermore, body measurements in a size chart can vary from one manufacturer to another, and even within the same manufacturer (Pandarum et al., 2011). Salusso-Deonier (1989) attests that clothing sizing has been unreliable for over 50 years, and that manufacturers are relying on an 'absolute database' to manufacture garments for a fast-changing and growing population.

In an attempt to standardise and understand body types within a country's population, numerous studies have been conducted. A limited number of these studies conducted worldwide focus on different clothing types and demographics of the population, for example Gupta and Gangadhar (2004) study in India sought to establish a statistical model for body size standards, with a focus on standardising the size charts to reflect variations in body types. Chen (1998) developed a sizing system for 7800 Taiwanese "elementary and high school students" to establish a statistical method for developing standard size charts. Laing et al. (1999) undertook a study to develop a sizing system for firemen's protective clothing in New Zealand and McCulloch et al. (1998) formulated a novel approach for developing an apparel sizing system of dress shirts for the military in the United States. Such studies indicate not only the underlying problem of outerwear garment sizing and fit within the different apparel categories but also the scale and the fragmentation needing to be addressed at a basic level in the clothing sector.

Additionally, since 1972, large-scale national anthropometric studies have been conducted in countries such as Sweden, China, Japan, Germany, America, the United Kingdom and Mexico using 3D body scanners. In South Africa, limited data have been collected since 2004 using TC^2 (Keiser and Garner, 2003) NX12 and NX16 full 3D scanners (Pandarum, 2009) in focused and generic studies. A national South African body-sizing survey is currently being conducted under the auspices of the Department of Trade and Industry. As previously stated, the sizing system in South Africa operates with a legacy adopted from the American, mainland European and British sizing systems. This system has been adapted over the years to accommodate a population that is slowly but continuously changing in size and shape (Pandarum, 2014). This is the only known study to date being focused on the African continent to collect the body measurements using 3D scanning technology, as other studies have used the traditional manual methods.

7.5.2 Intimate apparel

In the intimate apparel manufacturing industry the sizing system is also established. For example, the bra sizing system established in 1926 is still in use today. The system was originally proposed by Warner (Ewing, 1971) in America, to classify breast shapes into analogous types, with the breast volume being incorporated into bra size specifications in 1935. However, the United Kingdom only adopted this system in the 1950s (Yarwood, 1978). This, then, formed the basis for the alphabetically notated sizing system for bras, and for the basic modern bra sizing charts in use today. Although manufacturers and bra designers of different brands use different sizing charts for specific target markets. The over-bust circumference (1) and the under-bust measurement (2), as indicated in Figure 7.4, are the basic measurements used in bra sizing (Lynn, 2010). There are various international standards, including the International Organization for Standardization (ISO), European, Japanese and Chinese bra sizing standards in existence today, which use the difference in the horizontal linear measurements of the under-bust and over-bust measurements (Anon, 2013d). Interestingly enough, different countries also use different systems for bra size notation.

The United Kingdom and the United States use the imperial sizing system with the alphabetical bra cup size notation, while most European countries use the metric sizing system and the alphabetical bra cup size. Italy uses a sizing system that is based on numbers, with the bra cup size denoted by a letter of the alphabet; while in the Australian sizing system, the band size is based on dress size and the bra cup size denoted alphabetically. In the United States and the United Kingdom, bras with smaller band sizes of 28 and 30 cm, respectively, are available, whilst the smallest band sizes available in other countries include 70 cm in Europe, 85 cm in France, and a dress size 10 in Australia. The bra cup sizes show more consistency in the

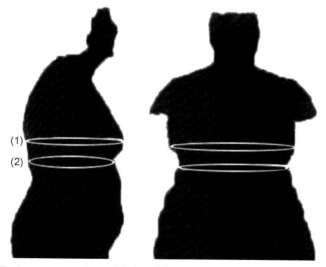

Figure 7.4 Basic measurements used in bra sizing.

denominations used in the United Kingdom, United States and France, with the smallest bra cup size of AA, while in Italy and Australia the smallest available bra cup size is an A (Lynn, 2010). In South Africa, bra manufacturing companies use a band and cup size notation similar to that of the United Kingdom (Pandarum, 2009). Consumers, however, still select their bra sizes by trial and error, by trying on bras of different styles, sizes and brands, until they find a bra that they believe fits them well.

7.6 Sizing and fit systems

Clothing size refers to the garment label found on ready-to-wear or off-the-rack clothing sold in retail stores around the world. This garment sizing system is currently specific to a country's figure types and garment types (Chen-Yoon and Jasper, 1993). Individual countries define figure types, garment size and garment types differently and use different key body dimensions to develop the sizing systems. Hence, the establishment and use of sizing systems vary worldwide, and body measurements do not necessarily translate into the garment measurements (SANS 1360-2, 2008). In an attempt to standardise sizing systems, either within the country or for export purposes, various countries have acquired new technologies, such as 3D body scanners, to collect large amounts of anthropometric data from their population to develop a statistically valid database from which a viable sizing system can be established. The garment size is communicated to the customer by the information on the ready-to-wear garment label in retail stores. According to Milliam, 'retailers, brands and manufacturers use garment size labels as a marketing tool, as a means to convey information to the customer, while consumers use them as a guide to determine the fit prior to purchasing the garment' (Milliam, 2014).

Various studies (ISO 3635 Standard, 1981; Ashdown, 2003; Doustaneh et al., 2010; Ming-Jung et al., 2007; Chen-Yoon and Jasper, 1993) reinforce the notion that the true value of a sizing system is the extent to which, with a limited number of sizes, it caters to a population's sizing and fit requirements. According to these studies, a system that caters well for its population sizing and fit requirements is said to have the advantage of influencing production numbers, the number of sizes produced and hence the fabric usage and production planning in a garment manufacturing environment. Garment size labelling methods vary in different countries; for example, in the United States, women's sizes are classified by figure types, of Women, Misses, Half-sizes while the junior sizes were alternatively labelled by age (Chen-Yoon and Jasper, 1993). According to Tyrrell (1968) the garment labelling system can also use numerical codes such as size 6 or size 8. The junior and the missy sizes are sometimes labelled 1 or 2. Hence numeric codes of garments on labels do not characterise a woman's age or her body measurements. Ironically, the sizing system developed in Germany in 1981 also did not indicate body measurements, but the garment sizes were coded as 18,018 and 18,518. According to Chen-Yoon and Jasper (1993), a size code 18,018 and 18,518 represented a garment that should fit a woman with a bust girth of 84 cm; while the numbers 0 and 5 designated suitability for a woman with slim or full hips, respectively;

and the number 18 indicated a short height and an increasing number within a size group indicated larger body frame sizes.

In 1991, ISO suggested the use of body measurements to describe garment sizes, using key body dimensions in the form of pictograms (Chen-Yoon and Jasper, 1993; SANS 1360-2, 2008 — South African National Standard). Figure 7.5 gives an example of a label for a woman's jacket. This is for a woman with a body height of 164 cm with a bust and hip girth of 96 and 104 cm, respectively.

Since then, countries such as Japan, the United Kingdom, South Korea, Hungary, South Africa and Kenya have adopted or are adopting ISO standards (Pandarum, 2014) developed under the technical committee titled ISO/TC133 — Clothing Sizing Systems — size designation, size measurement methods and digital fitting.

To date, the ISO/TC133 committee consists of 19 international participants and 22 observing member countries. Countries active in the committee convene yearly to deliberate over updating the current and proposed new standards. It is noted that the majority of these ISO standards are currently under review in an attempt to accommodate the global trade markets and to standardise the information on garment labels (Pandarum, 2014).

Although most sizing systems have differences and similarities with respect to the use of the key dimensions for the different garment types, there is a conscious move

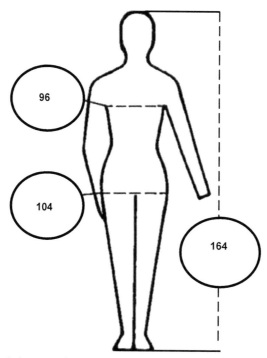

Figure 7.5 Adapted pictogram for a woman's jacket.
From SANS 1360-2 (2008).

internationally towards standardisation spearheaded by global trade and new and emerging global markets.

7.6.1 Fitting models and target markets: representing the concept of the 'ideal customers' body dimensions

Companies use mannequins or live-fitting models to test their base size pattern block as fashion silhouettes continuously change. According to Workman (1991) the fitting model's body measurements are said to resemble the commercial pattern measurements established by the company for a particular sample size. Vogt (2007) adds that these measurements generally consist of the height, bust, waist, hip circumferences, the leg and arm lengths and shoulder width, depending on the garment type. This, then, forms the base pattern size for either a size 8 or a size 10 garment with the other sizes graded proportionally from this (Chen-Yoon and Jasper, 1993), depending on the brand and the retailers target markets. Live-fitting models are preferred or used in conjunction with fitting mannequins as the live-fitting models are said to provide valuable information on the drape and the tactile aspects such as the feel of the garment to improve the designing process as opposed to using a static mannequin. However, at any given time, a particular company might produce several clothing lines or brands that may require different fitting models possibly of different ages with different body dimensions. Additionally, manufacturer's sizing systems may differ over time and the live-fitting model's body dimensions may also change with time; this further complicates the issue of the sizing and fit of garments that are produced for specific brands or target markets.

7.6.2 Benefits and challenges of the standardisation of garment sizes

Garment sizing and labelling systems that are easily understood worldwide, would create less confusion for customers and manufacturers alike. As a consequence, the global standardisation of garment sizes would encourage free trade between countries and hence stimulate a growing clothing market sector within countries. As previously mentioned, many countries are making a concerted effort towards adopting a standardised garment sizing system worldwide; however, the industry is either slow to adopt these standards in manufacturing or has insufficient data on the shapes and sizes of their populations to be able to add significant value towards a standardised system. Attempts at standardisation have been largely initiated in the First-World countries with the Third-World countries being slow to follow. Global standardisation is further complicated by the demographics of a country's population, which may vary considerably between and within countries. Therefore, the successful adoption of a standardised system will be determined by the following factors: "how consensus is reached by the different countries on the classification of garment types; how realistic the key dimensions selected within and between manufacturers in a country will be; whether a new standardised system will be used and adopted by customers worldwide; and whether customers will know and understand a garment sizing system using body measurements," or even know their own body measurements at any given time. Even so, Workman (1991) notes that the reality is that manufacturers may prefer not to support a standardised sizing system, because sizing is used as a selling or a marketing tool by retailers to distinguish themselves from other retailers and attract customers and hence enhance profits within their target markets.

7.7 Three-dimensional (3D) body scanning — current and potential future applications in clothing manufacture and retailing

Obtaining body measurements using a 3D body scanner is a rapidly evolving technology aimed at reducing the problem of ill-fitting clothes, because it can readily take many more body measurements than the manual methods to enable custom-fitting apparel (Workman, 1991). Since the 1980s there have been many types and models of body scanners developed (Vogt, 2007) for different clothing applications. Countries such as the United States, the United Kingdom, Sweden, France, Germany, Spain, Poland, South Korea and China have invested large sums of money since 1998 to 'size' their populations. Brooks Brothers, a dress-shirt retailer, has been using 3D body scan measurements successfully to make custom-fitted suits and dress shirts for men (Winsbrough, 2014; Istook, 2008). Levi's is another retailer that is reported to use 3D scan data to make custom-fitted jeans for their customers. In the intimate apparel industry, Victoria's Secret in the United States and DB Apparel in South Africa have conducted 3D studies in an attempt to address garment sizing and fit challenges for their retail markets (Pandarum, 2009). In addition, 3D body scanning has broad current and future applications in the following areas (Ashdown et al., 2004; Loker et al., 2005).

- Standardising sizes, in custom-fitting garments for an individual using a smart card to store an individual's body measurements and for the mass customisation of garments with a made-to-measure fit.
- Internet shopping, where customers can visualise themselves from all possible angles in the form of a personalised avatar, and in garment size recommendations also known as made-to-measure tailoring.
- Virtual-try-on systems, where the avatar can be virtually dressed with digital clothing, with the option to choose fabric design and colour, and even the accessories such as the trimming and buttons and with dynamic poses such as walking. This allows for the electronic exchange of information between the pattern-makers and designers without initially making a sample pattern, and hence fabric wastage is reduced.
- Production of anthropometric mannequins based on actual body sizes extracted from data collected during national surveys of gender, height and shapes used in garment production and in quality control in the manufacturing environment.
- Personalised mannequins, either for made-to-measure tailoring or for fitting models whose body dimensions are not static. Hence the manufacturer does not have to contend with the added challenge of those fitting models who are representative of the retailers' different target markets either gaining or losing weight.

7.8 Conclusions

Technologies such as 3D body scanners have become more affordable since their inception in the 1980s. Numerous studies (see Section 7.7) have demonstrated the significance and value of using 3D technology to address the challenges associated with garment sizing and fit worldwide. Manufacturers are now able to develop protocols for scanning large numbers of their population to collect body shape and size data, quickly and non-intrusively. This could ultimately lead towards a standardised sizing system both within and between manufacturers and retailers worldwide. Having said that, 3D body scanners will not solve an entire population's garment sizing and fit needs. They will accommodate the majority of the public needs including plus-sized, petite or tall subjects, thereby creating niche markets. However, there will still be a tendency for vanity sizing and data 'skewed towards a brand' with manufacturers catering to their target markets.

Nevertheless, 34 years after the inception of 3D technology for use in the fashion retail environment, the adoption of this technology has been largely in areas such as virtual try-on for online garment sales, including that of cloth definition and characterisation; dynamic poses and movement of the avatar; making garment size selection faster and simpler; and towards the development of realistic anthropometric mannequins for quality control and garment production, even extended to personalised mannequins (Loker et al., 2005). However, there is limited information available on the public demand for the virtual try-on services being offered for online garment shopping. Illustrating the continuing difficulties, at a presentation in London a representative of Fits.me, a developer of the virtual fitting-room platform, reported that '60 percent of the people will end up choosing another garment size other than the one the system recommends and this is dependent on the customer's preference that is difficult to define through data' (D'Apuzzo, 2009; Haldre, 2014). Therefore, until such time as a concerted effort is made by manufacturers and retailers towards not only collecting up-to-date shape and size data of their populations but to combine this with educating design teams, students entering the garment industry and target markets, garment sizing and fit problems will continue unabated.

7.9 Sources of further information

1. Shishoo, R. (2012). The Global Textile and Clothing Industry: Technological advances and future challenges.
2. Fan, J., Yu, W and Hunter, L. (2004). Clothing Appearance and Fit: Science and Technology. Woodhead Publishing Limited.
3. Ashdown, S.P. (2007). Sizing in Clothing; Developing Effective Sizing Systems for Ready-to-Wear Clothing. Woodhead Publishing Limited.
4. ISO 3635 – Size designation of clothes – Definition and body measurement procedure.
5. ISO 4416 – Size designation of clothes – Women's and girls' underwear, nightwear, foundation garments and shirts.

6. ISO 5971 – Size designation of clothes – Pantyhose.
7. ISO/TR 10,652 – Standard sizing systems for clothes.

Acknowledgements

The authors wish to thank Professor Phil Lyon for his guidance with editing this chapter and the Consumer Sciences Department of the University of South Africa, Johannesburg, South Africa, for providing the 3D scan data and images.

References

Adu-Boakye, S., Power, J., Wallace, T., Chen, Z., 2012. Development of a sizing system for Ghanaian women for the production of ready-to-wear clothing. In: The 88th Textile Institute World Conference. Selangor. Malaysia.
Anon. The Basic Figure Types. Chapter V. (Online). Accessed: http://haabet.dk/patent/Corset_fitting_in_the_retail_store/5.html.
Anon, 2013a. Human Body and Phi, the Golden Ratio (Online). www.goldennumber.net/human-body (accessed 14.11.13.).
Anon, 2013b. The Bailine Manual-figure Analysis v345p.Doc Page 52 (Online). http://www.doc.bailine.info/man/us/01 (accessed 01.11.13.).
Anon, 2013c. Body Proportions (Online) Available: en.wikipedia.org/wiki/Bodyproportions (accessed 27.11.13.).
Anon, 2013d. How to Measure Bra Sizes. Bra Cup Size Facts (Online). www.howtomeasurebrasizes (accessed 21.12.13.).
Anon, 2014a. Architecture in Renaissance Architecture (Online). http://en.Wikipedia.org/wiki/Renaissancearchiture (accessed 13.02.14.).
Anon, 2014b. Apollo www.wikipedia the free encyclopedia. (Online) (accessed 14.02.14.).
Ashdown, S., Loker, S., Cowie, L., Schoenfelder, K., Clarke, L., 2004. Using 3D scan for fit analysis. JTATM 4 (1), 1–12 (Summer).
Ashdown, S., 2003. Sizing up the Apparel Industry: Topstitch: Cornell Newsletter for the New York State Apparel and Sewn Products Industry. Spring.
Burns, L.D., Bryant, N., 1997. The Business of Fashion. Fairchild Publications, New York.
Chen, F.L., 1998. The research of Taiwan female anthropometric data. JCTI 9, 1–11.
Chen-Yoon, J., Jasper, C., 1993. Garment sizing systems: an international comparison. IJCST 5 (5), 28–37.
Cooklin, G., 1999. Pattern Grading for Women's Clothing. Wiley, Blackwell.
Dale, C., 2013. The Shape of Things to Come. Uniquely Woman (Online). http://www.uniquelywoman.co.uk (accessed 12.10.13.).
D'Apuzzo, N., 2009. Recent Advances in 3D Full Body Scanning with Applications to Fashion and Apparel (Online). www.hometrica.ch (accessed 10.02.14.).
DesMarteau, K., 2000. Let the fit revolution begin. Bobbin 42 (2), 42–56.
Doustaneh, A.H., Gorji, M., Varsei, M., 2010. Using self organisation method to establish nonlinear sizing system. World Appl. Sci. J. 9 (12), 1359–1364.
Eberle, H., Hornberger, M., Menzer, D., Hermeling, H., Kilgus, R., Ring, W., 2002. Clothing Technology…from Fibre to Fashion, sixth edition. Beutg-Verlag GmbH. BurggratenstaBe.

Elicher, J., Evenson, S., Lutz, H., 2000. The Visible Self: Global Perspective in Dress, Culture and Society, second ed. Fairchild Publications, New York, p. 101.

Ewing, E., 1971. Fashion in Underwear. Taylor, Garnett, Evans Ltd, Great Britain.

Gupta, D., Gangadhar, B.R., 2004. A statistical model to for establishing body size standards for garments. IJCST 16, 459–469.

Haldre, H., 2014. Your virtual self will try on clothes before your buy. In: Presentation at Decoded in London (Online) Accessed. http://www.wired.co.uk/news/archive/2014-05/13/fit-tech fashion.

ISO 3635 Standard, 1981. Size designation of clothes. In: Definition Body Measurement Procedure.

Istook, C.L., 2008. Three Dimensional Body Scanning to Improve Fit. Advances in Apparel Production. Woodhead publication.

James, T., 2005. The British Museum Concise Introduction to Ancient Egypt. University of Michigan Press, Anne Arbor, Michigan, ISBN 0-472-03137-6.

Kaiser, S., 1998. The Social Psychology of Clothing: Symbolic Appearance in Context, second ed. Fairchild Publications: ABC Media, Inc.

Keiser, S., Garner, M., 2003. Beyond design. In: Kontais, O. (Ed.), The Synergy of Apparel Product Development. Sizing and Fit. Fairchild Publication, New York, pp. 301–324.

Kelly, C., 2005. The beauty of fit: proportion and anthropometry in chair design. Unpublished Master Dissertation in Industrial Design in the College of Architecture.

Kidwell, C., 1979. Size variation in women's pants. Clothing Textile Res. J. 21 (1), 19–31.

Knight, L., 2012. Dress-making to Flatter Your Shape. Bloomsburg Publishing Plc, London.

Laing, R., Holland, E.J., Wilson, C.A., Niven, B.E., 1999. Development of sizing system for protective clothing for adult males. Ergonomics 42, 1249–1257.

Laitala, K., Grimstad, I., Hauge, B., 2011. Materialised ideal: sizes and beauty. Culture Unbound 3, 19–41.

Le Corbusier, 1980. Modular 1 and 11. Harvard University Press, Cambridge, MA.

Lehner, M., 2013. The Complete Pyramids. Thames and Hudson, New York, p. 84 (Online). http://en.wikipedia.org/wiki/Egyptianspyramid. http://www.pinterest.com/pin/92394229829063665 (accessed 10.12.13.).

Livio, M., 2002. The Golden Ratio: The Story of Phi, the World's Most Astonishing Number. Broadway Books, New York.

Loker, S., Ashdown, S., Schoenfelder, K., 2005. Size-specific analysis of body scan data to improve apparel fit. JTATM 4 (3), 1–15.

Lynn, E., 2010. Underwear: Fashion in Detail. V&A Publishing, London.

McCormack, H., 2005. The shape of things to wear: scientist identify how women's figures have changed in 50 years. The Independent (Fashion Articles and News. November). (Online). http://www.independent.co.uk/news/uk/this-britian/the shape-of-things-to-wear-scientists-identify-how-womens-fgures-have-changed-in -50-years-516259.html.

McCulloch, C.E., Paal, B., Ashdown, S.P., 1998. An optimization approach to apparel sizing. J. Oper. Res. Soc. 49, 492–499.

Milliam, R., 2014. Figure Forms. Personal Communication. 22 May, 2014.

Ming-Jung, C., Hai-Fen, L., Moa-Jiun, J.W., 2007. The development of sizing system for Taiwanese elementary high-school students. Int. J. Ind. Ergon. 37, 707–716.

Morton, G.M., 1964. The Art of Costume and Personal Appearance, third ed. John Wiley and Sons Inc, New York, p. 31.

Pandarum, R., Yu, W., Hunter, L., 2011. 3-D breast anthropometry of plus-sized women in South Africa Ergon. J. 54 (9), 866–875.

Pandarum, K., September 2009. Size Predication for Plus-sized Women's Intimate Apparel Using a 3D Body Scanner. Unpublished dissertation.
Pandarum, R., 6 February, 2014. In: ISO/TC133. Chairman.
Papa, S.D., 2010. Corporate Identity for the Young Fuller Figured Women. Peninsula University of Technology, Cape Town.
Pheasant, S., 1986. Bodyspace: Anthropometry, Ergonomics, and Design. Taylor C Francis, London; Philadelphia.
Rasband, J., 1998. Fabulous Fit. Fairchild Publications, New York.
Roache, M.E.E., Eicher, J.B., 1973. The Invisible Self. Prentice-Hall Inc., Englewood Cliffs, New York.
Salusso-Deonier, C.J., 1989. Gaining a competitive edge with top quality sizing. In: ASQC Quality Congress Transactions. American Society for Quality Control, pp. 371–378.
SANS 1360-2, 2008. Size designation of clothes - Part 2: Women's and girls' outerwear garments.
Simmons, K., Istook, C.L., Devarajan, P., 2004. Female identification technique for apparel. Part 1: describing female shapes. JTATM 4 (1), 1–16.
Stamper, A., Sharp, S., Donnell, L., 2005. Evaluating Apparel Quality, second ed. Fairchild Fashion Group: Capital Cities Media, Inc.
Thirty, M.C., 2008. The power of womenswear. AATCC 8 (5), 18–26.
Tyrrell, B., 1968. Tribal Peoples of Africa. Gothic Printing Co, Cape Town.
Vogt, P., 2007. Career Opportunities in the Fashion Industry (Online) (accessed 10.01.14.). http://books.google.co.za/books?id=MTJeYljsCmYC&printsec=frontcover&source=gbs_ge_summary.
White, E., 2011. What Is a Size Chart? (Online). www.techexchange.com (accessed 04.02.14.).
Winsbrough, S., 2014. An Insight into the Design, Manufacturing and Practical Use of a 3D Body Scanning System (Online). http://www.wwl.co.uk/images/design%26manufacture.pdf (accessed 12.02.14.).
Woolman, M.C., 2003. Standardisation of Women's Clothing: Short History of Ready-made Clothing (Online). http://museum.nist.gov/exhibits/apparel/rde/htm (accessed 04.02.14.).
Workman, J.E., 1991. Body measurement specification of fit models as a factor in clothing size variation. Clothing Textile Res. J. 10 (1), 31–36.
Yarwood, D., 1978. History of Brassieres. The Encyclopedia of World Costumes. The Anchor Press Ltd, Great Britain.

Pattern construction

K. Kennedy
RMIT University, Melbourne, VIC, Australia

8.1 Introduction

The process of pattern construction for apparel design and production is complex and diverse. The traditions of patternmaking have evolved over the centuries, since the invention of the tape measure in the late eighteenth century allowed for a systematic approach to pattern-cutting methods and enabled garments to be mass produced (Breward, 2001).

Thus, today with complex and global supply chains, the techniques of patternmaking as a preproduction phase can vary from an in-house operation to a remote or virtual focus. The requirements of global sourcing have fundamentally changed "how and where patterns are developed" by placing more emphasis on outsourced vendor-controlled patternmaking. Virtual product design and development encompass an evolving technology in the three-dimensional (3D) computer-aided design (CAD) domain.

The principles and the rules of pattern construction for apparel development in particular have developed according to a range of trade methods such as tailoring, draping, block or sloper construction, or pattern drafting. While these approaches to pattern construction techniques may differ, the primary objective is to achieve the same outcome, that is, to provide the optimum two dimensional (2D) template shape to delineate where the material or fabric of choice is cut. The success of the pattern template is determined or proven according to how well the 3D constructed shape fits the garment or object's design intent.

Thus, the range of methods by which a pattern is constructed offers characteristics particular to the mode of construction and the nature of the design. The traditional approaches could be via the tailor's traditions of chalking cut lines directly on fabric, or by the more sculptural approaches of modelling or draping on a body form, or by constructing a flat pattern in a geometric 2D $x-y$ coordinate mode. The range is diverse and determined by the garment's design intent and patternmakers professional training and often tightly held trade knowledge. Is the patternmaker approaching the task from a qualitative instinctual slant or as a quantitative pattern engineer or from a combination of both methods? Patternmaking can therefore be seen as a trade craft that draws on a variety of methods and inputs. Problem-solving abilities and a capacity for creative thinking are required capabilities for an effective patternmaker. For flat pattern cutting, the ability to interpret a design or sketch into a 2D flat pattern shape to create an acceptable 3D garment is a professional requirement. The draped method, where fabric is moulded on a dress stand or mannequin that approximates the body, the process is reversed, that is the 3D draped form must be reverse-engineered back to a flat pattern template before the material can be cut. Thus, this chapter aims to analyse the features, or pattern construction inputs, according to a range of modes rather than the geometrics of pattern construction systems and to identify the components of pattern construction as a technical design process.

8.2 Pattern construction modes

Within the apparel industry the traditional role of the patternmaker is characterised as a technical creative collaboration between design and manufacturing. The patternmaker's role may function within a variety of conditions from an in-house patternmaker to a broader range such as a designer/patternmaker, a designer/entrepreneur, a bespoke or private order tailor or dressmaker. With trends in outsourced manufacturing, patternmaking may be performed by a full make vendor and the pattern development is instructed by a remote garment technologist who may never see an actual physical pattern. The employment conditions can vary as well, ranging from a company employee working within a design room or contracted or subcontracted as a freelance patternmaker.

The traditional approach to learning the craft of patternmaking tended to adhere to a didactic format where knowledge or skills were handed down according to the master/apprentice model. The preferential block or sloper construction method and the application of patternmaking principles are often culturally based according to a range of methods. For example, the method based on a proportional measurement system that applies formulae according to anthropometric correlations (Muller and Sohn, 1997; Davis, 2008), or a system that requires direct body measurement inputs (Aldrich, 2009; Bray, 1966).

8.2.1 Pattern design considerations

The design source for a pattern may be derived from a variety of conditions. This can range from original design to an appropriated source; that is the design a new creation or is it a knock off of an existing garment? The starting point for the pattern may evolve from a master block or sloper or an existing pattern, or be translated from numbers and words on a specification sheet. Conditions around the garment performance criteria must also be considered. Does the garment require the capacity for repeatability or is it a one-off bespoke creation? Will the pattern be tested or sampled locally or will it be made in an outsourced location? Have the necessary quantitative inputs such a measurement data, fabric performance specifications, wearer fit preferences, construction and make conditions been clarified? Is there a requirement to produce a physical pattern output (manual or CAD) or is a remote vendor to develop and own the pattern? Consideration must be given to the variable inputs such as body shape (rather than size), fabric structure (woven vs knits) and design preference (tight vs loose).

8.2.2 Patternmaking contemporary supply chain influences

The demands of the fast fashion model within a global supply chain have mandated a different mode of production and pattern formats. The patternmaking function may be a virtual process with emphasis on the requirements for preproduction management in quality assurance or garment technology. Thus the ability to develop technical packs that include detailed technical drawing and specification documents is changing the emphasis from creating physical patterns. As a preproduction technician, knowledge in the fundamental principles of pattern construction and garment fit is an essential requirement.

Within the outsourced supply chain is the role of patternmaking vulnerable to becoming the hidden production element? The fast fashion model with an emphasis on quick turnaround, shorter lead times, high volume and low profit margins jeopardises the optimum garment prototype development requirements. Higher volume styles provide greater opportunity for development cost amortisation than a low volume style. The cost of the patternmaker's labour also directly affects this cost equation. Thus there has been a trend to outsource the pattern construction function to lower labour cost countries as a necessary preproduction function.

Therefore, by approaching the practice of pattern construction from a methodology that considers the various data inputs as quantitative and qualitative by nature, it is possible to analyse and evaluate the parameters from a variety of operational scenarios. This methodology can be applied to a range of pattern development situations from a pattern for an original design, a pattern iteration of an existing style or as a direct copy of a physical garment. Is the pattern a first generation construction or is it an adaptation or copy of an existing garment? The format and conditions under which patterns are developed can vary greatly with a range of variable inputs. These conditions are separate to and should not be confused with the rules or instructions on how to construct a pattern, be it a master pattern or a design derivation.

8.2.3 Traditional and contemporary approaches

There are numerous traditional and contemporary masters in the craft of patternmaking such as Aldrich (2009), Bray (1966), Kunick (1967), Armstrong (2006), Muller and Sohn (1997), to name a few. Each has a particular approach, technical nuance and preferred set of technical instructions on how to best construct, draft or adapt patterns for apparels. A patternmakers preferred method, formula or process is similar to a favourite cooking recipe. It is a method that can be relied on to achieve a successful outcome. The ability to assess the variety of elements and components and combine a complex range of inputs to achieve a successful outcome is what determines an adept patternmaker.

Contemporary approaches using the pattern as the design source rather than the traditional sequential didactic pattern drafting approach offer a more experimental format. Describing Roberts (2014) Subtraction Cutting techniques, Lindqvist (2013) cites Roberts as 'designing with patterns instead of creating a pattern for a design' (p. 50). The zero waste philosophy of McQuillan (2014) and Rissanen (2014), Singo

Sato's Transformational Reconstruction (2011) techniques, and Nakamichi's Pattern Magic (2012) pattern puzzles alter the function of patternmaking as a design translation to a creative technical design source. The relationship to form and fit is also fundamentally skewed from Erwin et al.'s (1979) classic fit benchmarks of grain, set, line, balance and ease. These contemporary practitioners playfully disregard the traditional rules around grain, set, line, and ease to experiment with volume, drape, and kinetic interactions. Off-grain effects, wrinkles and seams not perpendicular to the floor become design features rather than indicators of poor fit as defined by Erwin et al. (1979).

8.3 Body, material and design

To expand on the previous suggestion, it is possible to consider the elements common to all pattern construction modes as a way of defining the pattern input variables. This in turn helps to establish a methodology that can best provide the capacity to identify, analyse and evaluate the optimum approach according to a particular design requirement or mode of production. For example, the preferred patternmaking system or construction method according to the Aldridge method or the Muller system is only one part of the process. It is the adoption of a methodology that broadens the scope of pattern construction beyond a didactic set of instructions on how to draft a pattern or how to pivot a dart. Being able to seek the best method and work in an efficient way to achieve a successful outcome means that the analysis of the relationship of the variables provides a greater scope for the patternmaker. The scope is therefore more than simply acquiring an understanding of the principles of pattern construction.

It is also the capacity to understand the range of conditions under which the principles are applied, that is how the body moves, how a fabric performs and what is the tolerance factor for any variable input according to the design requirement? Thus by categorising the elements or input factors according to the typological classification of body, material and design (Figure 8.1) a methodological framework that can be applied to any preferred pattern construction method can be created. For example, the customer's body measurements are essential to a made-to-measure garment as the tailor or patternmaker has a direct connection to the body, whereas at the other extreme a pattern can be made remotely by copying an existing garment and never fitted on a body. These elements can be described as prepattern data and can be considered as separate factors or variables that inform the pattern construction process.

This discussion therefore proposes a consideration of the relationship among the three separate domains of (1) the body, (2) the material and (3) the garment (Figure 8.1) as essential nonhierarchical variable inputs. This model is based on the design model proposed by Goulding (©2010), as the body, textile and fashion garment dependency matrix for design consideration (Townsend and Goulding, 2011). Although Townsend and Goulding highlight how each domain provides a particular design focus and a way of approaching garment design, they also 'have a mutual dependency and in many ways are inseparable' (p. 288). Thus this concept of the mutual dependency, directed by specific domain focus, for example the body, the design or the material, can be

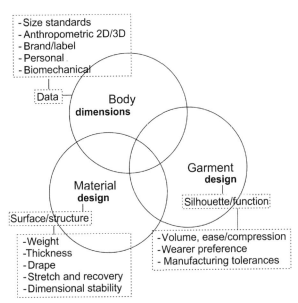

Figure 8.1 Diagram of the domains of the body, material and garment.

applied to the pattern design process as well. Zero waste pattern techniques place the primary emphasis on material/garment relationship with the body as a secondary site. The made-to-measure garment requires an approach with equal consideration of the body, garment and material. The knock-off pattern for the copied garment is driven by the garment domain with reference to the material domain and may never include the body. When constructing a block or foundation pattern, either manually or by CAD, the process is fundamentally driven by body dimensional data, as the block is a template of the static body. It is only when the block is used to create a style that the garment/material/body matrix is evoked.

By separating the functions into domains, it is likewise possible to consider the methods of pattern creation as an algorithmic process model, or one that facilitates parametric thinking (Table 8.1). Gill (2008) advocates the need to consider the matter of ease as a separate variable to allow quantification. Aldrich (2013) identifies the need for designers to understand fabric properties and develop 'fabric sense'; Lindqvist 'challenges the fundamental relationship between dress, patternmaking and body' to incorporate biomechanical points as dynamic references points for 'new expressions in dress' (Lindqvist, 2013, p. 5). The Grafis CAD system applies parametric thinking via a construction method to achieve a more efficient pattern outcome as it is based on a construction method that follows pattern construction protocols.

To further support the approach of identifying the elements of patternmaking as separate variable inputs, it is the use of 3D CAD technology from 3D CAD programs where such a model is being used to create virtual prototyping. Some computer programs include clo3D and Marvellous Designer by CLO Virtual Fashion Inc., V-Stitcher by Browzwear, and OptiTex where virtual fit is created by combining

Table 8.1 Table of body, material, garment input variables

Body	
Dimensions – data sources	
Standard measurements	From published national or international size standards
Anthropometric data	Measurements from data bases such as, CEASAR (http://store.sae.org/caesar/), SizeUSA (http://www.tc2.com/sizeusa.html), SizeGermany (https://portal.sizegermany.de/SizeGermany/pages/home.seam) Measured manually or by 3D body scanning
Brand or label size definitions	Developed from the above sources and or evolved from professional knowledge of customer market profiles. Can be formal or ad hoc.
Personal direct measurements	Taken directly from an individual by manual measurements or body scanning
Biomechanical inputs	Features to accommodate body movement
Material	
Design – surface and structure	
Weight	Test standards informal assessments, e.g. light, medium, heavy (Aldrich, 2013)
Thickness	Test standards or informal assessments
Drape	Test standards or informal assessments
Stretch and recovery	Test standards or informal assessments
Dimensional stability	As above plus pre- or post-garment washing factors
Garment	
Design – silhouette and function	
Volume and ease	The measurement value (+/−) calculated according to design aesthetics, fabric performance data and body asymmetry, i.e. the body to garment relationship
Wearer preferences	The tightness or looseness value (+/−) separate to the pattern ease requirement
Manufacturing tolerances	The value (+/−) to accommodate construction variables

quantified data inputs from: (1) an avatar body, (2) the material's physical properties and performance, (3) patterns from 2D vector files (Figure 8.2).

8.3.1 Measurement

Townsend and Goulding (2011) suggested that 'the relationship with the body is more direct in fashion than textile design' (p. 289). Therefore, an understanding of the body and measurement data is perhaps the most significant aspect of pattern construction.

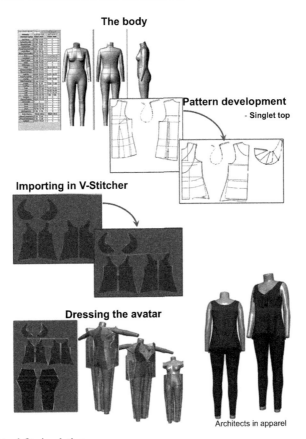

Figure 8.2 Virtual fit simulation.

Measurement data can be derived from a range of sources (Table 8.1), that is published measurement standards, anthropometric surveys (2D measurements or 3D data sets), brand or label size preferences or individual measurements. Such data can be viewed as a quantitative element to the pattern construction process. Measurement is a necessary input in creating a block, pattern or approved template according to the preferred construction method.

It is common practice in apparel design and development to create a base or sample size garment according to an accepted brand or house standard, tangibly represented by a mannequin and/or fit model. How to determine size profiles for sizes beyond the base size, in the absence of reliable anthropometric data, is a somewhat more speculative process. Published generic size standards have been expected to provide this information. The base size should be representative of the average of the population or the market demographic and as close as possible to 'the statistical average'. To be successful it should represent the 'dimensions and proportions' of the target market (Schofield, 2007, p. 158). In reality however, fit models are rarely truly representative of the statistical average in terms of both measurement and shape (Crawford and Kung, 2010, p. 371).

8.3.2 Anthropometric data

An understanding of anthropometric measurement and data is now a necessary part of apparel practitioner knowledge (Otieno, 2008, p. 63). How to measure the body, whether by manual measurement or by 3D body scanning, and how to apply data and develop appropriate fit profiles, according to market definitions is fundamental to developing a pattern to fit the sweet spot of a particular market definition. It is no longer feasible to expect that a standard measurement data set is sufficient to achieve a satisfactory outcome for the diversity of the global apparel market. Patternmakers working within a homogenous fashion market with a limited size range have a less challenging task than those operating in a complex broadly defined market as is required by the corporate apparel or uniform apparel for example. Therefore, a new area of skill development for pattern technicians is the ability to consider, gather and apply strategies around developing a valid sizing framework.

8.3.3 Size standards

Discussion of measurements standardisation is itself a varied and controversial topic, with validity, compliance and relevance common concerns. There is an expectation from both apparel manufacturers and consumers that apparel size standards are necessary for the efficient production and distribution of mass market apparel. However, with the increasingly globalised fast fashion product and anthropometrically diverse customers, retaining a standard size definition for the mass produced garment is virtually impossible. Anthropometric changes to the physical size of our population because of ageing and or the so-called 'obesity crisis' cannot be ignored. Understanding the correct anthropometric profile of the target customer is vital to success.

When considering the elements within the domains of material structure and performance against garment silhouette and function (Figure 8.1), there is a complex matrix of variables for pattern makers to consider. This complexity is causing a shift from standardisation to a more customised approach to size definition. Ashdown and Loker (2010) have highlighted the trend away from standardisation toward a target approach to size designation.

Maintaining the connection to a body by the essential human factor; a fit model and representative fit mannequin are preferential tools and ideally should represent the target market. Thus the ideal scenario to create a successful pattern includes a suite of test tools, that is an anthropometrically representative target market fit mannequin accompanied ideally by a corresponding human fit model. However, in the globalised supply chain with offshore production, there is possibly less emphasis on the human factor as part of the preproduction pattern function. A garment technician can direct a full-vendor supplier in the pattern/garment development process via a technical pack that specifies garment details, fabrication and finished garment measurements. A reference sample may accompany the pack, or the new garment development may be derived from an iteration of a previous garment. The demands of volume, reduced lead time, quick turn around and mass produced clothing are the dominant factors. The remote vendor patternmaker may have little or no direct connection to the body and garment fit may not be a priority.

For a patternmaker, knowing or defining 'the body' measurement points versus finished garment measurement points is vital data input. The International Organization for Standardization (1998): ISO 8559:1998 'Garment construction and anthropometric survey—Body dimensions' standard, specifies anthropometric measurement points, is a valid reference source for body measurement. However, it is not simply having a list of measurements to accompany and direct the specific pattern construction method; it is also an understanding of body shape and posture. As suggested by W. H. Hulme (1944) in his opening chapter of *The Theory of Garment—Pattern Making*:

> *The student of clothing should possess a good working knowledge of the parts of the body, the various types of figures, physical proportion, the effects of movement, and the relation of various fitting and draped garments to the body. (Hulme, 1944, chapter 1)*

Aldrich (2013) emphasises that it is important for students and designers to develop intuitive skills in fabric characteristics and performance. The ability to identify 'the relationship of fabric to pattern cutting' is a key driver in determining 'the pattern cutting method or the block chosen...' (Aldrich, 2013, p. 8). It is possible therefore to consider fabric performance as both quantitative and qualitative input for pattern development. Understanding fabric characteristics such as weight, drape, stretch and recovery or dimensional stability are necessary factors for a successful conversion from design to pattern, pattern to garment. This is the textile/garment integration design equation (Townsend and Goulding, 2011).

8.4 Pattern construction tools

Patternmaking requires a range of equipment including hardware and software to incorporate both the manual and the CAD domains. For manual pattern construction the components of a patternmaker's tool kit consist of an array of essential equipment items to accurately draft the flat pattern shape. If starting by constructing a basic block or foundation pattern according to a preferred construction method as previously discussed (Muller and Sohn, 1997; Davis, 2008; Aldrich, 2009, 2013) the starting point is a set of measurements, measured either from a body directly or from a set of standard measurements. Thus, a nondistortable tape measure to take accurate measurements is the foundation of any patternmaker's tool kit. For mass production, having a mannequin and or fit model that represents the required size and shape of the target market sample size is an essential resource for verifying garment fit (Figure 8.3).

8.4.1 Manual pattern tools

When manually drawing or drafting a pattern, the primary substrate used as the basis for pattern construction is commonly brown or white wide width craft paper ($80-120$ g/m^2). A light-weight transparent tracing paper or tissue paper is useful to transfer a shape to prove or validate the geometric correctness of a dart, check the

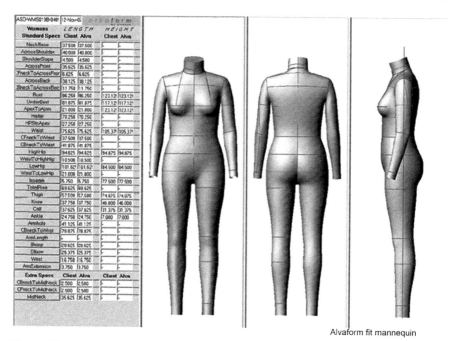

Figure 8.3 Fit model mannequin with measurement table.

flow through from front to back of a seam line, or transfer a design line. Tracing paper can also be used as an inexpensive way to trace a pattern for a prototype first sample garment. Once the garment fit has been approved then the final correct pattern shape can be transferred to cardboard as a robust pattern template for production patterns (Figure 8.4).

To create an accurate pattern, a sharp hard black pencil or mechanical pencil assists with fine line definition. Developing the capacity to draw a confident accurate line guided by the eye, known in tailoring as the 'rock of the eye', is an aspirational skill for a novice patternmaker.

Manual drafting according to a variety of construction protocols as previously discussed adhere to a geometric 2D $x-y$ coordinate grid construction format. Tools to draw accurate straight lines, such as straight edged rulers, grid or grading rulers and cutting mats, help keep construction lines parallel and right angles square. An adjustable set square, protractor and flexible curves assist with defining angles and curves with accuracy.

Equipment for making pattern notations, such as pattern notches or nickers for seam matching points, a hole punch to mark drill hole positions to indicate sewing lines for darts, pattern hooks to hang cardboard patterns and paper scissors for pattern cutting, is useful in the manual patternmaking process. A sturdy tape dispenser, glue stick and pattern weights are handy additions to the pattern makers tool kit (Figure 8.4).

Pattern construction

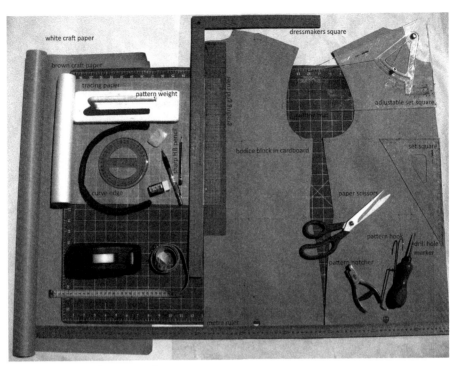

Figure 8.4 Manual pattern tools.

8.4.2 CAD pattern tools

When moving into the CAD domain, the most elementary way to transit manual patterns to a CAD environment is to digitise or trace physical patterns via a digitising table or pattern scanner into vector files. Many preproduction work methods prefer to use a mixture of manual and CAD pattern construction processes. For example, a first pattern for a new style may be derived from a combination of methods of flat patternmaking combined with some shape modelling on a mannequin to create the pattern template for the first prototype garment. Once the prototype garment fit has been approved, the pattern template can be manually digitised or scanned into a CAD system. The capacity to attribute various grading rules at strategic coordinate points (x/y) on the pattern in turn defines the required increase or decrease ($+/-$) to form a set of graded pattern pieces (Figure 8.5). This is also known as a 'graded nest' of patterns. Once the pattern is digitised with the CAD system as a vector, the CAD file can be used to make subsequent style changes and amendments.

The debate within the patternmaking profession on the merits of manual versus CAD patternmaking is greatly influenced by the patternmaker's preferred professional practice and access to technology. The capacity for a patternmaker to have affordable access to CAD software packages is a necessary CAD pattern skill enabler. Proprietorial apparel CAD programs that require specific hardware and software licence fees

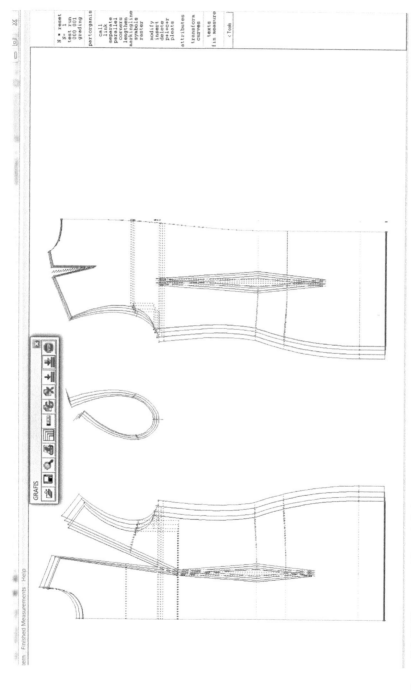

Figure 8.5 CAD pattern with graded size nest.

Pattern construction

have in the past created a barrier adoption for both patternmaking professionals and students.

However, the process of becoming a digital patternmaker is one that is becoming increasingly easy with access to generic CAD tools such as Adobe Illustrator, Corel Draw and AutoCAD. While Gerber Technology and Lectra pattern design software systems are considered the dominant CAD platforms, it is possible to emulate manual pattern construction techniques via vector-based programs such as Adobe Illustrator (Figure 8.6) or AutoCAD to create digital pattern files. Files can be saved or exported in the shareable.dfx format for importing into multiple CAD environments. The growing body of knowledge in how to use common CAD tools for pattern construction is well documented and shared within the *YouTube* community. Pattern

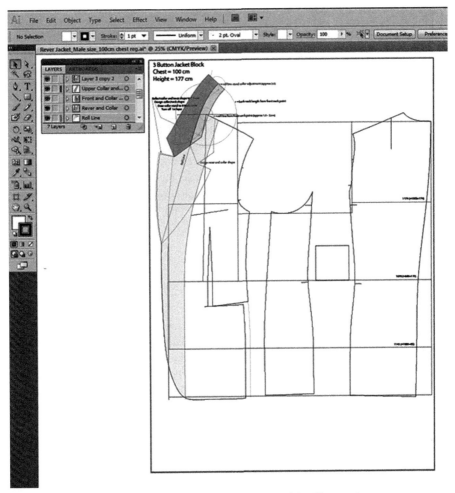

Figure 8.6 Jacket pattern constructed in CAD software (Adobe Illustrator).

Figure 8.7 Architects in Apparel CAD studio.
Architects in Apparel Web site (http://integral-t.com/index.php/services/training/).

printouts can be inexpensively printed in black and white wide width formats from commercial print shops. Thus a computer is the essential tool for the patternmakers of the future.

8.4.3 Future technology tools

The transition from 2D CAD to 3D CAD technology in programs was previously noted; clo3d and Marvellous Designer, V-Stitcher, and OptiTex are creating new methodologies for apparel design and development. The transition to the 3D mode increasingly offers the capacity for virtual prototyping, textile print and pattern simulations to extend the patternmakers domain beyond a 2D mode. The 3D view also creates the interface to technologies such as engineered pattern print design, 3D printing and seamless knitting. This further supports a broadening of the craft of patternmaking to a new digital trade craft where a greater range of design variables can be factored into the pattern development process. The laptop computer (Figure 8.7) is replacing the traditional manual patternmaking tools (Figure 8.4) to develop the next generation of digital patternmakers.

8.5 Conclusions

Within the contemporary context, pattern construction is a complex process dependent on a range of design sources and manufacturing modes. Thus, being able to develop and evolve according to the interaction of the three domains of body, design and material allows for consideration of the most appropriate method for creating a successful pattern shape according to the desired outcome. The patternmaker is no longer simply a design interpreter or convertor, working in support of a creative designer. The process of patternmaking can itself be a creative technical design source. Therefore, the capacity to identify and dissect the elements and components and consider them as variables allows for a methodological approach to creating patterns for apparel. In addition, consideration of these elements is also embedded within the 3D CAD future, which in turn is developing a new digital trade craft for patternmaking.

8.6 2D and 3D CAD Web sites

Information on 2D and 3D CAD systems can be obtained from the following websites:

http://www.optitex.com/
http://www.browzwear.com/
http://www.lectra.com/
http://www.gerbertechnology.com/
http://www.grafis.com/
http://www.stylecad.com/
http://www.clo3d.com/
http://www.marvelousdesigner.com/

Acknowledgements

Figure 8.2: 3D simulation by *Architects in Apparel* from 2D Grafis patterns in V-Stitcher by Browzwear.
Figure 8.3: AlvaForm fit mannequin and spec.

References

Aldrich, W., 2009. Metric Pattern Cutting for Women's Wear. Wiley, Chichester.
Aldrich, W., 2013. Fabrics and Pattern Cutting. Wiley, Hoboken.
Armstrong, H.J., 2006. Patternmaking for Fashion Design. Pearson Prentice Hall, Upper Saddle River, NJ.
Ashdown, S., Loker, S., 2010. Mass-customized target market sizing: extending the sizing paradigm for improved apparel fit. Fash. Pract. J. Des. Creat. Process Fash. 2, 147–174.
Bray, N., 1966. Dress Pattern Designing: The Basic Principles of Cut and Fit. Crosby Lockwood, London.
Breward, C., 2001. Manliness, modernity and the shaping of male clothing. In: Entwistle, J., Wilson, E. (Eds.), Body Dressing. Berg.
Crawford, A., Kung, A., 2010. Commercialising size survey data size UK – the UK national size survey. In: Proceedings of the International Conference on 3D Body Scanning Technologies. Lugano, Switzerland.
Davis, B., 2008. Towards Pattern Making Technology as a Source of Competitive Advantage. Barbara Davis, Melbourne, VIC.
Erwin, M., Kinchen, L., Peters, K., 1979. Clothing for Moderns. Collier Macmillan Publishers.
Gill, S., 2008. Book review: sizing in clothing: developing effective sizing systems for ready-to-wear clothing. J. Fash. Mark. Manage. 12 (4), 579–581.
Hulme, W.H., 1944. The Theory of Garment-pattern Making – a Textbook for Clothing Designers, Teachers of Clothing Technology, and Senior Students. The National Trade Press Ltd, London.
Kunick, P., 1967. Sizing, Pattern Construction and Grading for Women's and Children's Garments: A Treatise and Standard Textbook for All Who Are Engaged in the Production and Distribution of Women's and Children's Garments. Philip Kunick, London.

Lindqvist, R., 2013. On the Logic of Pattern Cutting: Foundational Cuts and Approximations of the Body. Licentiate Thesis. University of Borås.

Mcquillan, H., 2014. Make It Zero Waste: Kimono Twist Dress (Online). (accessed 01.05.14.).

Muller, M., Sohn, 1997. Schnittkonstruktionen fur Rocke und Hosen. Deutsche Bekleidungs — Akademie, Munchen, Germany.

Nakamichi, T., 2012. Pattern Magic. Stretch Fabrics. Laurence King Publishing Limited.

Otieno, R., 2008. Approaches in researching human measurement MMU model of utilising anthropometric data to create size charts. EuroMed J. Bus. 3, 63—82.

Rissanen, T., 2014. Zero-waste Fashion Design (Online). Available: http://timorissanen.com/category/zero-waste-fashion-design/ (accessed 01.05.14.).

Roberts, J., 2014. Subtraction Cutting (Online). Available: http://subtractioncutting.tumblr.com/ (accessed 01.05.14.).

Sato, S., 2011. Transformational Reconstruction. Center for Pattern Design, St. Helena, CA.

Schofield, N.A., 2007. Pattern grading. In: Ashdown, S.P. (Ed.), Sizing in Clothing. Woodhead Publishing Limited, Cambridge.

Townsend, K., Goulding, R., 2011. The interaction of two and three dimensional design in textiles and fashion. In: Briggs-Goode, A., Townsend, K., Textile, I. (Eds.), Textile Design: Principles, Advances and Applications. Woodhead Pub. in Association with the Textile Institute, Oxford.

Fabric spreading and cutting

I. Vilumsone-Nemes
Technical Faculty 'Michael Pupin', University of Novi Sad, Novi Sad, Serbia

9.1 Introduction

A cutting room is a separate area in a production enterprise where garment components are cut out. Before cutting takes place, textile materials for cutting are stored, counted and inspected, the cutting process is planned, markers to cut components are created and fabric spreads are laid out for cutting. Some components may then be fused with interlinings. All components are then inspected, sorted and bundled for further processing in a sewing line.

Many of the activities in the cutting room were traditionally performed manually. The quality of the process was dependent on the skills and experience of the cutting room workers. Now most of the work processes, such as cutting process planning, marker making, spreading and cutting, are automated and can be performed much faster and with more consistent quality than manual processing. This chapter describes fabric cutting to obtain garment components before joining them in a united article. To ensure the best utilisation of raw materials, machinery, time and labour, the cut process is precisely planned using specialised software. Multiple spreading is performed by spreading machines, which provide fabric feeding and transportation, ensure tension free spreading and material cutting in needed length pieces. Cutting of textile materials is performed by automated cutting systems using different computer-controlled cutting devices: knife, laser, ultrasound or water-jet. Final operations are quality control, the recutting of faulty components, the marking, sorting and bundling, preparing the cut components for their further processing in the sewing lines.

9.2 Cut process planning

Cut process planning prepares and organises work in a cutting room and ensures the smooth flow of work to further production processes. It is performed taking into consideration several basic parameters influencing productivity and work efficiency in a cutting room. The principal ones are as follows: the number of articles ordered for each size of a style; the technological constraints of the spreading and cutting processes; maximising fabric utilisation and ensuring the best use of time and labour.

9.2.1 Automated cut process planning

Today, special cut process planning software is used to improve the work planning process. It links together ERP (enterprise recourse planning), fabric management systems, CAD (computer aided design) and CAM (computer aided manufacturing) and creates the best solutions to cut material for incoming manufacturing orders. The most important steps of the automated cut process planning are running of different planning scenarios, establishing the marker processing time, performing marker calculations, spreading planning and processing of manufacturing reports.

9.2.1.1 Running of different planning scenarios

After the cut process planning, operator fills up all necessary order data (order quantity for each fabric type, initial marker and spreading settings), the programme tries all possible size combinations in markers. Based on a model information and marker library (data from previously used, same or similar styles), the programme estimates length and efficiency of every marker, yet uncreated. Finally, the programme selects and displays the best marker combinations for a certain order.

9.2.1.2 Establishing the marker processing time

The yet uncreated markers are classified by their importance depending on the number of sizes in a marker, lays in a spread and garment pieces produced from the marker. More time to find the best fabric consumption is given to progress more important markers. Marker processing time can be determined automatically (the programme distributes the time for each marker considering the markers size) and semiautomatically (the user can indicate how much time he wants the programme to spend for each marker).

9.2.1.3 Performing marker calculations

Getting ready markers back from CAD, the programme obtains the exact length and the efficiency of every performed marker. Now the fabric amount needed to produce the order is calculated using data of markers' length. The available information is also used to calculate statistics regarding average fabric usage per product, per fabric type or in total, and efficiency.

9.2.1.4 Spreading planning

Trying to represent the maximum number of fabric plies in the lay, the programme generates all spreads. Fabrics with similar properties are grouped together for one spread to reduce spreading time. If a disproportionate number of layers (very small number) appears for separate spreads, the programme can perform automatic balancing.

9.2.1.5 Processing of manufacturing reports

During and at the end of the cut process planning several reports could be generated such as: marker making, spreading and cutting instructions as well as fabric usage reports. To avoid mistakes, barcode scanning can be used to deliver data for automated spreading. The barcode can also be used to load a marker for a cutting process. The fabric report is sent to the warehouse to know how much fabric must be taken to spreading. It can also go down to a roll level. Then planning is done for every roll separately, finding its best usage.

Cut process planning software is developed by various companies: Lectra[1] (Optiplan), Polygon Software[2] (Cut Planning), AMS[3] (CutPlan), Option Systems[4] (Cutting Room Planning), Optitex[5] (CutPlan), Assyst[6] (Lago), FK Group[7] (Future Cutplanner), Plataine[8] (Cut-Order Planning), Reach Technologies[9] (Reach Cut Planner), and others.

A complete description of the basic parameters influencing productivity and work efficiency in a cutting room, as well as a description of the manual cut planning process, can be found in the book *Industrial Cutting of Textile Materials* (Vilumsone-Nemes, 2012). The automated cut planning process using Artificial Intelligence is described in the book *Optimizing Decision Making in the Apparel Supply Chain Using Artificial Intelligence (AI): From production to Retail.*

9.3 Spreading of textile materials

9.3.1 Manual spreading

The manual spreading process is suitable for small-scale production. Manual spreading may be used for all kind of fabrics, including those with complex structures and intricate patterns. In large-scale production, manual cutting is often used for working with intricately patterned and high-cut pile fabrics. When compared to automated spreading, the cost of technical equipment in manual spreading is low, but the productivity is poor.

9.3.1.1 Characteristics of the manual spreading process

The manual spreading process is performed in sequential steps. They are marking of the spread data; spreading of the fabric plies; and fixing a marker on the top of the spread.

Marking of the spread data
The marker, which is printed on paper, is placed on the spreading table. It is fixed in the required position and the following spread data are marked on the both sides of the table: the beginning and end of a spread, splice marks (places in the spread where the fabric may be cut and laid double to deal with flaws without damaging the cut components) and size change places (marks used in performing step spreads).

---

[1] www.lectra.com
[2] www.polygonusa.com
[3] www.cutplan.com
[4] www.styleman.com
[5] www.optitex.com
[6] www.assystbullmer.co.uk
[7] www.fkgroup.com
[8] www.plataine.com
[9] www.reach-tech.com

Figure 9.1 The manual transportation of a fabric ply above a spreading table.

Fabric spreading

At the beginning of the spreading process, an underlay paper ply is laid on the table to ensure easy transfer of the spread along the table during the cutting process. The fabric spreading process is carried out by one/two workers at each side of the spreading table who move the fabric ply to the beginning of a spread (see Figure 9.1). The end of the fabric ply is placed precisely at the beginning of the spread and secured. Returning to the initial position (the place where the fabric roll is fixed) one worker aligns the laid down fabric ply with the edge of the table and the previously spread fabric plies with a permitted variant of ±0.5 cm. The second worker smoothes the surface of the ply, ensures an even tension in the fabric and prevents creases or folds appearing during the spreading process. The spreading process is repeated until the desired number of fabric plies are laid down.

The optimal length of a manually performed spread is 4–7 m. Short markers may be joined and laid as one spread, forming either a traditional or a step spread. The number of fabric plies in a spread depends on the size of the order, the fabric properties (thickness, slickness, friction between the fabric and a cutting device, etc.) and the technical limits of the manual cutting machines (the stroke size, shape of the blade, etc.). Narrow tubular fabrics and interlinings are spread by a single worker.

The fixing of a marker on the top of a spread

A marker printed on a paper is placed on the top of a prepared spread. Clamps are placed around the edges of the spread to hold it in position. If the marker is printed on the paper with glue on its reverse side, it is lightly fixed to the top ply of the spread by using a special large base iron.

9.3.2 Automated spreading

Automated spreading systems have significantly increased the productivity of the spreading process, but have not altered its main work principles. Similar operations are performed in both the manual and the automated spreading processes.

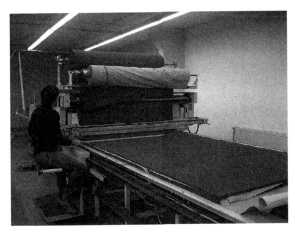

Figure 9.2 Automated spreading of a fabric.

Several companies produce automated spreading machines for a variety of textile products such as: clothing, furniture, car interiors, technical textiles and several other industries. The best known of these are: Lectra[1] (France), spreading machines of Progress-Brio series; Gerber[10] (USA), XLs series; Kuris[11] (Germany), SHATTLE and A series; Assyst-Bullmer[6] (USA); Eastman[12] (USA); Cosmotex[13] (Spain), APOLO series; FK group[7] (Italy); B.K.R. Italia[14] (Italy); and Oshima[15] (Taiwan).

9.3.2.1 Spreading machine and its main parts

Multiply spreading is performed by a spreading machine that provides fabric feeding and transportation over the spreading table and ensures tension-free spreading (see Figure 9.2). Spreading machines can work with materials of varied types and quality, as well as with woven, nonwoven or knitted materials. Fabrics can be kept folded and rolled.

Spreading machines are designed to work with fabrics of differing weight and are categorised according to the maximum weight of the spread fabric roll:

- Rolls up to 60 kg – spreading machines for light fabrics (roll diameter 300–500 mm).
- Rolls of 100–200 kg – spreading machines for medium heavy fabrics (roll diameter 500–700 mm).
- Rolls of more than 200 kg – spreading machines for heavy fabrics (roll diameter more than 700 mm).

---

[10] www.gerbertechnology.com
[11] www.kuris.de
[12] www.eastmancuts.com
[13] www.cosmotex.net
[14] www.bkritalia.com
[15] www.oshima.com.tw

Special spreading machines are available for: napped and pile fabrics (machines with a turntable); technical materials and denim fabrics (for long and high lays, large and heavy rolls); highly elastic fabrics (for production of lingerie) and tubular fabrics. Spreading machines are also characterised by the maximum height of a fabric lay, the working width and spreading speed.

The automated spreading process is performed on special tables. The main parts of a spreading machine are a fabric spreader truck, a fabric feed system, an automatic cutting device, an end-catcher, an operator stand panel, an encoder system, and a control panel.

Spreading table

The automated spreading process is performed on tables that can withstand the load of a fully equipped spreader. A table surface may be perforated and equipped with an air flotation system that lifts the lay while it is moved or with a vacuum system to hold the lay in place.

Fabric spreader truck

A spreader truck ensures the transportation of a fabric roll above a spreading table in the lengthways and transverse directions. It consists of two main parts — a body and a turret.

The body of a spreader truck is fixed on wheels. It ensures lengthways transportation of a fabric roll above a spreading table (see Figure 9.3). After laying the fabric to a specified length the spreader stops and then moves in the opposite direction. The speed of the spreader can be adjusted according to the fabric type and properties.

A fabric roll is placed on the turret of the spreader truck. The turret ensures transverse transportation of the roll during the spreading process (see Figure 9.3). It can move up to 15 cm laterally to adjust the fabric roll in order to achieve a perfect alignment of a fabric edge on the table. This movement is controlled by a double sensor that reads the position of the fabric during the spreading process and moves the whole turret as required. The body of the truck also carries several special spreading machine devices: a cutter (to cut the laid fabric ply), a zig-zag spreading device (for spreading in the 'face to face in both directions' mode), a tubular fabric spreading device

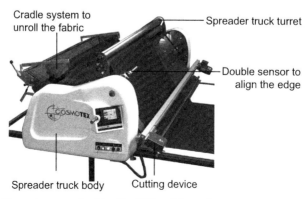

Figure 9.3 A Cosmotex spreader 'Apollo 100' and its main parts.

(to spread tubular fabrics in the zig-zag mode), and a fabric tearing device (for tearing light fabrics).

Fabric feed system
The loading, unloading, threading and rewinding of fabrics are controlled by a fabric feed system that rolls material from a fabric setting bar or a special cradle (see Figure 9.3) and moves it to the spreading table. A cradle feed system is synchronised to the speed of the spreader and can ensure automatic or manual fabric tension control to provide tension-free spreading.

The end section of the fabric on the body of the truck and the attached cutting device are lifted after each automatically laid ply. These are driven by the sensor of a lifting device to avoid contact between preceding spreads and newly laid plies. During the spreading process, the fabric feed system automatically adjusts the feeding speed of the material and measure its length.

Automatic cutting device
During the spreading process, the cutting device moves together with the spreader truck (see Figure 9.3). When the lay of fabric is fully spread, a round knife automatically moves across the table and cuts the fabric off. The cutting process may be done in one direction or in both directions. The cutting device is sharpened automatically. It has an adjustable cutting speed. An automatic height detection sensor ensures the minimum distance between the cutting device and the table top, so avoiding wrinkles in the laying process.

End-catcher
The end-catcher fixes the end of a laid fabric ply and prevents it from moving during the spreading process. It is particularly useful when working with slippery fabrics.

Operator stand panel
A special platform allows the operator to ride alongside the table during the spreading process. It also can be equipped with an adjustable seat. When the spreading process is fully automated, the moving platform does not require an operator.

Encoder system
The fully automated spreading process is driven and controlled by an encoder. Using a special belt with metal denticles fixed to one side of the spreading table, the encoder system counts the number of denticles and recalculates them in the distance (metres inches) from the start point. This defines the placement coordinates of the spreader truck on the spreading table at a given moment.

Control panel
An interactive control panel is used to set up parameters and to programme the spreading process. The main parameters are as follows: the lay length, the number of plies, the start point of the spread, the spreading mode, the fabric tension, the spreading speed, the 'dead head' speed, the fabric cutting speed, the frequency of knife sharpening and the fabric cutting length.

9.3.2.2 Semiautomated and fully automated spreading processes

A layer of a firm perforated paper is placed on a spreading table at the beginning of the spreading process. It ensures easy transportation of the spread along the table and prevents deformation of the lowest fabric plies during this process. The perforation of the paper is necessary to ensure air penetration and to compress the spread lays together during the automated cutting process. The spreading process may be performed in semiautomated and fully automated modes.

Semiautomated spreading process
In the semiautomated spreading process, the operator moves along the spreading table (walking or riding on a stand panel) and follows the spreading process. The operator smoothes the surface of the lay, identifies faults in the spread fabric and decides whether to leave faults in the spread or to cut them out. The operator uses a manually operated speed control handle to change the spreading speed and to reduce the speed in problematic areas, or even to stop the spreading process if it is necessary to define the location of a fault and to cut it out.

Fully automated spreading process
The fully automated spreading process is used for high-quality materials that are easily spread. An operator sets the necessary parameters (the length of the lay, the spreading speed, the fabric tension, etc.). The spreading machine automatically performs the following operations: lays the fabric in the required length of the spread, cuts the material at the end of every ply, counts the number of plies and stops after laying the required number of plies.

9.3.2.3 Automated fabric fault registration and management systems

Spreading machines may be equipped with special fault registration and management systems. During spreading, faults are visually recognised by a spreading operator. Using a joystick and a laser beam, the operator marks the fault and determines its placement on a ply and also on its marker on the screen. This way the operator sees if the fault affects any cut component and can make a decision to leave a piece of fabric with the fault in the spread or not. The spreading machine can automatically cut off the unusable part of fabric, register it (for further use) and move to the place on the cutting table to continue the spreading process.

9.3.3 Fabric spreading modes and their applications

Depending on the pattern and other properties, a fabric may be spread in different ways. The spreading mode determines the placement of the face side of each fabric ply in a spread — up or down and the placement direction of each fabric ply in a spread — in one or both directions.

9.3.3.1 'Face up in a single direction' spreading mode

All the plies are spread with their face side up and in one direction (see Figure 9.4). This is the most commonly used spreading mode. There are several reasons for its wide application:

- Most types of textile materials can be spread in this manner.
- The rejection of fabric faults can be carried out during the spreading process.
- Styles with asymmetrical components can be cut as pattern pieces in a marker are also placed with their face side up.

The disadvantage of this mode is the necessity of spreading of each ply from one end of the spread. It is necessary to repeat 'dead heading' movement of a fabric ply or the entire roll over the table, thus increasing the spreading time and work required.

9.3.3.2 'Face up in both directions' spreading mode

All the plies are spread with their face side up and in both directions (see Figure 9.5). To ensure that the face side is upward in all plies, the fabric roll must be turned through 180° at the end of every ply.

In contrast to the 'face up in a single direction' spreading mode, there is no need for "dead heading" in this mode. However, additional time is needed for turning the fabric roll at the end of every ply.

This spreading mode cannot be used in the following situations:

- If the fabric has a nap or pile.
- If the fabric has a pattern in one direction only and its location in a style is strictly determined.

9.3.3.3 'Face to face in a single direction' spreading mode

The first material ply is spread with its face side up. After or during the "dead heading" movement, the fabric roll is turned through 180° and the next ply is spread in the same direction with its face side down (see Figure 9.6).

This spreading mode is used for materials with a short-cut pile (velvet, corduroy, plush, and artificial fur) to prevent the plies form slipping during the spreading and cutting processes.

Figure 9.4 The placement of fabric plies performing 'face up in single direction' spreading mode.

Figure 9.5 The placement of fabric plies performing 'face up in both directions' spreading mode.

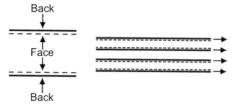

Figure 9.6 The placement of fabric plies performing 'face to face in a single direction' spreading mode.

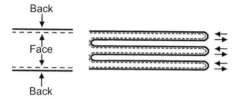

Figure 9.7 The placement of fabric plies performing 'face to face in both directions' spreading mode.

9.3.3.4 "Face to face in both directions (zig-zag)" spreading mode

The fabric is spread in both directions. At the end of each ply, the fabric is folded and spreading continues in the opposite direction (see Figure 9.7). 'Zig-zag' spreading may be used for different fabrics. It does not require that the fabric plies are cut, thus saving time. It is also possible to save material as there is no need for safety allowances at the end of a spread. If required, folded material plies may also be separated by shears or a knife. This spreading mode cannot be used in following situations:

- Where the rejection of fabric faults must be carried out during the spreading.
- Where a style has asymmetrical components.
- Where a fabric has an asymmetrical pattern and its location in the style is strictly determined.

A more complete description of manual and automated cutting processes, as well as individualities of spreading moods performed in manual and automated spreading, can be found in the book *Industrial cutting of textile materials* (Vilumsone-Nemes, 2012).

9.4 Cutting of textile materials

9.4.1 Manual cutting process

The manual cutting process ensures cutting of all kinds of textile materials. In comparison with automated cutting, its productivity is much lower, but the equipment is much less expensive and the repair and maintenance costs are small. For these reasons, the manual cutting process is widely used in small production units. The accuracy of cutting depends on the type of equipment used and on the skills and experience of the cutting operators. The greatest problem of the manual cutting process is its inability to eliminate displacement of fabric plies in a spread during the cutting process.

9.4.1.1 Characteristics of the manual cutting process

A fabric spread is processed by different cutting machines performing sequential cutting steps. These are:

dividing large spreads into smaller parts using movable cutting machines;
rough cutting of the components by movable cutting machines;
fine cutting of the components and their notches using movable and static cutting machines; and
placing of drill marks in the inner area of the components by specialised drilling machines.

Dividing a spread into smaller parts

A large spread is first divided into smaller parts for ease of movement and processing with various types of cutting machines. Then the following blocks of components are separated: blocks of large, similar length components (see Figure 9.8, white components); blocks of small size components (see Figure 9.8, dark grey components) and blocks of fusible components (see Figure 9.8, light grey components). A movable straight knife machine (see Figure 9.9) or a round knife cutting machine is used. These move during the cutting process while the spread or its parts are kept in a fixed position.

Rough cutting of components

During the rough cutting process, a specified fabric allowance is left around the contours of the components. The rough cutting is performed for fusible components, components with complicated shape and components from intricate patterned fabric. The rough cutting of components is done by movable cutting machines.

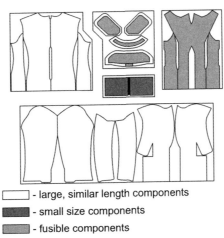

Figure 9.8 The separation of different component blocks.

Figure 9.9 A straight knife cutting machine.

Fine cutting of components

During fine cutting, the components are precisely cut along their original contours. The process can be carried out by a movable straight or round knife cutting machine (see Figures 9.9 and 9.10) or by a band knife machine (see Figure 9.11). The band knife machine is used to cut small and complicated shape components.

Drill marks placed in cut components

Drill marks — round holes in the area of cut components to show ends of darts and placement points of 'put on' components (patch pockets, flaps) — are made in bundles of cut components, using a special drill machine.

9.4.1.2 Manual cutting equipment

Manual cutting machines to cut fabrics with differing properties may be divided in two groups according to whether their operation is movable or static. The cutting process may be performed by straight, round or band knives.

Manual cutting equipment is produced by companies such as Kuris[11] (Germany), Eastman[12] (USA), Cosmotex[13] (Spain), Svegea[16] (Sweden), Hoogs[17] (Germany), Amppisani[18] (Italy), Suprena[19] (Japan), and others.

---

[16] www.svegea.se
[17] www.hoogland.de
[18] www.amppisani.com
[19] www.suprena.co.jp

Fabric spreading and cutting

Figure 9.10 A round knife cutting machine.

Figure 9.11 A band knife cutting machine.

The straight knife cutting machine

A straight knife cutting machine (see Figure 9.9) is used to cut components of differing sizes. They are moved along the cut contours while the fabric spread remains in a fixed position. Small capacity production units may use only straight knife machines. The main functional part of the machine is a vertically oscillating straight knife with a sharp blade. The knife stroke determines the maximum height of the spread which may be cut. In more powerful machines, the knife and its stroke are longer and higher spreads can be cut.

The round (rotary) knife cutting machine

A round knife cutting machine (see Figure 9.10) is used to cut large and medium size components of simple shape from low fabric spreads. The machine is moved along the cut contours while the fabric spread remains in a fixed position. The cutting device is a circular knife with a blade sharpened along one side. Round blades are used to cut light fabrics but polygonal blades are used to cut thicker and harder materials. As the cutting device is round, the machine cannot be used to cut notches and the height of the cut spread can be only approximately $\frac{1}{4}$ of the knife diameter.

The band knife cutting machine

A band knife machine has a working surface and a knife that forms a moving circle during the cutting process (see Figure 9.11). During the work process the fabric is moved while the knife is in a fixed position and performs a continuous downward movement. The band knife is characterised by its length, width and thickness. The work surface of the machine is smooth and may be equipped with a special air blowing system to facilitate maneuverability of cut fabrics. The band knife machines can ensure the highest level of accuracy in manual cutting.

The fabric drilling machine

Special fabric drilling machines are used to make drill marks straight through ply spreads. These are made after the components have been cut by placing the drilling machine onto the component bundle. The marking drills are done by a rotating steel needle that is fixed to a stand on the machine. The machine may provide the additional option of heating the needle.

9.4.2 Automated cutting process

Currently cutting of textile materials can be performed by automated cutting systems using different computer-controlled cutting devices: knife, laser, water-jet, or ultrasound. Automated knife cutting systems are most often used to manufacture garments. There are several companies which produce automated knife cutting systems for textiles: Gerber[10] (USA), GERBERcutter® Z1, and others; Lectra[1] (France), LECTRA Vector series; Kuris[11] (Germany), TexCut series; Topcut-Bullmer[20] (Germany), PremiumCut, TURBOCUT, PROCUT series; Eastman[12] (USA), Saber-C series; Tecno-Systems[21] (Italy), different series; FK group[7] (Italy), TopCut series; B.K.R. Italia[14] (Italy), BCUT series; Aeronaut[22] (Australia), different series, etc.

9.4.2.1 Automated knife cutting systems and their main parts

The main parts of the automated knife cutting system are a cutting device and a carriage in which the cutting device is fixed, a crossbar (beam), which carries the carriage

---

[20] www.topcut-bullmer.com
[21] www.tecno-systems.com
[22] www.aeronaut.org/

Fabric spreading and cutting

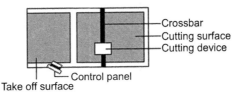

Figure 9.12 The schematic of an automated cutting system.

across the cutting surface, a cutting surface and a control panel to control the cutting process (see Figure 9.12) and nesting and cutter control software.

Cutting tools

The cutting process is done by multitool cutting heads that contains knives, notch tools, drill punches and markers. The choice of a knife depends on the configuration of the required contours and cutting operations:

- Rotary blade knife ('Pizza Wheel' type cutter) — is circular and rolls over the material (see Figure 9.13). Rotary blade knives are used in single-ply cutters to cut the contours of the components.
- Drag knife — its blade is angled (see Figure 9.14). During the cutting process, the knife is dragged along the profile of the cut component. Drag blades are used in single-ply cutters to cut detailed shapes, small circles or notches.
- Oscillating (reciprocating) knife — has a reciprocating action and it is used to cut multi-ply spreads (see Figure 9.15). Because of the oscillating blade, the surface of the cutting table must be soft enough for the blade to penetrate.

Other operations carried out during the cutting process can be performed using punches and drills. Marking tools are used to identify cut components.

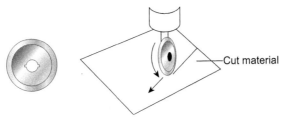

Figure 9.13 A rotary knife in a cutting process.

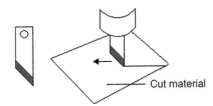

Figure 9.14 A drag knife in a cutting process.

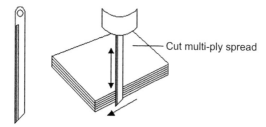

Figure 9.15 An oscillating knife in a cutting process.

The cutting device drive system

Two synchronised servo motors move the beam (crossbar) along the length of the cutting table (the X axis), while the third servo motor moves the carriage on the beam across the width of the cutting table (the Y axis) (see Figure 9.16) and a fourth rotates (the C axis) the cutting tools during the cutting process. If the cutting tool (oscillating knife) also moves vertically (the Z axis), an additional motor is used to complete the cutting process. The cutting process can be carried out at variable speeds, depending on the properties of the material and the specific requirements of a cutting operation.

The cutting table

The cutting process is carried out on horizontal cutting tables. Depending on their shape and the mode of operation, they are categorised as follows:

> Static tables: the cutting process is completed in one step on the fixed surface of the cutting table.
> Conveyorised tables: the surface of the cutting table is movable, which ensures a continuous cutting process and increased productivity.

Cutting tables are often equipped with a powerful vacuum system to secure the material during the cutting process and to ensure a high degree of accuracy in the cut components.

The control panel

The operator is provided with information on the cutting data on a computer screen from which all the processing functions are controlled by a mouse or a touch screen

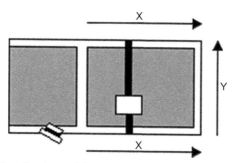

Figure 9.16 The driving directions of the servo motors.

display. A cut marker is placed on the screen of a control panel and is visible during the cutting process. The operator can distinguish the components that have already been cut (these are marked in different colours from those as yet uncut) and the location of the cutting device at any given moment of the cutting process. The operator sets the main parameters of the cutting process (depending on the properties of the material) such as the cutting speed, the oscillation frequency of the knife, the knife sharpening frequency, the vacuum level, etc. before starting the process.

The nesting and cutter control software

Nesting and cutter driving software is used to make the connection between CAD drawings and the cut output. The workflow in automated cutting involves the following steps:

- Nesting (marker making): arranging the pattern pieces in the correct orientation to fit into the most economical amount of material.
- Setting the optimal tool and the optimum speed for cutting components.
- Sending the work task to the cutting system.

Special software is used to assign different cutting tools to certain shapes in CAD drawings (different tools are used to cut contours, sharp corners, notches or holes). A complete tool setup also includes the cutting pressure of the blades, the cutting speed, the acceleration of the knife, and the time delay settings for moving the tool up and down.

Each job may be cut in several ways. The markings are normally done first for the entire work zone. This is followed by any necessary punch or drill marks. Next, all the contours of the components are cut using the chosen cutting tools. The cutting process is carried out in separate work zones. All the cutting operations are performed in a specified work zone. The material is then moved (working with conveyorised cutting surface) or the cutter by itself moves to the next cutting zone and the cutting of the components is continued. The width of the work zone is adjustable and depends on the size of the cut components (a smaller work zone is used for smaller components and vice versa). The cutting control software controls the starting and stopping of a job, the pausing and moving of the cutter on the table and its restarting at any point.

9.4.2.2 The single-ply cutters

The cutting process is carried out on a static or conveyorised cutting table by a multi-tool cutting head. The material may be fed straight from the roll. A porous cutting surface and regulated aspiration vacuum system are used to fix the material on the table during the cutting process without using clamps. If the cut material is air permeable, it must be covered with polyethylene film to ensure the formation of a vacuum.

The single-ply cutting systems can ensure cutting of single fabric ply and also low spreads (the maximum thickness of the compressed material is 25 mm). They are used for sample making, made-to-measure manufacturing, cutting intricate patterns (striped, checked, others) fabrics, preproduction runs and small orders. Single-ply cutters ensure a higher cutting speed and accuracy than multi-ply cutters.

9.4.2.3 The multi-ply cutters

Multi-ply cutting can only be used for materials that are porous and air permeable. Several material plies are spread on top of each another (using a spreading machine) and covered with a polyethylene film. Using a vacuum system, all the plies are fixed to the cutting table and pressed tightly to form an airtight seal that ensures the material is held together during the entire cutting process (see Figure 9.17).

As the cut material, consisting of several plies pressed together, is much thicker and not homogeneous, many of the tools used for single-ply cutting cannot be used here and the normal cutting time is two to three times longer than that of single-ply cutting. The cutting process is carried out by a vertical high oscillation knife (Z axis) and a drill or a punch. Deflection of the knife during the cutting process is prevented by steering it through a narrow gap in the feed of the cutting head (see Figure 9.18). The system ensures automatic sharpening and cleaning of the blade. The knife cooler reduces fusing between the cut parts and keeps the cutting head components clean.

Because of the vertical movement of the knife and the necessity of ensuring accurate cutting from the top ply to the lowest ply, the cutting surface is covered with bristles

Figure 9.17 The fabric plies on a cutting surface compressed by a vacuum system.

Figure 9.18 The feed of a cutting head of Gerber Cutter XLc7000.

that allow the knife to pierce all the plies easily and without causing any damage (see Figure 9.19). During the cutting process, a bristle cutting surface automatically advances the material to the cutting zone and simultaneously moves the pieces that are already cut to a take-off table where a cutting operator manually unloads the cut components.

A variety of multiply power cutting systems are available. Their power levels are determined by the maximum thickness of the compressed material to be cut: low ply cutters, 25, 30 mm; middle ply cutters, 40, 50 mm; high ply cutters, 80, 90 mm. Low ply cutters work at higher speeds than high ply cutters. They are used for both low and high volume production, but are more suited to industries in which high accuracy is required, and in which short runs, just-in-time production or one-offs are required. High ply cutters are used for high volume manufacturing, comparatively low cost goods such as garments and upholstery where a very high degree of accuracy is not required.

9.4.2.4 Other automated cutting methods to cut textiles

Automated laser cutting systems
Textiles may be cut by carbon dioxide (CO_2) lasers. Single-ply cutting is the most widely used. Laser is good for cutting components with complicated shapes and very light fabrics (silk, cotton) where the use of bladed tools might move or drag the fabric during the process. During laser cutting synthetic fibres melt and cut edges cannot fray. This is an advantage of working with technical materials. Cutting garments melting of synthetic materials is a problem as hard edges can irritate the skin. Often laser cutting is faster than knife cutting. Laser cutting of textiles can be performed by the cutters of 'Electron Laser' series developed by Aeronaut[22] and others.

Automated ultrasound cutting systems
Ultrasonic cutting is common in the garment manufacturing where melted and sealed edges free from fraying are required. Ultrasonic cutters are also good for thick fabrics or fabrics with an uneven weave or variation in thickness, where laser cutting often gives a poor edge finish. Ultrasonic cutting ensures a high level of accuracy in both simple and complex shapes and notches. It also enables a single-ply cutting process. Ultrasound cutting of textiles can be performed by the 'PremiumCut US' cutting system developed by Topcut-Bullmer[20] and others.

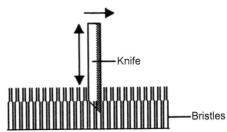

Figure 9.19 The bristle surface of a cutting table.

Automated water-jet cutting systems

Water-jet cutting does not generate levels of heat that would harm textile fibres or fabrics. Depending on the diameter of the jet, it is possible to process very narrow and sharp corners and difficult contours. There is no deformation as the cutting forces are very small. The use of water-jet cutting for textile materials may cause wet edges. Water-jet cutting of textiles can be performed by an 'Ecocut Standard Premium' cutting system developed by Tecno-Systems[21] and others.

Multipurpose cutters

A number of dual or multipurpose cutting systems are also available on the market today. They are developed to cut wider ranges of materials than with traditional cutting systems and perform additional operations during the cutting process. Usually blade cutting and laser (e.g. 'Matercut', 'Genesis' and 'Aristomat' series by Tecno-Systems[21]) or ultrasonic cutting (e.g. Electron B1 and B2 Ultra cutters by Aeronaut[22]) are combined in one cutting system. The cutter can have also special printing tools to print barcodes, logos and certain patterns directly on the fabric during its cutting.

Wider description of manual and automated cutting processes can be found in the book *Industrial Cutting of Textile Materials* (Vilumsone-Nemes, 2012).

9.5 Fusing of cut textile components

Fusing is a process in which cut components or separate parts of a lay (blocks of components) are fused with interlinings that are coated with thermoplastic resin. Fusing certain components creates strength and stability and improves the shape and crease resistance of a garment. Fusing of the components is usually performed in a cutting room as this process must be performed after material cutting and before the final operations in a cutting process (quality control, sorting and bundling of cut components).

9.5.1 Fusing presses and their characteristics

The fusing process is performed by special fusing presses. They can be divided into two groups according to the way they perform the work process: discontinuous work process (flat) fusing presses and continuous work process fusing presses.

Discontinuous work process fusing presses realise sequential, separated from each other fusing process. They are less productive and are suitable for small and medium production units. Continuous work process fusing presses enable an ongoing process by moving the components on a conveyor belt. They offer a higher level of productivity and are more energy efficient. Because of these advantages, continuous fusing presses are designed for different power production units and used more often.

9.5.1.1 Discontinuous work process fusing press

The discontinuous fusing press is generally flat bed, which has a heating zone with two work surfaces. The other arts are handle, head and buck. Depending upon the

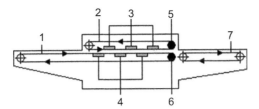

Figure 9.20 The schema of a continuous fusing presses: 1, loading and feed (lower) conveyer belt; 2, upper conveyer belt; 3, upper heating zone; 4, lower heating zone; 5 and 6, pressure rolls; 7, exit conveyer.

construction, there are two ways in which the work surfaces may move to open the press such as: an upper surface which folds sideways (fusing plate press) and an upper or lower surface, which is moved whilst maintaining its parallel position. The discontinuous fusing press is flexible for a range of materials, needs low maintenance, which are designed for low to medium sized production requirements.

9.5.1.2 Continuous work process fusing press

The operation of this type of press (see Figure 9.20) depends on a continuously moving conveyor belt, which moves the face fabric and its interlining components in and out of the heating chamber. The heating chamber of an advanced continuous fusing press consists of several (up to 7, 9 or 12) individually controlled separate heating zones for the even distribution of heat. The temperatures for the upper heating zones (which heat the interlining) and the lower heating zones (which heat the face fabric) can be adjusted separately and precisely using a special heat control system. The long heating chamber with its several heating zones extends the heating time and ensures a gradual temperature increase so that the cut components can be fused perfectly even under lower temperatures, avoiding fabric shrinkage and fading. The heating capacity and the geometry of the heating chamber determine the material to be fused. Lighter fabrics require lower heating capacity and shorter heating chambers with fewer heating zones. The main heat should come from the face fabric side, in order to have the melting resins flow towards the face fabric. For heavier fabrics higher heating capacity and a longer heating chamber with several heating zones are needed. Heating from both top and bottom sides, with separate thermo control, ensures the correct temperature balance.

The fusing presses are produced by companies: Meyer[23] (Germany), VEIT-Group[24] (Germany), Reliant Machinery[25] (UK), The Martin Group[26] (Italy), Macpi[27] (Italy), Konsan[28] (UK), Hashima[29] (Japan), Oshima[15] (Taiwan).

---

[23] www.meyer-machines.com
[24] www.veit.de
[25] www.reliant-machinery.com
[26] www.martingroup.it
[27] www.macpi.it
[28] www.konsan.co.uk
[29] www.hashima.co.jp

9.5.2 Determination of the optimal fusing parameters

The most important factor for performing qualitative fusing is temperature. It must correspond to the glue line temperature, required for the certain interlining. However, the temperature setting or reading on the control panel of a fusing press indicates its belt temperature, not the temperature applied to the interlining. The real temperature that will be delivered to the interlining through the face fabric can be determined only during the test. It can be performed putting special temperature control tape in-between the face fabric and its interlining sample and fusing them. The colour of the tape will show the real temperature applied to the interlining. The optimal fusing temperature must be found for every fabric of the order to avoid its shrinkage or damaging during the fusing process.

A more complete description of the fusing process, discontinuous work process (flat) fusing presses and continuous work process fusing presses, as well as fusing quality problems and their prevention can be found in the book *Industrial Cutting of Textile Materials* (Viļumsone-Nemes, 2012).

9.6 Final work operations of the cutting process

Final operations complete the cutting process and prepare the cut components for their further processing in the sewing lines. The operations are as follows: quality control of the cut components, the recutting of faulty components, and the numbering, sorting and bundling of the cut components.

9.6.1 The quality control of cut components

A quality control operation is performed after the components are cut and fused. The replacement of faulty components is more convenient at this stage as it does not affect the previously finished cutting process, or the forthcoming sewing process. All the cut components are inspected and the following quality parameters are controlled: fabric quality; the conformity of the size and shape of cut components to their pattern pieces; the quality of notches and drill marks.

9.6.1.1 The control of fabric quality

Fabric quality of the cut components is inspected visually. If textile faults or faults arising from the cutting or fusing processes are found, the component is taken out of the bundle and sent to be recut.

9.6.1.2 The control of size and shape

After the manual cutting process, the size of cut components is controlled by comparing the components with their pattern pieces. The dimensions of the cut components may have certain admissible tolerances. These depend on the importance of the dimensions in the garment (smaller tolerances are admissible in the horizontal dimensions of components), all dimensions of a component (the smaller its size, the smaller the admissible tolerances) and the application of the component (larger tolerances are admissible in interlining components). After automated cutting the control of size and shape of components is not performed as the process ensures high cutting precision.

9.6.1.3 The control of notches and drill marks

After the manual cutting process, the quality of notches and drill marks is inspected. Their placement is compared with the corresponding marks in the pattern pieces. If inaccurately placed notches or drill marks are found during inspection, the components are taken out of the bundles and sent to be recut. After automated cutting the control of notches and drill marks is not performed as the process ensures high cutting precision.

9.6.2 The recutting of faulty components

The components are recut from the fabric that remains from the spreading process. A piece of fabric with the correct colour shade must be found and the component is recut using its original pattern piece. All notches and drill marks are also performed and if necessary, the interlining is cut and the component is fused. The new component is marked and replaced in the bundle at the same place (the original sequence of the components in the bundle must not be altered).

9.6.3 The numbering of cut components

During the numbering process, every component is given a sequential number. This corresponds to the number of the fabric ply from which the component was cut. In the subsequent sewing process, the components with the same sequence number will be joined to make one article. This numbering helps to eliminate any confusion of the components and prevent the joining of components with different colour shades. The numbering is done manually, using a special hand-held numbering device (a sticker gun) with small labels. If a bar code system is used to monitor the production process, the bar codes are fixed on the bundle tags.

9.6.4 The sorting and bundling of cut components

The sorting and bundling of cut components are the final work operations during which the following actions are taken: a full set of components of each cut style is formed; the cut articles are grouped by size, and also by colour if several different coloured fabrics were laid in one spread.

9.6.4.1 The formation of a full set of components

The full set of all the component bundles of a style is placed on a table. After, they are tied with fabric strips to fix the components together and to prevent their loss during transportation to the sewing line. The bundles of components that will later be joined are combined in larger bundles (e.g. one large bundle may contain bundles of an outer sleeve, an under sleeve and cuffs). If the sewing line has a separate section for processing small components, their bundles are sorted separately and transported directly to this part of the sewing line.

9.6.4.2 The grouping of cut components by size

A full set of component bundles for each size is completed separately. This ensures the concurrent processing of one size of garment in the sewing line, the fixing of the correct sizing labels and the separate packing and recording of completed goods according to size.

9.6.4.3 The grouping of cut components by colours

The grouping of garments by colour is necessary to process different colour articles with the corresponding colour threads and matching accessories such as buttons and zippers. It is also the usual practice to pack and record garments according to colour.

9.7 Future trends

The development of cutting room technologies is still a target for maximum productivity, optimum quality and maximum effective material utilisation. The following tasks are needed in the near future to achieve these targets:

- To improve marker making algorithms to ensure more effective placement of pattern pieces in a marker and shorter computation time.
- To improve fabric defect elimination software to reduce spreading time and fabric loss.
- To develop automated pattern matching methods for multi-ply spreading of fabrics with intricate patterns.
- To increase cutting speed, to improve cutting tool technology and to reduce cutting time.
- To develop universal spreaders and cutters to process the most diverse materials with the same machinery.
- To develop multipurpose cutters to cut wider ranges of materials and perform additional operations during fabric cutting.
- To simplify operation of automated systems and improve work process monitoring and control systems to operate machinery without special knowledge, experience or additional training.
- To improve Smart Service for automated systems – online support to an operator, remote technical assistance, predictive maintenance, and anticipation of breakdowns; and
- To improve production planning and control systems to plan and schedule the work process more effectively and reduce fabric loss.

9.8 Conclusions

Well planned, efficient work processes in a cutting room ensure the maximal material utilisation and smooth flow of work to downstream operations in sewing lines. The significance of the cutting department in the whole production process is the main reason for continued innovations in the cutting room during the past decades. Due to advanced software, high-tech equipment and intelligent services, it has become the most advanced department in an apparel manufacturing enterprise. Its management and manufacturing processes can successfully react on the latest tendency of the global economy — increase of labour, energy, raw material costs — and changes in the apparel market — increasing diversity of styles and materials, decreasing quantities, and short manufacturing cycles. In the nearest future, because of the very high competition in the apparel industry, manual work in a cutting room will have to be fully eliminated by advanced technologies. Furthermore, new ways to reach the best productivity, quality and material utilisation in a cutting room must be found.

9.9 Sources of further information

Bowers, M.R., Agarwal, A., 1993. Hierarchical production planning: scheduling the apparel industry. Int. J. Clothing Sci. Technol. 5 (3/4), 36−43.

Degraeve, Z., Gochet, W., Jans, R., 2002. Alternative formulations for a layout problem in the fashion industry. Euro. J. Oper. Res. 143, 80−93.

Gersak, J., 2013. Design of Clothing Manufacturing Processes: A Systematic Approach to Planning, Scheduling and Control. Woodhead, Cambridge.

Gutauskas, M., Masteikaite, V., 1997. Mechanical stability of fused textile systems. Int. J. Clothing Sci. Technol. 9 (5), 360−366.

Gutauskas, M., Masteikaite, V., Kolomejec, L., 2000. Estimation of fused textile systems shrinkage. Int. J. Clothing Sci. Technol. 12 (1), 63−72.

Hui, C.L., Ng, S.F., Chan, C.C., 2000. A study of the roll planning of fabric spreading using genetic algorithms. Int. J. Clothing Sci. Technol. 12 (1), 50−62.

Jevšnik, S., Jelka Geršak, J., 1998. Objective evaluation and prediction of properties of a fused panel. Int. J. Clothing Sci. Technol. 10 (3/4), 252−262.

Jevšnik, S., Jelka Geršak, J., 2001. Use of a knowledge base for studying the correlation between the constructional parameters of fabrics and properties of a fused panel. Int. J. Clothing Sci. Technol. 13 (3/4), 186−197.

Kim, S.J., Kim, K.H., Lee, D.H., Bae, G.H., 1998. Suitability of non-woven fusible interlining to the thin worsted fabrics. Int. J. Clothing Sci. Technol. 10 (3/4), 273−282.

Lai, Sang-Song, 2001. Optimal combinations of face and fusible interlining fabrics. Int. J. Clothing Sci. Technol. 13 (5), 322−338.

Nayak, R., Khandual, A., 2010. Application of laser in apparel industry. Colourage 57 (2), 85−90.

Ng, S.F., Hui, C.L., Leaf, G.A.V., 1998. Fabric loss during spreading: a theoretical analysis and its implications. J. Text. Inst. 89 (1), 686−695.

Ng, S.F., Hui, C.L., Leaf, G.A.V., 1999. A mathematical model for predicting fabric loss during spreading. Int. J. Clothing Sci. Technol. 11 (2/3), 76−83.

Tyler, D.J., 2003. Carr and Latham's Technology of Clothing Manufacture. Blackwell Publishing, Oxford.

Vilumsone-Nemes, I., 2012. Industrial Cutting of Textile Materials. Woodhead, Cambridge.

Walter, L., Kartsounis, G., Carosio, S., 2009. Transforming Clothing Production into a Demand-driven, Knowledge-based, High-tech Industry. Springer-Verlag, London.

Wong, W.K., Chan, C.K., Ip, W.H., 2000. Optimization of spreading and cutting sequencing model in garment manufacturing. Comput. Ind. 43 (1), 1–10.

Wong, W.K., Guo, Z.X., Leung, S.Y.S., 2011. Applications of artificial intelligence in the apparel industry: a review. Text. Res. J 81 (18), 1871–1892.

Wong, W.K., Guo, Z.X., Leung, S.Y.S., 2013. Optimizing Decision Making in the Apparel Supply Chain Using Artificial Intelligence (AI): From Production to Retail. Woodhead, Cambridge.

Yoon, Soon Young, Park, Chang Kyu, Kim, Hyeong-Seok, Kim, Sungmin, July 2010. Optimization of fusing process conditions using the Taguchi method. Text. Res. J. 80, 1016–1026.

Sewing, stitches and seams

10

G. Colovic

The College of Textile – Design, Technology and Management, Belgrade, Serbia

10.1 Introduction

Sewing is the most important task of making a garment or other similar product. It can be defined as the craft of fastening or attaching objects using stitches made with a needle and thread. It is a term used to describe the process used in factories to mass-produce a wide range of garments and other goods that are created by joining different components together along the course of a structured process. Sewing is done by putting parts together and joining into a whole garment. The basic sewing parameters include:

- stitches,
- seams,
- a method of sewing.

According to Kunz and Glock (2004), sewing garment pieces together with thread forming stitches and seams is the most used method of garment production. Seam and stitches are related to each other as seam can not be held without a stitch. Seam is the join between two or more plies of pieces of material, whereas a stitch is formed by one or more threads or loops of threads. Both the seam and stitch type affect the quality of a sewn garment which is characterized in terms of strength, durability, elasticity, security and appearance. The selection depends on the end use application and the relative importance of these characteristics.

10.1.1 Stitches

Depending on the formation, stitches can be made by hand or machine. According to ISO 4915 standard, a stitch is one unit of conformation resulting from one or more strands or loops of thread based on the following three principles (Figure 10.1):

1. Intralooping – passing a loop of thread through another loop by the same thread.
2. Interlooping – passing a loop of thread through another loop formed by a different thread.
3. Interlacing – passing a thread over or around another thread or a loop of another thread.

10.1.2 Seams

Seams may be created with thread by hand or machine, or with fusion through chemical bonding. A seam is a method of joining two or more pieces of materials together by a row of stitching. Ready-made-garment seams have utility as well as aesthetic

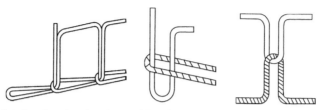

Figure 10.1 Intralooping, interlooping and interlacing. From ISO 4915.

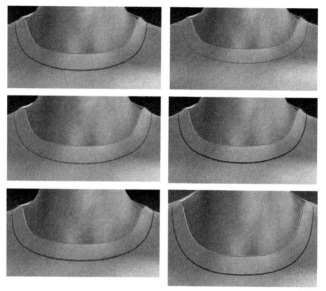

Figure 10.2 Utility and aesthetic value of seams.

value (Figure 10.2), and certain characteristics are necessary for the seams such as strength, elasticity, durability, security and appearance.

10.2　Stitch classes

A sewing stitch may be formed without any material, inside a material, through a material and on a material. All stitch types are formed by a needle penetrating into a fabric while transporting a sewing thread. It is essential to understand how the unique characteristics of every stitch type are dependent upon the mechanical actions of the sewing machine and fabric type, which determines the stitch properties.

The stitch properties are stitch size and stitch consistency. Stitch size includes stitch length (SPI − stitches per inch, Figure 10.3), stitch width (distance between the outermost lines of stitches) and depth (distance between the upper and lower parts of the

Figure 10.3 Stitches.

surface stitch). Stitch consistency is the uniformity with which each stitch is formed in a row of stitches. There must be a compatibility of fabric, stitch and seam type, needle, thread and machine setting to achieve a good quality sewn product. Stitch lengths of about 2.5 mm are used for light- to medium-weight fabrics, 3 mm for medium- to heavy-weight fabrics and 3.5−4 mm for heavy and thick fabrics.

Stitch classes are based on the type of loop formation by a sewing machine. Textile stitch types, classification and terminology are catalogued into six classes (as per BS3870/ASTM D-6193/ISO 4915:1991), which are identified by the first digit of the three digit numerals. Each class is further divided into several types which are identified by the second and the third digit. The international organization for standardization uses stitch classification based on numbering, as shown in Table 10.1.

10.2.1 Class 100: chain stitches

Chain stitches are formed using one or more needle thread(s) and are characterized by intralooping. One or more loops of thread are passed through the material and secured

Table 10.1 Stitch classes

	Stitch classes	Type
100	Chain stitches	101, 102, 103, 104, 105, 107, 108
200	Hand stitches	201, 202, 204, 205, 206, 209, 211, 213, 214, 215, 217, 219, 220
300	Lockstitches	301, 302, 303, 304, 305, 306, 307, 308, 309, 310, 311, 312, 313, 314, 315, 316, 317, 318, 319, 320, 321, 322, 323, 324, 325, 326, 327
400	Multi-thread chain stitches	401, 402, 403, 404, 405, 406, 407, 408, 409, 410, 411, 412, 413, 414, 415, 416, 417
500	Over-edge chain stitches	501, 502, 503, 504, 505, 506, 507, 506, 509, 510, 511, 512, 513, 514, 521
600	Cover-seam chain stitches	601, 602, 603, 604, 605, 606, 607, 608, 609

From ISO 4915.

by intralooping with a succeeding loop or loops after they have been passed through the material. Chain stitches are elastic, thicker than lockstitches (Class 300). Applications of chain stitches include basting, tacking, button sewing and label setting.

According to the Mihailovic process of forming a chain stitch, there are several phases of stitch formation as follows (Figure 10.4). At first the needle thread slacks at the position where the needle slightly goes up from its lowest position, and the looper catches the needle thread, which has become like a loop. Then needle comes off the cloth and the cloth is fed. The looper rotates and removes the thread that the looper caught before while pulling in the needle thread. The needle bar continues going up and the needle thread take-up lever lifts the thread. The looper continues rotating and pulls in the thread in the centre of the looper, and the thread take-up lever tightens the thread which the looper removed before.

Cloth feed is finished and a stitch is formed. The needle penetrates into the cloth to continue to the next stitch.

The stitch class 100 includes stitch types 101, 102, 103, 104, 105, 107 and 108. One of the simplest of all chain stitch types is 101, which is formed from a single thread. It is used for basting operations in manufacturing garment because it can be easily removed (in positions such as edges, flaps, collars, and so on, it is a temporary stitch). Figure 10.5 shows stitch type 101.

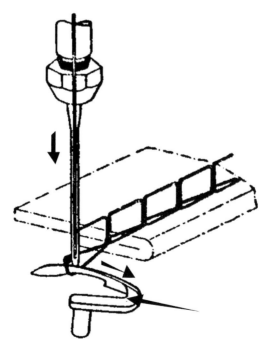

Figure 10.4 Formation of stitch class 100.
Mihailovic N. (1986). Masine I Uredjaji U Konfekciji. VTTS, Beograd.

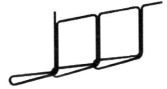

Figure 10.5 Stitch type 101.
From ISO 4915.

Figure 10.6 Stitch type 103.
From ISO 4915.

Type 103 is formed with one thread and a curved needle that passes from the left to the right, entering and exiting from the same side of the material carrying a needle thread that is intercepted by a blind looper (Figure 10.6). It can be used for hemming, belt loops, and felling operations.

10.2.2 Class 200: hand stitches

The stitch class 200 consists of stitches done by hand with the exception of 205, which simulates a hand-running stitch but is formed by a special machine (pick stitch). Hand stitches formed by a single thread are passed from one side of the material to the other in successive needle penetrations. Thread passes through the material as a single line, and the stitch is secured by the succeeding formation of loop that passes in and out of the material or intralooping of the loops with themselves. When more than one thread is used, the threads pass through the same perforations in the material. As an example, hand stitch type 205 is shown in Figure 10.7.

Hand stitch type 209 (Figure 10.8) is formed with one thread. It is a machine-made version of traditional hand stitching, sometimes referred to as a saddle stitch.

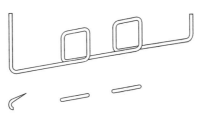

Figure 10.7 Hand stitch 205.
From ISO 4915.

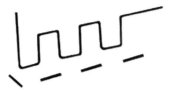

Figure 10.8 Hand stitch 209.
From ISO 4915.

Hand stitching is used at the expensive end of garment production. For example, these stitches are used in the lapel of jacket, coat, edge of the jackets, designed bed sheet, designed pillow cover and some luxurious items according to the consumer's choice.

10.2.3 Class 300: lockstitches

The lockstitch class 300 is the most commonly used, as a complete garment can be sewn on a lockstitch machine. These stitches are formed by a needle thread or threads, introduced from one side of the material, interlacing with an under-thread supplied from a bobbin on the other side.

Applications of lock stitches include seaming operations on all types of garments and run stitching. Lock stitch is extensively used for joining fabrics collar, cuff, pocket, sleeve, facing etc. Lockstitch type 301 is the simplest, which is shaped from the needle thread and the bobbin thread (Figure 10.9).

Type 304 (zigzag version) is commonly used for attaching trimmings (attaching lace and elastic on lingerie) where a broad row of stitching but no neatening is needed, and to produce forgetting. Forgetting is a decoration stitch used to connect two pieces of fabric by allowing space (width of stitch) between the pieces. Figure 10.10 shows stitch type 304.

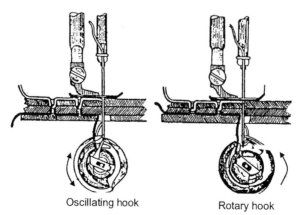

Figure 10.9 Formation of stitch type 301.
Mihailovic N. (1986). Masine i uredjaji u konfekciji. VTTS, Beograd.

Figure 10.10 Stitch type 304.
From ISO 4915.

Other types of zigzag lockstitches are class 308 and 315, which form a longer and a wider zigzag by using several stitches before changing direction. They are used for seaming, top-stitching, cover stitching and knits or wovens where wide coverage or greater stretch is required. Stitch type 306 is a lockstitch used as blind stitch. This is more secure than the traditional type 103 and its typical uses include hemming or seaming lining to shell fabric.

10.2.4 Class 400: Multi-thread chain stitches

The stitch types in class 400 are formed with two or more groups of threads, and their general characteristic is the interlooping of the two groups (Figure 10.11). Loops of one group of threads are passed through the material and are secured by interlacing and interlooping with loops of another group. One group is normally referred to as the needle threads and the other group as the looper threads.

According to the Mihailovic process of forming a multi-thread chain stitch, there are six phases, as shown in Figure 10.12.

The needle passes through the lowest position upwards. From the side of a short needle groove, the loop of the upper thread is formed into which the top of the looper gets in. The needle gets out of the material, the material moves for the length of a stitch. The looper moves to the left, shifting the loop of a needle thread on itself and moves in the direction of forming a stitch. The needle moves down, goes through the material,

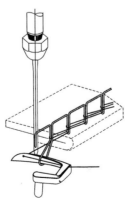

Figure 10.11 Formation of stitch class 400.
Mihailovic N. (1986). Masine i uredjaji u konfekciji. VTTS, Beograd.

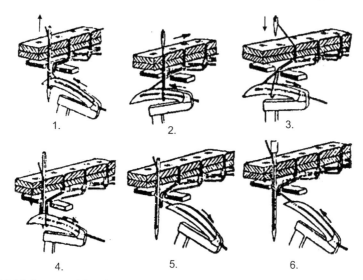

Figure 10.12 Process of forming a multi-thread chain stitch.
Mihailovic N. (1986). Masine i uredjaji u konfekciji. VTTS, Beograd.

and with its top gets into the triangle between the loop of the upper thread, the lower thread and the top of the looper. Then the needle moves down and the looper to the right, thereby tightening the lower thread. In the lower end position the needle tightens the previous stitch and keeps the looper thread on its body. The looper moves and takes the starting position. The needle moves for a certain length and the process of forming a stitch continues.

Stitches in this class are sometimes referred to as double-locked stitches, because the needle thread is interconnected with two loops of the under-thread. Applications of multi-thread chain stitches are seaming operations on all types of garments.

The most common stitch types in this class are type 401, 402, 404, 406 and 407. Type 401, or the two-thread chain stitch, has the appearance of a lockstitch in the top but has a double-chain effect formed by a looper thread on the underside. The chain generally lies on the under surface of the material, with the needle thread being drawn through to balance the stitch. The loop formation of the chain stitch elongates when extended; thus it is used for seams that require elasticity, such as setting sleeves and attaching elastic materials. Stitch type 401 is shown in Figure 10.13.

Figure 10.13 Stitch type 401.
From ISO 4915.

Type 402, the cording stitch, is used primarily for stitching permanent creases. It uses two needle threads that produce two parallel rows of stitching on the face of the fabric. A looper thread travels between the two needle threads on the back of the fabric, creating a ridge or crease between the needle threads on the face. Type 402 is used on sportswear where a crease needs to be maintained e.g. at the back of gloves.

Stitch type 406 is formed with two, three or more needles. The needle from the lower end moves up and the looper starts moving to the left. It is on the needles that loops are formed into which the looper gets carrying the lower thread. Needles come into the lower end and go through the triangle that is formed between the top of the looper, the previous needle loop and the looper loop.

Types 406 (two-needle version) and 407 (three-needle version) use needles and have a looper thread covering the fabric between them on the underside. They are known as bottom-cover stitches (Figure 10.14). These stitches are used to cover seams or unfinished edges on the innerside of garments and to keep them flat, for attaching lace and braid trimmings to garments and for attaching elastic edging to briefs.

10.2.5 Class 500: over-edge chain stitches

This class of stitch may be formed from one or more needle and/or looper threads with at least one thread passing round the edge of the material being sewn. Loops of one group of thread are passed through the material and are secured by interlooping with loops of one/more interlooped groups of threads before the succeeding thread loops of the first group are again passed through the material.

These stitches are often called over-edge, overcast, overlock, serge or merrow. Applications of over-edge chain stitches include bolt end, seaming and serging operations.

Stitch types 502 and 503 are formed by two threads, a needle and a looper thread. The 502 type is a tight stitch that is used primarily for seaming the outer edge of bags, while 503 is used for blind hemming (e.g. T-shirts) and serging (the process of finishing a single ply of fabric to prevent raveling), especially menswear. Stitch types 504 and 505 are three-thread over-edge stitches that are formed with one needle thread and two looper threads. Type 504 is a highly extensible but a secure stitch that makes an excellent seam for knit garments, such as sweater seams.

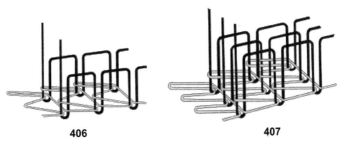

406   407

Figure 10.14 Stitch types 406 and 407. From ISO 4915.

Figure 10.15 shows stitch types 503, 504 and 505.

Stitch types 512 and 514 are also called mock-safety stitches. These are four-thread over-edge stitches that are formed with two needle threads and two looper threads. Type 514 is stronger and more elastic than 512, but both may be used for seaming knits and wovens (Figure 10.16).

Types 515, 516 and 519 are a combination of an over-edge stitch and a 401 chain stitch. This type is called a safety stitch (the chain stitch closest to the seam is backed by a row of another tight over-edge stitches), and is used in manufacturing shirts, jackets, blouses and jeans. Figure 10.17 shows stitch types 515 and 516.

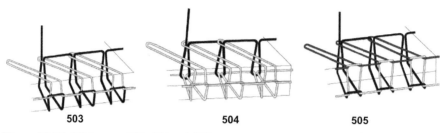

Figure 10.15 Stitch types 503, 504 and 505.
From ISO 4915.

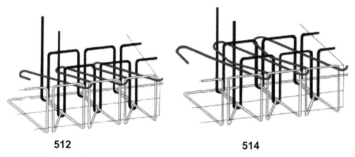

Figure 10.16 Stitch types 512 and 514.
From ISO 4915.

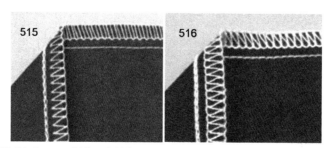

Figure 10.17 Stitch types 515 and 516.
From ISO 4915.

10.2.6 Class 600: cover-seam chain stitches

Stitch types in class 600 are formed with three groups of threads, and their general characteristic is that two of the groups cover both surfaces of the material. Loops of the first group of threads (the needle threads) are passed through loops of the third group already on the surface of the material, and then through the material where they are outer-looped with loops of the second group of threads on the underside of the material. The second and third groups are usually referred to as the top-cover threads and the bottom-cover or looper threads, respectively. Figure 10.18 shows the formation of stitch class 600.

Cover-seam chain stitches are used in binding operations in knitted undergarments, athletic shirts and infant wear. These stitches are very strong, and elastic, which are used extensively by manufacturers of knit garments to cover raw edges and prevent raveling. They may be used for attaching flat knit or ribbed knit collars.

Cover stitches 602 and 605 are very strong and elastic stitches. Stitch type 602 is formed with four threads (two needle threads, one looper thread and one top-covering thread) and used for covering stitch or seaming knit. Class 605 is a similar stitch but formed with three needle threads, one looper thread and one top-covering thread. This class is used for covering stitches or butt-seams. Figure 10.19 shows stitch types 602 and 605.

The most complicated stitch type in this class is type 606, known as flat lock, which can be used to join fabrics that are butted together in what used to be called a flat seam. Type 606 is used on knitted fabrics, especially underwear fabrics.

The flat-seaming stitch type 607 with nine threads (four needle threads, four looper threads and one top-covering thread) is used for trims and seams simultaneously, producing the flat, butted seams as on infant panties, men's briefs and other knitted garments (Figure 10.20).

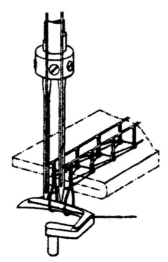

Figure 10.18 Formation of stitch class 600.
Mihailovic N. (1986). Masine i uredjaji u konfekciji. VTTS, Beograd.

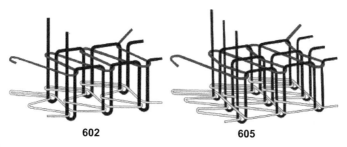

Figure 10.19 Stitch types 602 and 605. From ISO 4915.

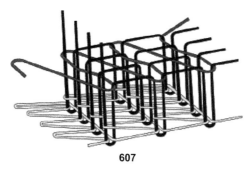

Figure 10.20 Stitch type 607. From ISO 4915.

A combination of stitch types is often used in the production of a garment. They are formed simultaneously in one operation as two or more rows of stitch classes, denoted by using the two individual stitch type designations, for example: 401.504.

Figures 10.21–10.27 show the stitch types on different garments, such as blazer, suit, shirt, dress, boilersuit, sport jacket and underwear.

10.3 Seam types

According to ASTM D 6193 (American Society for Testing and Materials), seam classification relates directly to the positioning of fabric sections at the junction where these sections are sewn. Seams are divided into six classes. Each class is subdivided into types and designated by symbols as follows:

- Class of seam: two or more upper-case letters.
- Types of the class: one or more lower-case letters.
- Number of rows of stitches: one or more Arabic numerals.

The seam classes are as follows.

Sewing, stitches and seams

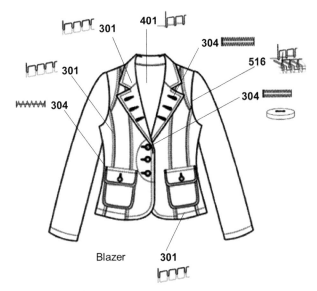

Figure 10.21 Stitch types on blazer.

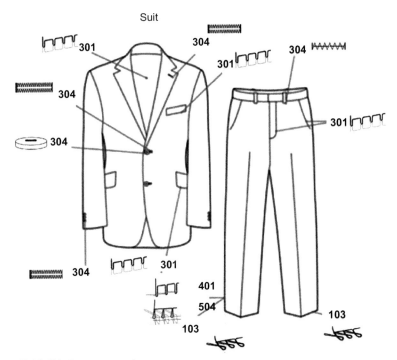

Figure 10.22 Stitch types on suit.

260   Garment Manufacturing Technology

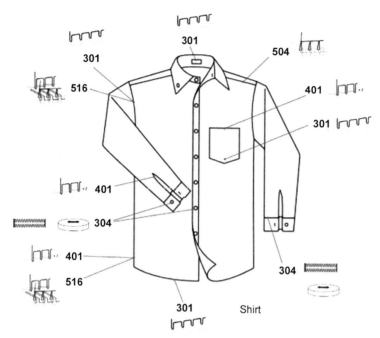

Figure 10.23 Stitch types on shirt.

Figure 10.24 Stitch types on dress.

Sewing, stitches and seams

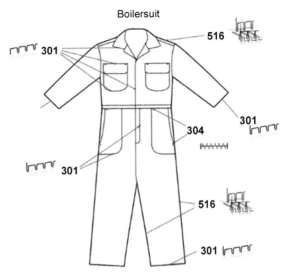

Figure 10.25 Stitch types on boilersuit.

Figure 10.26 Stitch types on sport jacket.

10.3.1 Superimposed seam – class SS

Superimposed seam is achieved by two or more separate fabric pieces put together. This is one of the most recognized methods of seaming. Superimposed seam is sewn with a lockstitch, chain stitch, over-edge stitch or safety stitch.

The most basic superimposed seam is the SSa used for in-seam and side seams in garments. One ply of fabric is stacked upon another with thread stitching through all

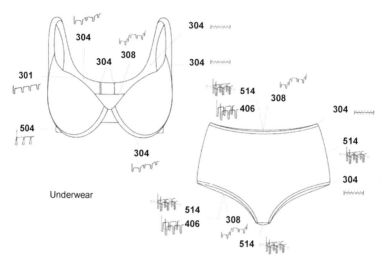

Figure 10.27 Stitch types on underwear.

Figure 10.28 Superimposed seam.

plies of fabric. Superimposed seam SSa (Figure 10.28) joins two plies using Class 100, Class 300, Class 400 or Class 500 stitches.

Seam SSb (for finishing belt ends and attaching elastic to waistline), seam SSc (for ends of waistbands on jeans), seam SSd (for seaming (but not widely used)) and seam SSe (for collars or cuffs seamed and topstitched) are two plied seam with one edge folded, which uses Class 100, Class 300, or Class 400 stitches.

10.3.2 Lapped seam – class LS

This class of seam requires that the plies of material are lapped and seamed with one or more rows of stitches. A lapped seam is a seam formed by lapping two pieces of material commonly used in joining garment parts such as yoke, gusset and other garment parts (Figure 10.29).

The most common lapped seams are LSb and LSc overlap seams of two plies at the edges using Class 100, Class 300 or Class 400 stitches. LSb is used for attaching curtains and to attach the waistband of men's dress slacks and LSc for side-seams of dresses, shirts and jeans. Long seams on garments such as jeans and shirts

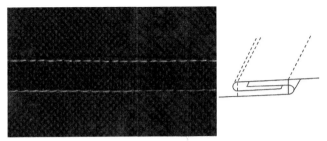

Figure 10.29 Lapped seam.

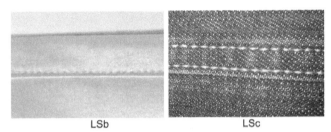

Figure 10.30 LSb and LSc seams.
From ASTM D 6193.

(lap-felled seams) are sewn with two rows of stitches on a twin-needle machine equipped with a folding device. This provides a very strong seam in garments that will be worn a lot, although there is a possibility that the thread on the surface may suffer abrasion in areas such as inside leg seams. Figure 10.30 shows the LSb and LSc seams.

10.3.3 Bound seam − class BS

The BS class of seam is formed by folding a binding strip over the edge of one or more plies of material and seaming the binding strip to the material with one or more rows of stitches. This bound seam is always used in edge neatening or for decorative purposes, mainly on high-quality garments (finish edges, necklines, short sleeves on T-shirts, sleeveless tank tops with binding).

Seam BSa is used for edges bound with ribbon or braid, BSb for T-shirt necklines or sleeve edges with knot trim, BSc for neckline or front edges bound with bias-woven material, and BSd and BSe for seaming and binding, respectively.

Bound seams deliver extra durability and strength (Figure 10.31) to the sewn product.

10.3.4 Flat seam − class FS

This class of seam is formed by sewing the abutted edges of material together in such a manner that the stitches extend across and cover or tend to cover the edges

Figure 10.31 Bound seam.

Figure 10.32 Flat seams.

of the plies joined. The flat seam class is the smallest class with only six different types.

Flat seams are constructed to remain flat through care and maintenance of clothes. A flat durable seam is used on men's sports shirts, work clothes, children's clothes and pyjamas. For example, FSa is used for raglan seams of sweatshirts; FSb, FSd and FSe are used for sweatshirts and underwear.

Figure 10.32 shows the flat seams.

10.3.5 Ornamental stitching – class OS

This class of seam requires a series of stitches to be embodied in a material either in a straight line, a curve or following a design, for ornamental purposes. For example, OSa is used for decorative stitching on jeans pockets (Figure 10.33), OSb for decorative

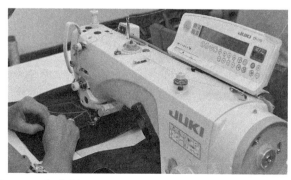

Figure 10.33 Ornamental stitching.

stitching with cording insert, OSc for raised stitching without cording insert for backs of gloves and OSe for pin tucks on the front of blouses.

10.3.6 Edge finishing – class EF

This class of seam requires the edge finishing to be accomplished by either sewing a series of stitches at or over the edge of a material while the edge may or may not be folded as specified, or the edge of the material is folded and stitched to the body of the material with a series of stitches. The most common seams are EFa (single-fold hem), EFb (double-fold hem), EFc (T-shirt hem), EFd (edge finish, serging) and EFe (ornamental edge finish). Figure 10.34 shows EFb seam.

10.3.7 Seam classes as per ISO/BS

According to the ISO 4916 or BS 3870 standards, seams are divided into eight classes (1–8). Each stitched seam is designated numerically by five digits, with the following configuration:

- The seam class (1–8).
- The material configurations (01–99).
- Needle penetrations, material configurations (01–99).

Class 1 – Plain seam and French seam, formed by superimposing the edge of one piece of component over the other. These seams are produced with a minimum of two pieces of components. Each component is limited on the same end. An example of seams in Class 1 is shown in Figure 10.35.

Class 2 – Welt seam, formed by lapping two pieces of component and produced with a minimum of two pieces of component. One component is limited on one end and the other is limited on the other end. The limited edges of these two components are put in opposite directions (Figure 10.36).

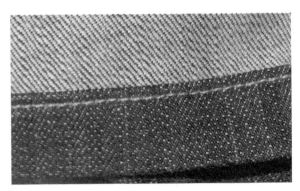

Figure 10.34 EFb seam.
From ASTM D 6193.

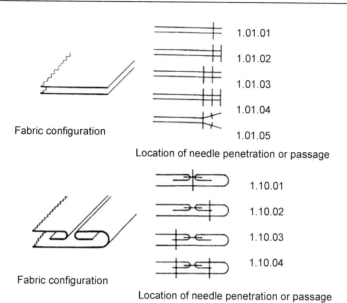

Figure 10.35 Example of Class 1 seams. From ISO 4916.

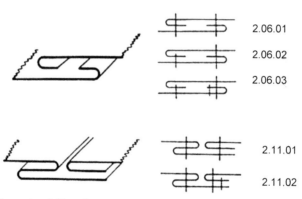

Figure 10.36 Example of Class 2 seams. From ISO 4916.

Class 3 — Bound seam, constructed by binding one component edge with another narrow component. The seam is produced with a minimum of two pieces of component. One component is limited on one end and the other is limited on both ends (Figure 10.37).

Class 4 — Channel seam, formed by two pieces of fabric that are laid flat with their edges closing each other without overlapping. Seams are produced with a minimum of

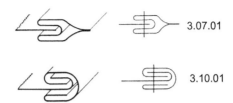

Figure 10.37 Example of Class 3 seams. From ISO 4916.

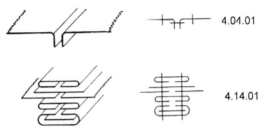

Figure 10.38 Example of Class 4 seams. From ISO 4916.

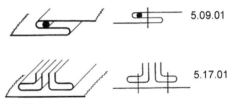

Figure 10.39 Example of Class 5 seams. From ISO 4916.

two pieces of component. Both components are limited on one end but they are placed opposite to each other on the same level (Figure 10.38).

Class 5 — Ornamental seam, produced with a minimum of one piece of component, with unlimited edges on both ends. An example of Class 5 seams is shown in Figure 10.39.

Class 6 — Turned hem seam, produced with one piece of component. This seam has one limited edge on one end. An example of Class 6 seams is shown in Figure 10.40.

Class 7 — Edge-stitched seam, produced with a minimum of two pieces of component. One component is limited on one end, and the other narrow component is limited on both ends. Figure 10.41 shows Class 7 seams.

Class 8 — Enclosed seam, produced with a minimum of one piece of component with a limited edge on two ends. Figure 10.42 shows examples of Class 8 seams.

Figure 10.43 shows examples of different seam types on necklines and shirts.

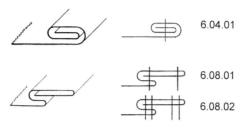

Figure 10.40 Example of Class 6 seams. From ISO 4916.

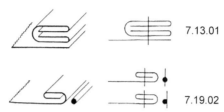

Figure 10.41 Example of Class 7 seams. From ISO 4916.

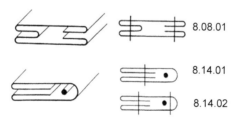

Figure 10.42 Example of Class 8 seams. From ISO 4916.

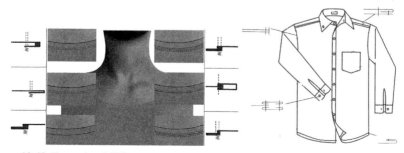

Figure 10.43 Example of different seams on necklines and shirts.

10.4 Seam-neatening

Seam-neatening is finishing of raw seam edges. It is done to protect and strengthen the raw edges of seamed fabric. The method of neatening is chosen depending on the type of fabric (e.g. in a knitted fabric, the neatening might be omitted), the character of a garment and seam type.

A seam finish is a treatment that secures and neatens the raw edges of a plain seam to prevent ravelling, by sewing over the raw edges or enclosing them in some sort of binding. A seam is finished by neatening the edges using an overlock or zigzag stitch. The seams can be finished together on lightweight fabrics. On medium- to heavy-weight fabrics, seam openings are pressed and each seam allowance is neaten separately.

Overlock stitches are extremely versatile, as they can be used for decoration, reinforcement or construction. Overlocking is also referred to as overedging, merrowing or serging. Although serging technically refers to overlocking with cutters, in practice these four terms are used interchangeably.

Overlock stitches are typically used as follows:

- End-to-end seaming of piece goods for textile finishing (one-thread).
- Finishing seam edges, stitching flatlock seams, stitching elastic and lace to lingerie and hemming (two-thread).
- Sewing pintucks, creating narrow-rolled hems, finishing fabric edges, decorative edging, and seaming knit or woven fabrics (three-thread).
- Decorative edging and finishing, seaming high-stress areas (four-thread).
- Seam construction in apparel manufacturing (five-thread). Some of the overlock stitches used for seam-neatening are discussed below.

Overlock stitching with three threads is shown in Figure 10.44.

Zigzag stitches are sewn as a single seam or hem using zigzag-shaped stitches. Zigzag stitches are mainly used to sew strips of elastic to fabrics, particularly jersey, swimwear and sportswear (Figure 10.45).

Sewing tape comes in several different forms, and serves a variety of functions. The three most common types of sewing tape are hemming tape, bias tape and stitch tape. Hemming tape is frequently used to hem a garment when a perfectly flat hem is necessary (pair of cuffed pants).

Bias binding is used as a decorative finish for the raw edges of a fabric and is used either for hemming or as a form of a bound seam where the ends of the fabric are tucked within a tape, so that no raw ends are exposed (Figure 10.46).

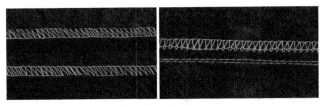

Figure 10.44 Overlock stitches with three threads.

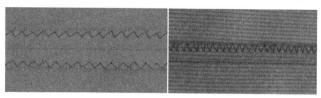

Figure 10.45 Zigzag stitch.

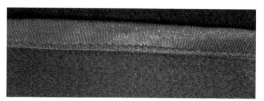

Figure 10.46 Seam with bias tape.

Hong Kong style (bound) is ideal when the inside of the garment might show, as the seam allowances are neatly bound. The binding used can be bias-cut strips of lining, organza or a special seam-binding tape. The difference between a bound and Hong Kong seam finish is that the edges of a binding strip are turned under both front and back.

10.5 Future trends

Seam and stitch types affect the quality of a sewn product to a large extent. Careful selection of both these factors needs much skill and depends on the nature of the product and end-use application. If the garment is made from non-porous materials (such as those that are waterproof or fire or chemical resistant), then the perforations caused by a conventional sewn seam will compromise the integrity and performance of the garment. However, Welding, bonding and heat sealing are other options of sewing methods, which can avoid these problems. In garment manufacturing, welding and bonding have been used for several years in the past to apply pockets and zippers, make darts and pleats, hang pocket bags, hem garments, make stiffer plackets or panels and apply waistcords.

According to Kathlyn Swantko (2004), there are two basic methods for applying bonded and welded seams. The bonding method uses an adhesive (thermoplastic) film for bonding two fabrics together. During the process, the adhesive is tacked or laminated to one of the substrates and the second substrate can be laid on top. Heat and pressure activate the adhesive, and the film melts and penetrates into the fabrics and creates a bond.

The welding method involves welding two fabrics with radio frequency or ultrasound. Ultrasonic technology involves the creation and channeling of high-frequency vibratory waves that cause a rapid buildup of heat in synthetic fabrics. The heat can be used to weld, bond, cut or slit the materials, as desired. There are no thread holes with ultrasonically welded seams. The material is welded together creating a superior particle and fluid barrier.

A welded seam is formed where two pieces of material are fused together by heat and pressure alone, creating an adequate bond on thermoplastic coatings such as polyvinylchloride, polyurethane, polyethylene and polypropylene. Welded seams are also being used increasingly for women's bras and babies' pants and nappies, as they are comfortable. The finished seams rub the skin less, as they are flatter and lighter than the conventional seams. Chemical protective clothing is another area where welded seams provide improved penetration resistance, as there are no stitch holes for the chemicals to penetrate.

In the search for automated processes for seaming of fabrics, companies have looked to welding and adhesive bonding, particularly for synthetic fabrics. Laser welding offers the potential for both automation and improved seam performance (Nayak and Khandual, 2010). The principal advantages of this method include high welding rates; restriction of heating to joint surfaces without affecting external texture; single-sided access, so that welds can be performed beneath other layers of fabric; retention of seam flexibility; and gas-sealed or water-sealed seams where required.

Seam tape (heat-sealing tape) is a multilayered adhesive film that is applied to the sewn seams to prevent water from leaking through those seams on garments of waterproof fabric. It can be applied by using hot-air taping machines, laser or ultrasonic machines. Seam tapes have been proven in many applications, including rainwear and outerwear; work wear; military garments; and wear for skiing, snowboarding, cycling, sailing, climbing, paddling sports, hiking, hunting, fishing, golfing etc.

10.6 Conclusions

The kind of garment to be sewn and the shape of the cutting parts define the types of operations. Seam length and the shape of seam contour influence the number and the type of technological operations. The technological sewing operations differ according to the:

- form and the length of the seam;
- function of the seam and the place where it is positioned on the article of clothing;
- numbers of fabric layers to be sewn;
- length of the sewing stitch to be made;
- quality requirements;
- material.

For making a quality garment, it is necessary to know all types of stitches and seams. The strength of the seams in garments depends on various technical parameters, such as the type of fabric, type of sewing thread, sewing needle, sewing stitch, density of sewing stitches, type of sewn seam etc. In some places on clothes, a heavy strain in transverse direction can lead to the deformation or tearing of seams, so it is necessary to adjust the characteristics of the base material to the characteristics of the sewing thread and stitch type. When wearing a garment or maintaining it (cleaning, dry cleaning), strain occurs that can damage or tear the seams because of friction (Nayak and Padhye, 2014):

- Of a seam with other parts of a garment or machine components.
- Of a thread.

When sewing, multiple errors may occur due to the differential strains of upper and lower threads, variable sewing speeds, shrinkage of seams, skipping of certain stitches etc. The biggest mistake is the formation of stitches that are not of the same length, and often this error cannot be easily seen except when sewing with the thread in contrasting colours in relation to the subject of work.

10.7 Sources of further information and advice

To get perfect stitch sewing, sewing thread selection must be in accordance with the properties of materials that are being sewn. This provides for the optimal strength, flexibility, durability, safety and appearance of the seam, as well as the appearance of garment. Therefore, it is necessary to use standards for stitches and seams, such as:

1. ASTM D 6193 – 11, Standard Practice for Stitches and Seams.
2. ASTM D 7722 – 11, Standard Terminology Relating to Industrial Textile Stitches and Seams.
3. BS 3870-1: 1991, Stitches and Seams. Classification and Terminology of Stitch Types.
4. BS 3870-2: 1991, Stitches and Seams. Classification and Terminology of Seam Types.
5. ISO 4915:1991 Textiles – Stitch Types – Classification and Terminology.
6. ISO 4916:1991 Textiles – Seam Types – Classification and Terminology.

The ASTM D 6193 standard also includes sewing applications for buttons/snaps/hook and eye attachments, buttonhole stitching, bartack stitching, tacking, specialty stitching and seams and guide to count stitches per inch.

The book *Stitches and Seams* (Laing and Webster, 1998) provides a detailed analysis of the fundamentals of different types of stitches, seams and sewing threads; explains how they are produced; and examines their optimum design, performance and modes of failure.

In *The Technology of Clothing Manufacturing,* Carr and Latham (2009) show that much of the application of technology to garment manufacturing is concerned with the achievement of satisfactorily sewn seams. They describe the British Standards (BS 3870: Part 1: Classification and Terminology of Stitch Types and BS 3870: Part 2: Classification and Terminology of Seam Types).

On the website of American & Efird (www.com/technical-tools/stitches-and-seams/amefird), there are good instructions for use of stitches and seam types and SPI recommendations for wovens and knits, because using the correct number of SPI can greatly enhance the strength, appearance and performance of the seam for a given fabric type and application.

References

ASTM D 6193—11. Standard Practice for Stitches and Seams.
ASTM D 7722—11. Standard Terminology Relating to Industrial Textile Stitches and Seams.
BS 3870—1: 1991. Stitches and Seams. Classification and Terminology of Stitch Types.
BS 3870—2: 1991. Stitches and Seams. Classification and Terminology of Seam Types.
Carr, H., Latham, B., 2009. The Technology of Clothing Manufacturing. Blackwell Publishing.
ISO 4915:1991. Textiles — Stitch Types — Classification and Terminology.
ISO 4916:1991. Textiles — Seam Types — Classification and Terminology.
Kunz, G.I., Glock, R.E., 2004. Apparel Manufacturing: Sewn Product Analysis. Prentice Hall.
Laing, R.M., Webster, J., 1998. Stitches and Seams. Woodhead Publishing Limited.
Mihailovic, N., 1986. Masine I Uredjaji U Konfekciji. VTTS, Beograd.
Nayak, R., Khandual, A., 2010. Application of laser in apparel industry. Colourage 57 (2), 85—90.
Nayak, R., Padhye, R., 2014. The care of apparel products. Elsevier, pp. 799—822.
Swantko, K., 2004. Forming a New Bond, Fabric trends, pp. 12—14.

Sewing equipment and work aids

P. Jana
National Institute of Fashion Technology, New Delhi, India

11.1 Introduction

Sewing is the process of stitch formation involving needle and thread. Although hand sewing was prevalent during the Paleolithic era, most sewing processes could be performed by machines by about 1900 (Glock and Kunz, 2000). This chapter will discuss different features and functions of sewing machines and of work aids that assist in the sewing process. Industrial sewing machines differentiate themselves from home sewing machines by varieties of bed and feed type, computerised functions and the possibility of integrating the work aids. While the bed types of sewing machines are important for optimised handling, the feed types enable uncompromised quality, computerisation eliminates non-value-added elements and work aids de-skill the operation. Complete automation in sewing is limited due to difficulty in handling the limp dimensionally unstable material in 3D space (Jana, 2004). The cyclic sewing machines and sewing automats are primarily batch processes and concentrate on effective utilisation of operators by deploying a lower person-to-machine ratio (Jana, 2014a). The sewing thread and needle are highly specialised consumables in the sewing process, and basic knowledge is necessary for appropriate selection. Although no significant/fundamental change happened in sewing machine kinematics since 1900, electronics and computerisation (Jana, 2004) have driven the technology advancement in sewing machines during last three decades, and energy efficiency and sustainability will drive the development in future. The migratory nature of the industry has shifted apparel manufacturing to Asia and shifted the machine manufacturing base to China, affecting the research and development initiatives and the nurturing and decimation of knowledge.

11.2 Different bed types in industrial sewing machines

The bed of a sewing machine is that part of the sewing machine on (or against) which the fabric rests while it is being sewn. The classification of sewing machine bed types or shape types is done based on the manner in which the fabric falls, behaves and travels with respect to the bed during the course of sewing (Solinger, 1988), to enable easier movement of materials around the machine (Carr and Latham, 1999). Solinger (1988) divided the bed types into horizontal bed and vertical bed based on the plane of fabric sewing. While the horizontal beds are common in use and mentioned by other authors like Carr, Cooklin, Jana, Glock and Kilgus, the vertical bed type is uncommon and further classified into open vertical bed and closed vertical bed (Solinger, 1988).

Table 11.1 Sewing machine bed type classification

Solinger	Carr and Latham	Jana	Glock	Cooklin	Kilgus
Flat	Flat	Flat	Flat	Flat	Flat
Raised	Overedging	Raised	Raised	Raised	Raised
Cylinder	Cylinder	Cylinder	Cylinder	Cylinder	Cylinder
Feed-off-arm	Feed-off-arm	Feed-off-arm	Feed-off-arm	Feed-off-arm	Feed-off-arm
Post	Post	Post	Post	Post	Post
		Feed-up-the-arm	Feed-up-the-arm		
	Blind felling	Monoblock		Submerged	Side bed

In the open vertical-bed machine, the fabric is suspended vertically while being sewn. Bag-closing machines are open vertical-bed machines. In the closed vertical-bed machine, the vertically suspended fabric is surrounded by sections of the frame while being sewn, thus limiting the fabric size (Solinger, 1988). Horizontal bed types are further classified into five types (Solinger, 1988); six types (Carr and Latham, 1999; Glock and Kunz, 2000, pp. 487–488; Cooklin et al., 2006; Kilgus, 1996); and seven types (Jana, 2014b) by different authors (Table 11.1). Common types identified by all authors are flat, raised, cylinder, feed-off-arm and post bed. While feed-off-arm and feed-up-the-arm are sub-classifications of cylinder beds (Glock and Kunz, 2000; pp. 487–488), mono block and side-bed machines are modified raised-bed types without space at the ride of the needle, which are common for overlock stitch types. Blind felling is actually a repetition of the cylinder-bed type where material moves around the cylinder axis. The submerged bed type is actually a raised-bed machine converted to a flat bed.

11.2.1 Flat bed

A flat bed is a sewing machine frame that permits sewing of a flat 2D shape of item. The flat bed of the machine is therefore usually mounted on a sewing machine table, with the working surface of the bed flush with the table top (Figure 11.1; Solinger, 1988). The flat bed is used in the majority of sewing, where a large and open garment part can easily be handled past the needle and provides a suitable surface for use of markers to control the position of garment parts, for example a patch pocket in a shirt front.

11.2.2 Raised bed

A raised bed working surface is not flush with the table top; rather, it is generally raised up to 4 inches above the table top in the form of a plinth (Figure 11.2; Kilgus, 1996). The raised bed actually facilitates threading of loopers in 400, 500 and 600 class of

Sewing equipment and work aids

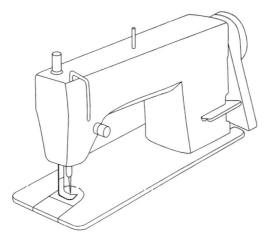

Figure 11.1 Flat bed.

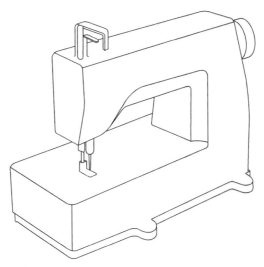

Figure 11.2 Raised bed.

stitch machines. It facilitates perimeter sewing of large-diameter tubular shaped goods, for example hemming of T-shirts (Jana, 2014b).

11.2.3 Cylinder bed

A cylinder bed is a cylinder-shaped sewing machine frame that permits sewing of a cylindrically shaped item along the circumference of the item. Here the cylinder axis is parallel to the direction of sewing (Figure 11.3). This bed type is used where the parts to be sewn are small, curved or otherwise awkward in shape (Carr and Latham, 1999). While hemming of hollow cylindrical shaped garment parts like trousers leg or T-shirt sleeves, the sewing direction is along the circumference of the

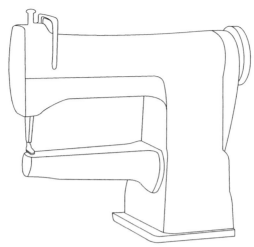

Figure 11.3 Cylinder bed.

cylindrical shape; therefore, a cylinder-bed machine is used for these operations. While sewing along the circumference of cylinder-shaped items using a cylinder-bed machine, the imaginary axis of the cylindrical item coincides with the axis of the cylinder bed and the item is self-supported against the machine bed, thereby providing ease of handling during sewing (Jana, 2014b).

11.2.4 Feed-off-arm

A feed-off-arm bed is a cylindrical sewing machine frame that permits joining of two opposite edges of a flat-shaped fabric to create a hollow cylindrical-shaped item. The axis of the cylindrical bed is parallel to the direction of sewing

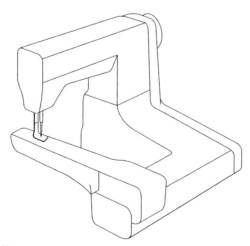

Figure 11.4 Feed-off-arm.

(Figure 11.4). Here the operator wraps the part to be sewn around the machine bed, two opposite edges are joined together by flat and fell seam and it is fed away from the operator, off the end of the bed, as the operator sews (Carr and Latham, 1999). While one end of the cylinder-shaped arm is attached to the machine frame, the sewing happens at the other end, which is open and away from the operator. Here the sewn fabric travels away from the cylinder bed; thus the sewn cylinder-shaped item automatically comes off the tubular/cylindrical-shaped arm. Shirt sleeve inseams and trouser leg inseams are operations where the flat shape of fabric is converted to a hollow cylinder/conical shape, and a feed-off-arm is the appropriate bed type for these operations.

11.2.5 Post bed

A post bed is a sewing machine frame in which the bed is the top surface of a pillar or post. The pillar or post is mounted vertically on the machine table (Figure 11.5). The sewing area is the horizontal cross-sectional area of the pillar or post, which is typically less than or equal to four square inches. Such a bed type permits sewing of concave- and convex-shaped 3D items, such as brassiere or girdle parts, with ease because the sewn part can follow its geometric inclination to encompass this bed shape as it is sewed (Solinger, 1988). This bed type is used where the parts to be sewn are small, curved or otherwise awkward in shape (Carr and Latham, 1999).

11.2.6 Feed-up-the-arm

A feed-up-the-arm bed is very similar to a feed-off-arm, where a tubular/cylindrical sewing machine frame permits joining of two opposite edges of a flat-shaped fabric to create a hollow cylindrical-shaped item. Here the open end of the cylinder is towards the operator and the attachment to the machine frame is away from the operator; the

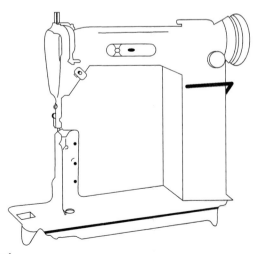

Figure 11.5 Post bed.

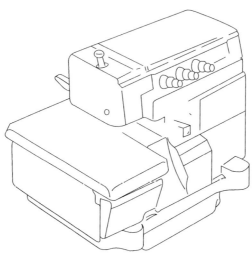

Figure 11.6 Side bed.

axis of the cylindrical bed is parallel to the direction of sewing. Here the direction of sewing is opposite to that of the feed-off-arm. In feed-up-the-arm, the tubular sewed items are accumulated in the cylinder-shaped machine arm, and do not come off automatically. The operator must lift the presser foot and take out the sewed items from the cylinder after completing a batch. Left and right shoulder joins of T-shirts or vests are done using this machine.

Apart from above six types, there are also monoblock or side beds (Figure 11.6), which are actually variations of raised-bed machines without any space on the right side of the needle.

As the sewing machine bed type is linked with the continuous movement of the fabric, which is guided by the operator, the bed-type classifications are relevant only for continuous sewing machines (Jana, 2014b). Cyclic sewing machines, like button-sewing, buttonhole-sewing and pattern-tacking machines, where fabric movement is guided by a clamp (and not by the operator), and the movement of fabric is neither linear nor continuous, are not covered under the bed type classification.

11.3 Different feed types in industrial sewing machines

Feed system, feed types and feed mechanisms are interchangeably used terminologies in different literature. Although Glock (Glock and Kunz, 2000) described only three parts — presser foot, throat plate and feed dog — responsible for feeding to take place, the needle also plays an important role in some of the feed types. Needles penetrate the fabric for stitch formation, and intermittent feeding of fabric is necessary for ensuring stitch formation takes place in continuous longitudinal form (Solinger, 1988; Juki, 1988). The universally used feed system in sewing machines for fashion and commodity apparel manufacture is called drop oscillation feed or simply the drop feed from the bottom, which accounts for more than 90% of sewing machines installed in factories (Jana and Khan, 2014). The machine parts responsible for feeding from the bottom are drop oscillating feed dog and belt, while parts responsible for feeding from the top are needle, puller, rotary presser foot, walking (oscillating) toothed presser foot and belt. Solinger explained five basic categories of prime feed mechanism (Solinger, 1988) — needle feed, presser foot feed, rotary feed, drop oscillation feed and platform feed — and how two or more prime feed components are combined to give one type of multi-feed action. These resultant multi-feed actions in sewing machine are known as feed type, feed system, transport system or primary feed type (Jana, 2014b). Carr and Latham (1999), Juki (1988) and Brother Industries Limited (2000) mentioned six different feed types available in commercial sewing machines: drop feed, variable top and drop feed, needle feed, unison feed, differential drop feed, variable top and differential drop feed. Kilgus (1996) classified five types, renamed the unison feed 'alternating compound feed', separately classified variable top and bottom feed into before-the-needle and behind-the-needle categories, and did not mention variable top and differential drop feed. Pfaff (1983) and Jana (2014b) mentioned an additional seventh feed type called bottom feed and alternating top feed (Table 11.2). Recently an eighth type of feed system was developed by Typical Corporation, called X-feed (Jana and Khan, 2014), and all eight different feed systems are available commercially.

11.3.1 Drop feed (bottom feed)

The three sewing machine parts that together constitute the drop-feed mechanism are presser foot, throat plate and feed dog (Carr and Latham, 1999). In a drop-feed system, the feed dog feeds the fabric from the bottom while the needle is up (Figure 11.7). As only the bottom ply is in contact with the teeth of the feed dog and the top ply is expected to slide (along with the bottom ply) below the presser foot with negligible friction, there is no positive binding force between both the plies. Depending on the surface characteristics of fabric, the drop-feed system is subjected to severe inter-ply slippage (for slippery fabric or fabrics with very low frictional value) or negligible inter-ply slippage (for very high surface frictional value). When the needle is up and stitch tightening is taking place, the feed dog is also up and the fabric is being supported from the top by the presser foot and at the bottom by the feed dog; therefore, there is no possibility of fabric being bulged by the tension of sewing thread, and drop feed is considered universally as the most suitable for sewing lightweight fabrics.

When two or more thicknesses of fabrics are being sewn, regardless of whether they are separate fabrics or folded sections of the same fabrics, the problem arises that friction between the bottom ply and the feed dog is greater than that between intervening plies (Carr and Latham, 1999). The tendency of the lower ply satisfactorily moving ahead with the feed dog and the upper ply being retarded by the presser foot is known as inter-ply shift, differential feeding pucker or just feeding pucker (Carr and Latham, 1999).

Table 11.2 Sewing machine feed type classification

Pfaff and Duerkopp Adler	Carr and Latham, Juki, Brother	Jana	Cooklin	Kilgus
Drop feed	Drop feed	Drop feed	Drop feed	Drop feed
Needle feed	Needle feed	Needle feed	Compound feed	Compound feed
Unison feed	Unison feed	Unison feed	Unison feed	Alternating compound feed
Differential feed	Differential feed	Differential feed		Differential drop feed
Variable top and bottom feed	Adjustable top feed	Variable top and bottom feed	Variable top and bottom feed	Variable top and bottom feed (before the needle)
Variable top and bottom differential feed	Variable top and bottom differential feed	Variable top and bottom differential	Variable top and bottom differential feed	Variable top and bottom feed (behind the needle)
Needle feed and alternating top feed		Needle feed and alternating top feed X-feed		

11.3.2 Needle feed (compound feed)

A compound feed is the combined simultaneous feed action of a needle feed and drop oscillation feed and is useful for sewing with a low inter-ply frictional co-efficient (Solinger, 1988). In a compound feed system, the needle and feed dog together feed the fabric while the needle is down and inside the fabric (Figure 11.8). The needle moves in the direction of the stitch in a pendulum action (the tip of the needle is actually moving in an arc). As the fabric is transported by the feed dog from the bottom while needle is inside the fabric, there is controlled inter-ply slippage. As the needle movement arc length is small (0.083 inches for 12 SPI) and the fabric layer thickness being sewn is also negligible, the resultant inter-ply slippage is negligible to be noticed in sewn material. However, when the needle is up and stitch tightening is taking place, the feed dog is down and fabric is not being supported from the bottom by the feed dog, and therefore there is a possibility of fabric being bulged/buckled by the tension of sewing thread in

Sewing equipment and work aids

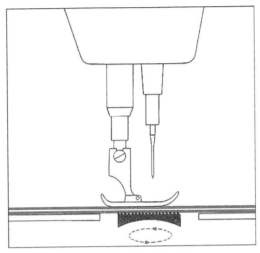

Figure 11.7 Drop-feed system.
From Pfaff special service catalogue.

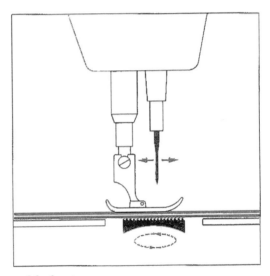

Figure 11.8 Compound feed system.
From Pfaff special service catalogue.

case there is not enough stiffness in the fabric plies. Although compound feed can control inter-ply slippage in thick material sewing (car seat cover, shoes, upholstery, luggage, etc.), it cannot be used for lightweight fabrics (Carr and Latham, 1999).

Compound feed is used to eliminate inter-ply slippage in multiple-ply sewing of thick materials, negotiating the corner of acute-angle-shaped components while top stitching at the edges. Generally double-needle lockstitch machines have a needle-feed system.

11.3.3 Unison feed

The feed dog, needle and presser foot all move together to feed the fabric while the needle has penetrated inside the fabric plies. The presser foot is split into two parts; one part has teeth beneath and moves in direction of feed while the other part moves only up and down and holds the fabric in between feed strokes, that is during stitch formation (Figure 11.9). Unison feed is used for sewing multiple plies of thick materials like tarpaulin fabric, polyvinyl chloride (PVC) and leather. It helps to do consistent sewing over cross-over seams of all types of thick materials without pitch errors (no stitch gathering or stitch elongation). It is used for sewing car seat covers, luggage, furniture using leather, PVC or other spongy thick fabrics, particularly in piping and binding-tape attachment processes.

11.3.4 Differential drop feed

Here two feed dogs are arranged in a series differential position in which the feed dog lengths follow one another in a straight line, one behind the needle and another in front

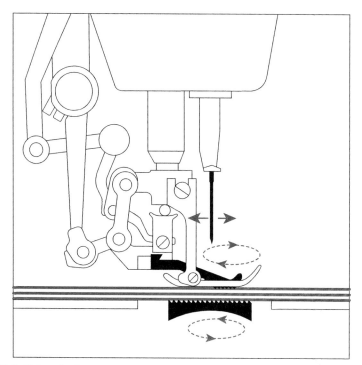

Figure 11.9 Unison feed.
From Pfaff special service catalogue.

Sewing equipment and work aids

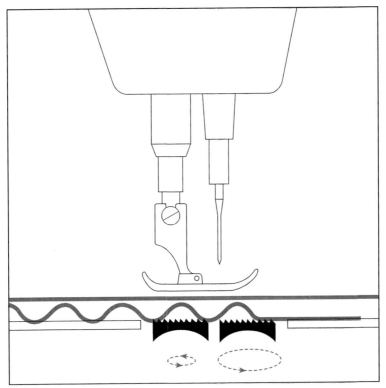

Figure 11.10 Differential drop feed.
From Pfaff special service catalogue.

of the needle (Figure 11.10). The rear feed dog (away from the operator) is called the main feed dog while the front feed dog (closer to the operator) is called the differential feed dog. When the differential feed dog takes longer or faster strokes than the main feed dog, the fabric will receive a linear compression force (Solinger, 1988), and vice versa. The amount of feed by two feed dogs can be independently adjusted. Generally all overlock and/or coverstitch sewing machines have a differential feed system. In all areas where fabric is prone to stretch during sewing (diagonal and curved edges), differential feed ensures forced compression of fabric, which counterbalances the stretch and ensures correct shape and size of the seam.

11.3.5 Variable top and bottom feed

One feeding foot at top and one feed dog below the throat plate feed the fabric plies together. The bottom of the feeding foot has a toothed surface. Feeding takes place while the needle is outside the fabric plies. Here, the top feed and bottom feed amounts can be independently adjusted to create inter-ply slackness or tension in the fabric being sewn. The presser foot is in two sections, one holding the fabric in position while the stitch is being formed and the other having teeth on the lower side and moving or

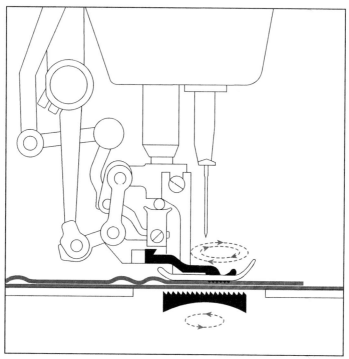

Figure 11.11 Variable top and bottom feed.
From Pfaff special service catalogue.

walking in such a way that the top ply is taken along positively while the needle is out of the material (Figure 11.11). Positive control of the top ply allows for adjustments so that the fabric plies will either be fed through exactly together or, if required, the top ply is gathered into the bottom ply (Carr and Latham, 1999). The feeding foot works before the needle to create a gather in the top ply, for example giving fullness while sleeve setting in the armhole of a blazer, and works behind the needle to create smooth seams, for example centring the back seam of blazer (Kilgus, 1996). The feed difference that can be created between top and bottom ply at best create fullness but cannot create visible gathers in one ply.

11.3.6 Variable top and bottom differential feed

One feeding foot at the top and two feed dogs (like differential feed) below the throat plate feed the fabric plies together. The bottom of the feeding foot has teeth underneath. Feeding takes place while the needle is outside the fabric plies (Figure 11.12). This mechanism enables sewing two plies with the top ply gathering while the bottom ply remains flat. Here the feed difference that can be created between the top and bottom plies at the best creates visible gathers in one ply. This feed system is available with an overedging machine; a long panel join in skirts with a smooth flat seam

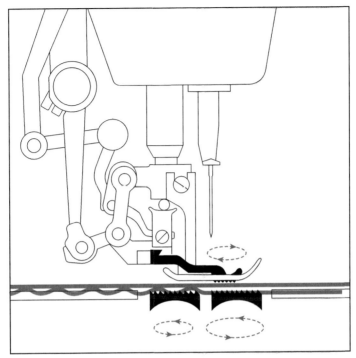

Figure 11.12 Variable top and bottom differential feed.
From Pfaff special service catalogue.

without inter-ply slippage is possible, as well as a waistline join in a dress with a bottom ply gathered and sewn with a smaller bodice block.

11.3.7 Bottom feed and alternating drop feed

One feeding foot at top and one feed dog below the throat plate moves in synchronisation to feed the fabric plies together. The bottom of the feeding foot has teeth beneath. Feeding takes place while the needle is outside the fabric plies. The presser moves up and down alternatively with the feeding foot. This feed system is used for difficult-to-feed materials and multilayered portions of medium and heavyweight material like synthetic leather products, vinyl coating products, leather products, rubber clothing and sponge clothing, with great precision. There is no irregular feed pitch due to positive feeding from both top and bottom plies.

11.3.8 X-feed

X-feed from Typical Corporation is characterised by needle feed and alternating drop feed. The feed dog is split into two independent feed dogs; the central one has a needle hole and acts like a needle feed, and the side one acts like a drop feed. During the first half of stitch length, the needle enters the material and also through the hole of the

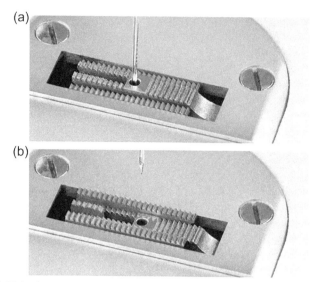

Figure 11.13 X-feed system. (a) Central feed dog (gold colour) is up and transporting the fabric while needle is down. (b) The side feed dog (sky blue colour) is up and transporting the material while needle is up.
From Typical Corporation product catalogue.

central feed dog (which is in 'up') and transports the fabric just like the compound feed mechanism; the side feed dog is below the throat plate level (Figure 11.13(a) and (b)). When the needle is leaving the material and going up, the central feed dog goes down and the side feed dog comes up and transports the material in the sewing direction just like the drop-feed mechanism. As the side feed dog is transporting the material in the sewing direction, the thread take-up lever is pulling the thread, an effect that stretches the stitch like on a puller machine. Due to the alternating movement of both feed dogs, material transport is continuous and calm in comparison to the intermittent feeding of drop feeding (Stitchworld, 2012).

The X-feed claims to sew with 35% lower thread tension (Stitchworld, 2012), which results in a flatter and pucker-free seam appearance, and the occurrence of ply slippage is also significantly lower in X-feed than in drop feed (Jana and Khan, 2014).

Apart from these eight different feed types, there are auxiliary feed types (Jana, 2014b) as well as modified feed types. In modified feed types, the shape of the presser foot and/or feed dog is often changed to wheels or belts for better material movement. While wheels help in negotiating acute curves, belts have a better grip and control on the fabric. For example, a needle-feed machine for sewing shoe uppers will have a wheel feed (wheel replacing the feed dog), needle feed and roller presser foot (roller replacing the presser foot). Similarly, a unison feed may have wheel feed, needle feed and driven roller presser foot, and variable top and bottom feed may have wheel feed and driven roller presser foot or top belt and drop feed. A computer-controlled top and bottom belt drive can adjust the speed of the top belt compared with the bottom belt. A secondary top belt to the right of the main one can move faster on curves and assist in taking in the excess material in the seam allowance compared with that along the stitch line (Carr and Latham, 1999). Auxiliary feed types are to enhance or emphasise the qualities of a primary feed-like puller feed in combination with drop feed. Clamp feed systems enable multi-directional stitching and are usually available in cyclic sewing machines.

11.4 Cyclic sewing machines

Cyclic sewing machines complete the sewing work in a short automatic cycle (Carr and Latham, 1999), thus the name cycle, cyclic (Jana, 2014b) or automatic (Kilgus, 1996). Glock categorised these machines as semiautomatic special-purpose machines, where the operator places the materials or garment part, activates the machine and the machine completes the cycle of a pre-programmed number of stitches (Glock and Kunz, 2000). While the fabric is held by clamps, sewing guides or templates, the movement of clamps is controlled by a cam-follower mechanism or electronics. Typical applications of this type of machine are button-sewing, buttonhole-sewing and pattern-tacking machines.

In these machines, the needle moves only up and down to form the stitch, and the clamp feed moves the fabric horizontally (x-axis) or vertically (y-axis), creating the required shape. The movement of the clamp is controlled by a cam-follower mechanism or electronics. In a cam-follower mechanism, the specific pattern of button sew (e.g. N or X), specific length and width of buttonhole or specific shape of tack (e.g. I or D) can be made by one specific cam type, and for any change of pattern and/or shape of tack, the cam needs to be replaced and involves machine downtime. In an electronic machine, the movement of the cam is controlled by a microprocessor, and thus a different pattern and/or shape of tacks can be sewn easily by changing the program in the microprocessor.

11.4.1 Button-sewing machines

These are either single chain-stitch (107 type) or lockstitch (304 type) machine heads where the fabric is placed over the cloth clamp and button is placed in the button clamp. Once the operator starts the machine the button clamp comes down and touches the cloth clamp; the button and cloth clamp together moves horizontally and vertically as per cam-follower or microprocessor movement and at the end of the cycle the thread is cut and the empty button clamp move up to the original position for sewing the next button. During the 1990s, there were chain-stitch button-sewing machines (Pfaff 3303) where the needle bar used to oscillate horizontally between two holes, in case of four-hole button, the button clamp used to move at appropriate moment to bring the second pair of holes in line of sewing (Kilgus, 1996). These oscillating bar button-sewing machines were preferred for sewing buttons with a reinforced washer at the back of the fabric (for button down collar shirt). Chain-stitch button-sewing machines were prone to unravelling of button and hence discouraged by most retailers. Since 2000, most button-sewing machines are lockstitch type and microprocessor controlled, therefore,

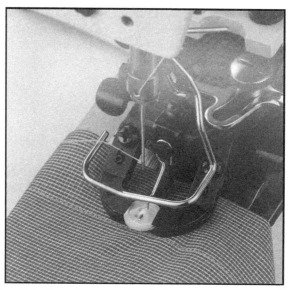

Figure 11.14 Thread shank can be created by placing a spacing finger over the button. From Brother catalogue.

easy to changeover between different stitch type (say from N to X type) with press of a button. Button with thread shank can be created by placing a spacing finger over the button during sewing (Figure 11.14), for shank button the button is clamped on its side and the needle passes alternately into the garment and then into the hole in the shank and into the garment (Carr and Latham, 1999).

11.4.2 Buttonhole-sewing machines

Buttonhole-sewing machines have either lockstitch zigzag (304 type) or double chainstitch zigzag (404) machine head where the fabric is placed over the cloth clamp. Once the operator starts the machine, the fabric-holding frame comes down and grips the fabric against the clamp and movement of the clamp starts. A buttonhole is formed by a series of zigzag stitches with varying degrees of needle throw-in synchronisation with $x-y$ axis clamp movement. The hole in the buttonhole is created by slashing/cutting fabric either after sewing or before sewing. In an ordinary straight buttonhole (shirts and blouses), once the sewing cycle is complete the knife comes down to slash the centre of the buttonhole — called the cut-after mechanism. In knitwear, often buttonholes get distorted after washing; the appearance can be improved by doing stay stitching around the proposed cut area at the beginning of the sewing cycle. A stay stitch acts as reinforcement and stiffens the area around the cut. Traditionally, knife size depends on the length of the buttonhole, and the knife needs to be replaced when changing the buttonhole length; however, nowadays small knives are mounted, and based on the buttonhole length, the knife calculates and makes multiple cutting strokes as required. No replacement of the knife saves time and increases flexibility in handling multiple styles. Cutting is generally done by slashing the fabric by a diagonal-shaped knife against a slit; however, for some loosely woven fabrics with stronger yarns, a flat-shaped knife with chopping action against a Teflon base is appropriate to get a clean cut.

While straight buttonholes are appropriate for most of the garments, shank buttons (common for thick garments like blazers or thick areas of garments like trouser waistband) require an eyelet buttonhole to accommodate the shank diameter. The shape of the button is like keyhole or eyelet, hence the name. Eyelet buttonholes are commonly sewn by double chain-stitch zigzag, with or without a thick gimp thread positioned at the edge of the buttonhole. The gimp thread provides stiffness and bulk to raise the purl effect of the buttonhole, and acts as reinforcement to preserve its shape during use. These buttonholes can be cut before or cut after; the advantage of cut-before is a neat appearance with the thread covering the raw edges of the hole effectively. The advantages of cut after is the edge of the fabric gives some protection to the thread and the fabric is more stable during sewing; here the cut after is preferred for loosely woven fabric and cut-before is ideal for densely woven fabric. Unlike a straight buttonhole, where cutting is done by slashing the fabric, here cutting is done by die-cutting eyelet-shaped fabric.

11.4.3 Pattern-tacking machines

Pattern-tacking machines have a lockstitch zigzag (304 type) machine head where the fabric is placed over the cloth clamp. The machine sews a pre-determined number of stitches in a particular shape or pattern using a combination of lockstitch and zigzag. The most common shape for cam-follower-driven machines is a small 'bar' or 'I', used to reinforce stress points like pocket openings and zip openings; thus the name of these machines was 'bar-tacking machines'. The microprocessor-controlled machine can create any shape and therefore the name of the machine changed to the more generic pattern-tacking machine. Figure 11.15 shows sewing of a cross-tacking in a pattern-sewing machine. The cost of these machines depends on the maximum size of the tack and corresponding clamp size. Machines are available from approximately 6 cm^2 to 4 m^2 stitch area.

11.5 Computerised sewing machines

Although no significant/fundamental change has happened in sewing-machine kinematics during the last five decades, electronics and computerisation have driven the technological advancements in sewing machines (Jana, 2004). Electronic control has introduced a new form of machine versatility that allows a general-purpose machine to become semiautomatic with specific programmable processes like counting of stitches, detecting bobbin run-out and trimming thread (Glock and Kunz, 2000). Computerised sewing machines or under-bed trimmer (UBT) sewing machines have memory banks and electronic control panels that the operator uses to program an operation that the machine will perform repeatedly (Solinger, 1988), and are reprogrammable in terms of stitch patterns, cycle times and operation of work aids (Glock and Kunz, 2000).

Figure 11.15 Cross-tacking in a pattern-tacking machine. From Brother catalogue.

Computerised sewing machines generally are a lockstitch (301 stitch type) sewing machine. An operator terminates a typical lockstitch sewing operation as following:

1. Slowing down the speed of the machine for accurate stop.
2. Performing a back-tack (BT) operation, if required.
3. Rotating the hand wheel to bring the needle up (during any stop there is 50% probability that the needle may be down or up).
4. Lifting the presser foot.
5. Pulling the sewn component away from the needle while continuously rocking the hand wheel to release both the needle and bobbin thread.
6. Picking up the scissors and cutting the thread.
7. Moving away the cut thread ends manually behind the presser foot.

The above non-sewing activities are non-value adding and involve wasteful motions. Computerised lockstitch machines are developed to reduce and/or eliminate these non-sewing wasteful motions. There were four different generations of development of computerised sewing machines, and every generation offered incremental benefits in features.

The fundamental principle behind a computerised sewing machine is the synchroniser, a device that monitors and counts the rotation of a hand wheel. While the synchroniser acts as the logic provider to the electronic circuit, the solenoids or pneumatic cylinders move the machine parts. With one complete rotation of the hand wheel, the needle bar completes one cycle of needle up/down movement. For example, when the needle is at topmost position, and the electronic marker is set at 90° position on the hand wheel, every time the electronic marker come to 90° position on the hand wheel the needle will be at topmost position. Using this logic, the synchroniser can locate the position of the needle simply by monitoring the rotation of the hand wheel. The electronic marker setting for the synchroniser comes pre-set for any computerised machine from the machine manufacturer, and the machine engineer at the garment manufacturing factory need not do anything. Starting and running these machines is same as with conventional machines; only when the operator performs 'back-heel' (pressing the foot pedal by the heel) the sewing machine gets the definitive signal to terminate sewing. The back-heel signal by the operator automatically performs a chain of events through a synchroniser. First the synchroniser gets a signal to lift the needle to the topmost position; then a movable knife cuts both the needle thread and the bobbin thread simultaneously. The operator simply needs to lift the presser foot and dispose of the sewn component.

As the synchroniser can monitor the rotation of the hand wheel, it can also count the number of rotations of the hand wheel. One complete rotation of the hand wheel equates to one stitch formation. To ensure that thread is not slipped out of the needle hole while starting the next sewing cycle, all generations of computerised machines have one thread catcher and thread wiper fitted near the needle point. While the thread wiper swipes away the thread tail towards the right side of the stitch line, the thread catcher holds the thread end by suction or pinch force to ensure that the thread is not slipped out of the needle hole while starting the next sewing cycle.

11.5.1 *First generation of computerised machines*

The first generation of computerised machine featured a needle positioner (NP) and UBT. The NP can definitively take/move the needle either up or down based on settings, when the machine stops; and the UBT automatically cuts the threads whenever the machine gets a signal to complete/terminate sewing. While using this first generation of the machine, only activities (3), (5) and (6) in the list in Section 11.5 are eliminated, but the operator still needs to perform the rest of the wasteful motions.

11.5.2 *Second generation of computerised machines*

The second generation of computerised machine featured an NP, UBT, and back-tack. While using a second-generation machine the number of back-tack stitches is programmed in a panel that is generally mounted on the sewing-machine head (Figure 11.16). The back-tack stitch can be of four types: single, double, triple and quadruple back-tack. The number of stitches in every step of the back-tack can vary and so does the back-tack type at the start and termination point.

The double back-tack is the most popular; however, all four types are available in all computerised machines. Any even back-tack is easier to sew because the starting point of the back-tack is the same as the starting point of the stitch. However, in odd back-tacking, the starting point of the back-tack is different from the starting point of the stitch, thus making it difficult to align the piece to be sewn. Even step back-tacks are used at the starting of a seam (as alignment of the piece is required at the start of seam), while odd-step back-tacks can be used at the termination of seam.

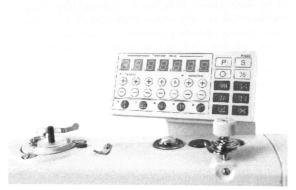

Figure 11.16 Control panel in a computerised sewing machine. From Typical Corporation product catalogue.

When an operator using a second-generation computerised machine presses the foot pedal, the start back-tack is completed automatically and the machine awaits beginning the seam at the starting point. The operator has to release the pressure at the foot pedal and press again to start the stitch. During continuous sewing, the operator has to guide, pivot the fabric manually while slowing down and speeding up the sewing at will. During the sewing process, whenever the operator releases the pressure from the foot pedal, the machine will stop with the needle automatically at the downmost position; this will help to align/pivot the piece as necessary. The operator can resume stitching by pressing the rear of the foot pedal by toe. Only when the sewing comes to an end and the operator wishes to terminate the sewing operation, the operator performs a back-heel. Once the back-heel is performed, the machine does a chain of events: perform a back-tack, raises the needle and the UBT cuts the thread. All three activities happen in sequence but automatically in a fraction of a second and no human intervention is required in between. If the operator performs a back-heel accidentally during the mid-course of sewing, the machine will perform same three activities in sequence to terminate sewing (with back-tack). The operator needs to be careful and vigilant not to back-heel by mistake.

The second generation performs (2), (3), (5) and (6) activities from the earlier list, but the operator still needs to perform the rest of the wasteful motions.

11.5.3 Third generation of computerised machines

The third generation of computerised machines features NP, UBT, back-tack and step programming (SP). The SP feature counts the total number of stitches that can be given in the continuous sewing operation between start back-tack and end back-tack. The total sewing length between start back-tack and end back-tack can be divided into multiple steps, and the number of stitches in each step can be programmed. Generally, there is a maximum of 9–13 steps possible, and each step can have 99 stitches maximum. For every step, the operator is required to press the foot pedal and release to switch to the

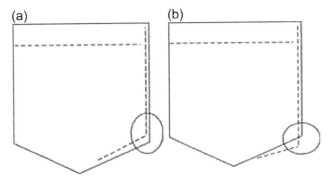

Figure 11.17 Step programming problem in sewing pocket corner. (a) Stitch pivot point is not reaching the corner of pocket. (b) Stitch line shoots off the corner of pocket.
From StitchWorld.

next step. The operator need not slow down to perform a precise stop, as the machine will stop automatically (even if the pedal is kept pressed) after sewing the pre-programmed number of stitches. SP can be used successfully for attaching pre-creased labels or pockets where creasing is consistently accurate up to a single stitch length. Such accuracy is not achievable while manually creasing pockets. Figure 11.17 shows that when the pocket crease measure varies, the pre-programmed number of stitches may fall short of the corner (Figure 11.17(a)) or may shoot off the corner (Figure 11.17(b)). These machines also feature an additional stitch button near the needle point; a single press of this button inserts one additional stitch without altering the SP.

11.5.4 Fourth generation of computerised machines

The fourth generation of computerised machines features an NP, UBT, back-tack, SP and fabric ply sensor (PS). Here the machine is not counting the number of stitches but sensing the difference of plies in the line of sewing. PSs can sense the number of plies and the machine can be programmed to run while sewing a pre-set number of plies (say three plies) and can come to stop when a different pre-set number of plies (say one ply) is sensed before the needle point. While attaching patch pockets, the PS can be programmed to sense the difference between a single ply and three plies. The sensor is mounted at the machine head at the line of sewing just after the needle point. While minor variation of creased pocket size can result in the programmed number of stitches shooting up or falling short of the corner, the PS feature is capable of stopping precisely by sensing the difference of plies. The operator only has to pivot at all corners, and press the pedal to start sewing.

11.6 Work aids

Work aids are labour-saving devices used to simplify an operation, reduce handling, increase productivity, improve work quality and reduce operator fatigue. The work aids may operate mechanically or by pneumatics or electronics, and can be modified machine parts (Jana, 2014b), separate devices, attachments and machine options (Glock and Kunz, 2000). Work aids are devices that are built into machines, added to them afterwards, attached alongside or made use of in whatever ways a resourceful engineer can devise to improve productivity, improve or maintain quality standards, reduce training time and minimise fatigue to the operator (Carr and Latham, 1999). The function of a work aid is to improve the qualitative and/or quantitative output of a sewing machine (Solinger, 1988) and may also de-skill the sewing operation (Jana, 2014b).

The typical sewing operation can be analysed into ply separation, pick up, orient, mate, control through sewing and dispose (Carr and Latham, 1999). The term 'handling' is normally used to describe those elements that are not sewing — like bundle handling, various aspects of machine attention and personal needs — and handling makes up 80% of the total time spent (Carr and Latham, 1999). The nature of clothing material makes ply separation (from a stack of fabric plies) and pick up difficult because the materials are limp and they slip or cling. The nature of clothing material makes machining also difficult; orientating, mating and controlling through sewing are slowed down because of the need to pause repeatedly to realign the parts or to fold into position (Carr and Latham, 1999).

While there is no clear categorisation of work aids available, they can be broadly categorised into machine options and separate devices. Machine options are varieties of built-in features in machines that can reduce the mental, visual and physical demands on an operator (Glock and Kunz, 2000). Presser foot, thread trimmers, and back-latchers are examples of machine options. Separate devices are positioned on or around the sewing machine for orderly stacking, orderly disposal and controlled feeding of materials. Attachments, stackers, elastic metering devices and bundle clamps are some of the examples of such devices.

11.6.1 Machine options

The presser foot can be categorised into solid (Figure 11.18) and hinged (Figure 11.19); solid is used for flat seams without cross-over seams, and the hinged type is used for universal application. Presser foots are generally double legged; however, a single-legged presser foot can be used for cording or piping where the stitch line should be closer to the cord or piping (Carr and Latham, 1999). Narrow leg presser feet are used for zip attachment. While sewing at the edge of thick material, one leg of presser foot runs on the thick material and another leg of the presser foot is actually hanging out without support from the bottom. A compensating foot can be used in such a situation where there is a difference in thickness to the left and right of an edge and stitching is required at a specific distance from that edge (Carr and Latham, 1999). These presser feet are classified as compensating right (Figure 11.20) or compensating left (Figure 11.21), followed by the amount of distance the stitch will be placed from the edge. The compensating presser feet are the hinged type and have spring-mounted separate legs that can compensate for variable thickness of material.

Top stitching the raglan line is by cover-stitch and is used to flatten out the overlock seam in case of medium- to heavy-weight fabrics for sweatshirts, whereas for

Sewing equipment and work aids 297

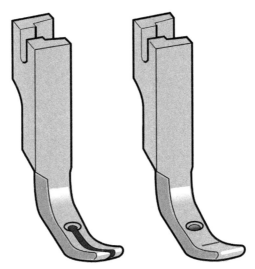

Figure 11.18 Solid presser foot.
From Suisei product catalogue.

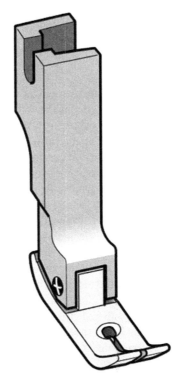

Figure 11.19 Hinged presser foot.
From Suisei product catalogue.

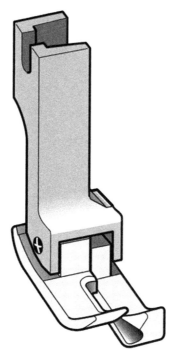

Figure 11.20 Right-compensating presser foot. From Suisei product catalogue.

lightweight fabrics in blouses and T-shirts, overlock is folded in one side and cover-stitched. The presser foot for a cover-stitch machine is designed accordingly.

The thread trimmer for a lockstitch machine trims both needle and bobbin thread together underneath the machine bed and is called a UBT; the same for cover-stitch machines can cut needle, looper thread and also top cover thread. The device operates either pneumatically or electrically. Thread trimmers for overlock sewing machines are called chain cutters. The device either sucks the chain and then cuts by scissor knife action or simply cuts by chopping knife action. The suction can be either vertical or flat type. In the vertical type, the suction inlet is positioned behind the needle and vertical to the cloth plate. These devices can be retrofitted in selected overlocks; there is a stationary knife and a moving knife that reciprocates while the machine is running by the moving part of the machine itself and requires no extra drive source. Here the chain has to be sucked sideways through the suction inlet. In the flat type, the suction inlet is positioned right behind the presser foot and flush with the cloth plate. The thread chain is sucked below the machine bed along gravitational pull and chain is cut by a moving knife. There is a separate drive source for the moving knife, either electrical (solenoid) or air (pneumatic) operated. The chopping knife action chain cutter is available for both overlock as well as flatlock machines and can cut both thread chain and/or cloth tape.

This device in a lockstitch machine trims all the plies of sewn material parallel to the stitch line at a pre-defined distance (e.g. collar run-stitch) and in a cover-stitch machine

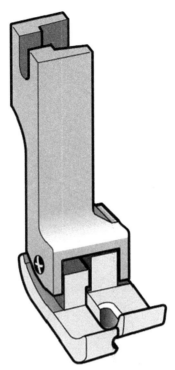

Figure 11.21 Left-compensating presser foot.
From Suisei product catalogue.

trims only the lower ply neatly during the felling operation (Figure 11.22), providing neatly finished products (e.g. bottom hem of T-shirts).

Stitch condensation devices in overlock (Figure 11.23) and cover-stitch machines prevent unraveling at the start and end of sewing by condensing the stitch length without stopping the machine.

A back-latching device in overlock is equivalent to back-tacking in a lockstitch machine. An overlock stitch is always left with a tail of chain at the start and end of a sewn piece. This is a device that back-latches the thread chain by sewing it back into the seams on the fabric at the start of sewing. In a manual version, the operator has to pull back the thread chain manually at the end of operation and clamp in the needle plate in front of the needle. In the automatic version, the thread chain is sucked (pulled back) by vacuum action through a hollow tube in the needle plate (Atlanta Attachment) or otherwise (CF ITALIA), resulting in a clean finish.

Tape cutters are photocell-sensing devices that detect the end of the sewing and automatically engage the cutter to cut the material behind the presser foot above the machine bed. A stitch-skip device in trouser waistband attaching skips a predetermined number stitches at the beginning and/or end of sewing to facilitate the next sewing process. This feature often works in conjunction with tape cutters.

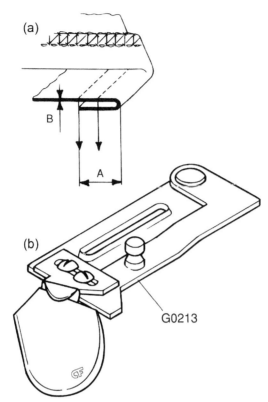

Figure 11.22 Felling operation: (a) Stitch formed and (b) Downturn feller. From CF Italia.

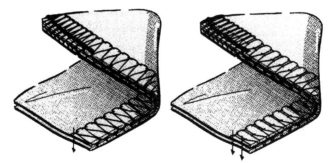

Figure 11.23 Condensed stitch in overlocks. From Pegasus catalogue.

11.6.2 Separate devices

These sewing machine attachments are mechanisms that are attached to sewing machines without cutting through or changing the original frame of the machine. The removal of such an attachment leaves the machine in its original condition (Solinger, 1988).

These attachments are fastened to the machines by screws or nuts and bolts. These are static, dynamic or synchronised; static attachments do not move during the sewing work-cycle; dynamic attachments are moved by the operator during the sewing work-cycle. Synchronised attachments have a link (or links) connected directly or indirectly to one of the driveshafts of the sewing machine. The moving link has a motion pattern that positions the fabric during the sewing element of the operation work-cycle (Solinger, 1988). This positioning pattern must be synchronised with the stitching or feeding action of the machine. Classification of attachments may be based on the function they perform: guide attachments, positioning attachments and preparation and finishing attachments (Solinger, 1988), or operations like straight flat seam, curved flat seam, hemming and binding (Garg, 2010). Attachments may be stationery, swivel, slide or pivotal in order to be engaged or disengaged from the area of operation (Glock and Kunz, 2000) and assist an operator in guiding, positioning, folding and regulating the material during sewing.

Edge guides can be simple magnetic metallic barriers fixed on the throat plate at a pre-determined distance from the right of the needle to complex zippy[1] where ply separator plates and vertical prongs are assisted by compressed air blow to align the edge of differently shaped fabrics and sew them together. Felling or hemming is a single or double turn of the fabric edge to finish the raw edge; fellers and hemmers can be upturned or downturned to enable sewing to happen from the reverse or front side, respectively. Binding is defined as covering of raw edges using a separate strip of fabric. Binding can be one-side finish, two-side finish or two-side raw edge. In one-side finish binding, the raw edge of a strip is visible at the back; in two-side finish binding, no raw edge of strip is visible either at the face or back; and in two-side raw binding, the raw edge of the binding strip is exposed in the face as well as the back (Garg, 2010). There are lap seam or flat and fell seam folders for the feed-off arm, two-piece folders for waistband attachment and yoke attachment with back, and downturn elastic hemmers that can accommodate elastic in between fabric folds. Pin-tucking attachments can be two types: knife tucking and air tucking. Knife tucking is done one at a time with a single-needle lockstitch machine, whereas air tucking is done multiply at a time with multi-needle chain-stitch machines (Solinger, 1988).

To ensure that all these folders, hemmers and binders can be fixed in the sewing-machine bed in such a manner that fabrics are fed to the needle point in the required shape, often the presser foot, throat plate and feed dog are reshaped and have to be changed with the respective attachment. These are called gauge parts, and all attachment catalogues mention the gauge parts along with specific attachments.

A run-stitch or profile-stitch operation can be done using a jig (Carr and Latham, 1999) or template (Solinger, 1988). It consists of two layers of rigid material such as aluminium or acrylic joined together by a hinge. A slot is cut out of both layers of material conforming to the exact shape and size of component to be sewn. The

---

[1] Zippy is a brand name of Profeel Inc. Italy.

inside surface of the jig is covered with a strip of non-slip material like emery sheet to have a firm grip on the fabric. The fabric plies are sandwiched in the jig and the jig is sewn in a specially adapted sewing machine where the throat plate has a protrusion to glide through the slot; the double roller presser feet and rubber-covered feed dog guides the jig along the slot. Sewing stackers and bundle clamps are additional devices attached to or outside the sewing table to organise the garment components before sewing, during sewing or after sewing. Elastic metering devices are motorised or non-motorised additional devices for controlled positive feeding of elastic to the needle point.

11.7 Sewing automats

Automation is a state of operating without external influence or control. Automated sewing systems are capable of feeding themselves cut parts from a stack, completing multiple sewing tasks and delivering finished parts. An automated sewing system eliminates human error but has a high cost of acquisition, installation and maintenance. Glock has defined three stages in the advancement of sewing technology: mechanisation, automation and robotics (Glock and Kunz, 2000). Robotics are the most advanced form of automation, which are computerised, reprogrammable, multifunctional manipulators designed to move materials, parts, tools or specialised devices through variable programmed motions for the execution of a variety of tasks (Rosenberg, 1983). Flexible reprogrammability is one of the hallmarks of robotic automation (Lower, 1987).

Sewing automats are engineered workstations, which are defined as a separate category (Glock and Kunz, 2000) as the combination of equipment and work aids that are needed to perform a designated operation. These include the machine table, power supply, compressed air supply, clamp stands, mobile work racks, pneumatic cylinders and programmable logic controllers (PLCs). Such workstations improve the ergonomics, efficiency and safety of operations and may be purchased off the self or custom developed by plant engineers or R&D departments (Glock and Kunz, 2000).

The early 1970s and 1980s saw intensive research and development activities in the United States, Europe and Japan, aiming towards a totally humanless sewing factory that is flexible as well as productive. The important research and development initiatives in the sewing area included pretreatment technology for stiffening and bending fabrics, temporary joining of pieces for efficient sewing assembly, sewing head movable-type automated sleeve mounting, work movable-type skirt waist belting, spatial clamp system shoulder pad wrap sewing, a mechanism that can grip flexible fabrics like a worker and technology for transporting fabric items being processed between different workstations (Jana, 2003). Any sewing operation can be broken up into three steps: loading, sewing and unloading (Jana, 2003). In commercially available workstations, loading is manual, while the sewing and unloading steps are completely automated without any human intervention (Jana, 2014a).

11.7.1 Loading of fabric components

Loading involves picking up the fabric components by one or two hands, placing/aligning/folding/matching, and sliding below the presser foot. While mechanising the loading operation, identifying the right/wrong side of a fabric will require a visual or touch sensor, and separating plies and picking up a single ply from a stack of fabric will require a ventura picker or Clue and Peabody[2] picker. Placing the component at the sewing table and sliding beneath needle point are achieved by a conveyor belt.

11.7.2 Sewing of fabric components

Sewing involves sewing needle movement and occasional stopping for guiding/aligning/matching/measuring/pivoting until the sewing cycle is complete. While start of sewing may be mechanised by proximity or an optical sensor, feed manipulation will require the tension-adjusted belt-feed system, dynamic edge alignment of dissimilar curves during sewing will require air-assisted multi-layer edge guides and pivoting the fabric while keeping the dimensional stability intact will require spatial clamp feed or template sewing (Jana, 2003).

11.7.3 Unloading of sewn components

Unloading or disposing of the sewn piece involves sliding/flipping/folding to arrange/stack/hang the piece in order once the thread is cut. Disposal involves severing the sewing thread as well as sliding/flipping the sewn component from the table to stack/hang in a stacker/tray to make loading for next operation less cumbersome and less time-consuming (Carr and Latham, 1999; Jana, 2003). Depending on the size of the sewed parts to be disposed off, the mechanism of disposal can be pick and place, slide and align, flipped to hang on fold, or concentrically arranged. There are two types of stackers: small part stackers and large part stackers.

Small part stackers are placed behind or on one side of the machine, as desired and controlled by a thread trimmer, chain or tape cutter, knuckle or knee switch. For a T-shirt sleeve hemming workstation, or cuff and collar band felling workstation, the flip-type small part stacker has a swivel bar for short parts folded up or folded down. Figure 11.24 shows a small part stacker from Pfaff Special Service catalogue. A turntable stacker is another type of small part stacker which stacks small parts concentrically like cards held in hand. The larger garment part after being sewn is slid off the back of the machine table over the horizontal bar of a stacker. Pneumatic force then moves the bar away from the machine so that the garment part falls astride it and the bar returns to its formal position to await the next part (Carr and Latham, 1999). Large part stackers are of different types, standard, with deflectors, flip type, etc. Freely stacked large part stackers are common in trouser panel serging workstations (Figure 11.25). The components are either stacked freely on a double-bar mechanism (for better stability) or stacked on a single bar with clamping bar. Rotary stackers are generally used for stacking closed-loop-type sewn

---

[2] Clue and Peabody is a trademark of Clue and Peabody Corporation.

304 Garment Manufacturing Technology

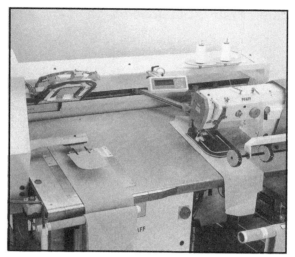

Figure 11.24 Small part stacker.
From Pfaff special service catalogue.

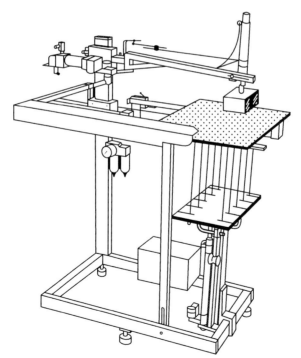

Figure 11.25 Large part stacker.
From Pfaff special service catalogue.

components, for example elastic loops made for closed waistbands. Elastic tapes are cut and sewn (butt join or overlap join) automatically before gliding onto the stacker, which rotates intermittently (Jana, 2003).

Some of the popular sewing automats commercially available include the following:

Automatic neck band creaser
Automatic cuff hemmer
Button-sewing indexer
Buttonhole indexer
Pocket attaching in shirt front
Yoke attaching in men's shirts
Back pocket attaching in jeans
Back pocket hemming in jeans
J-stitch making for jeans
Welt pocket making for formal trousers
Welt pocket making for formal jackets
Serging of formal trouser panel
Side seam joining of formal trouser panel
Dart sewing in formal trouser panel
T-shirt front placket making
T-shirt bottom hemming
T-shirt sleeve hemming

Figure 11.26 shows a shirt pocket attaching workstation with clamp feeding mechanism and large part stacker. While these types of automats are highly productive, they are highly inflexible, too. Generally simple mechanics, pneumatics or hydraulics actuate physical displacement, and electronics provides logic or command. A high level of complex mechatronics makes these workstations vulnerable to style and size changes (set up time) as well as to breakdown (Jana, 2003).

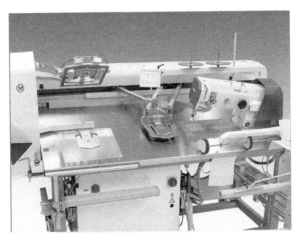

Figure 11.26 Shirt pocket attaching workstation.
From Ptaff industrial.

11.8 Sewing needles

The needle was one of humankind's first tools. Over the millenia it developed from a simple craft item to the precision tool for modern sewing machines. The most ancient (28,000 BC) sewing needle did not have an eye but a split end that gripped the thread to be sewn. Needles from 17,500 BC had an eye at one end and tapered at the other end (The Schmetz, 2001). Around 1800, Balthasar Krems from Mayen, Germany used, for the first time, a needle that had the eye moved close to the point; this eye-point needle paved the way for the mechanisation of sewing worldwide. The basic elements of a sewing needle, which are constantly modified to suit many fields of application, are the needle shank, the needle blade with one or two grooves and the needle point with eye. Figure 11.27 shows different parts of a sewing needle; it is an extract from DIN 5330-1. Table 11.3 also describes the different parts of a sewing needle.

The manufacturing of a needle involves almost 30–35 stages. Important stages are straightening of wire, grinding ends, swaging, stamping and punching, marking, groove milling, soft polishing, hardening and tempering, chemical deburring, optoelectronic straightening, polishing and hard pointing, electroplating and final inspection (The Schmetz, 2001; Groz Beckert, 2014). Though chromium plating is one of the most common surface finishes applied to sewing needles, other finishes like titanium nitride coated GEBEDUR needles from Groz Beckert (2014), titanium coated PD (Perfect Durability) needles from Organ Needles (2014) and diamond carboride needles from Schmetz (2001) are offering unique properties for special purpose use. There are three parameters used to classify sewing needles: needle system, needle point type and needle size.

11.8.1 Needle systems

Needle systems are classified based primarily on three measurements: eye to butt length, shank length and shank diameter, and specific to different stitch classes. Needle systems are expressed as an alphanumeric code with a multiplication sign within. For example, the most common needle system for lockstitch is DBX1, where eye to butt length is 33.80 mm, shank length is 16–10.5 mm and shank diameter is 1.62–2.02 mm. Another needle system, DAX1, is also used for lockstitch machine (for sewing lightweight fabrics) with a shorter eye to butt length of 29.60 mm, a shank length of 14.50 mm and a shank diameter of 1.62 mm. Similarly, a DCX1 needle is common for overlock, TQX1 for button sewing and so on. Although these alphanumeric codes are commonly used in industry, there are other parallel system code numbers used by different needle manufacturers. For example, DBX1 is also referred as 16X231 and TQX1 is also referred as 175X1. To date, there is no uniform categorisation of needle system numbers across major needle manufacturers, and according to Schmetz there are around 800 different types of frequently used needles (The Schmetz, 2001). The Organ catalogue lists as many as 28 different types of cloth-point needles only for lockstitch, and as many as 162 types of cloth-point needles across nine machine/stitch categories/classes (Organ Needles, 2014).

Sewing equipment and work aids

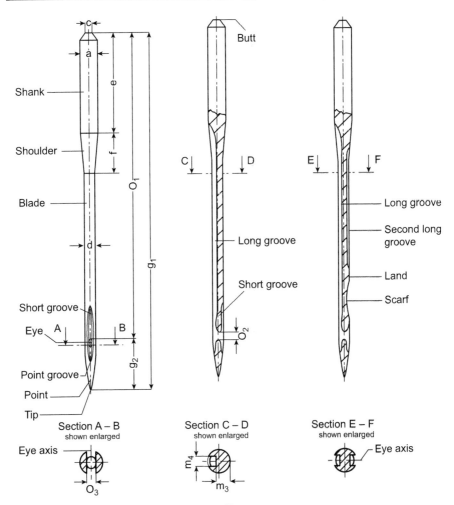

Figure 11.27 Different parts of a sewing needle. From Schmetz.

11.8.2 Needle point types

Needles used for woven and knitted fabrics are called cloth points. Cutting points are used for leather and other laminated materials.[3] Cloth points are generally of two types: round points and ball points. In general, round points are meant for woven fabric and ball points are meant for knitted fabric. While round points have sharp edges, ball points have rounded blunt edges to cause sliding action rather than penetrating action. While generally needle tips are concentric, there are instances of eccentric-point needles (The Schmetz, 2001). Depending on

---

[3] Cutting-point needles are not discussed in this chapter as their use is outside the scope of the chapter.

Table 11.3 **Measurement description of needle parts**

Legend	Description
a	Shank diameter
d	Blade diameter
c	Butt diameter
e	Length of shank
f	Length of shoulder
g_1	Length of needle
g_2	Length of point
m_3	Remaining thickness of blade
m_4	Width of long groove
O_1	Butt to eye length
O_2	Length of eye
O_3	Width of eye

From Schmetz.

sharpness and bluntness of tip there are further sub-classification, e.g. Groz Beckert lists a total of 36 point types (Groz Beckert, 2014), Schmetz a total of 13 point types (The Schmetz, 2001) and Organ 9 types (Organ Needles, 2014). Different point types are referred to by different alphabetical codes and there is fair bit of consensus between different brands. For example, normal round point is referred as R, light ball point as SES, special ball point as SKL and so on. There are differences, too; while Groz Beckert and Organ have an 'extra light ball point' (RG type), in the Schmetz classification there are no 'extra light ball points'. Table 11.4 shows the basic logic of cloth point types.

11.8.3 Needle sizes

The size of the needle is the diameter of the needle, which is decided based on fabric to be sewn. In 1953, German standard DIN 5325 standardised the needle sizing system (The Schmetz, 2001) with metric size designations. Number metric (under abbreviation NM) indicates the diameter of the needle blade in hundredths of a millimeter measured above the scarf or the short groove, but not at any reinforced part of the blade (The Schmetz, 2001). A sewing machine needle with a blade diameter of 0.80 mm therefore corresponds to NM 80. Apart from the number metric system there are many other sizing system that are in use like Singer, Union Special, System 292, etc. Generally in all needles, two size numbers are written today, that is, the NM and Singer system. For example, 90 NM is equivalent to size 14 of

Table 11.4 Sewing needle cloth point types

	Acute round point (SPI)	Normal round point (R)	Light ball point (SES)	Heavy ball point (SKF)	Special ball point (SKL)
Needle point top view					
Needle point side view					
Description	Tapered uniformly to sharp point	Convex tapered profile to sharp point	Straight tapered profile to small ball-shaped tip	Straight tapered profile to large ball-shaped tip	Double-stage tapering to small ball-shaped tip

Adapted from Groz Beckert and Organ Needle.

Singer and written as 90/14; similarly, 75 NM is equivalent to 11 size of Singer and so on. While 75/11 size needle is commonly used for sewing medium-weight fabric, 110/18 size may be used for 16-oz denim fabric and so on. There are a total of 24 sizes of needle available, with the thinnest size being 35 NM and thickest size being 200 NM.

11.8.4 Needle selection

Almost all major brands have some sort of ready reckoner for selecting the most appropriate needle point type for different standard fabric types. The logic to be followed for selection of the correct size is the lighter the fabric, the narrower the needle; however, due to continuous development of new fabric types (and continuous modification of point types by needle manufacturers), the process is often prone to error. All major needle brands offer to recommend the best point type and size of needle once a fabric sample is provided to them by the manufacturer. The selection of needle systems is a non-issue, as the machine catalogues always mention the needle system to be used for that machine and generally no other system would work.

11.9 Sewing threads

Sewing threads are special kinds of multi-yarns that are engineered and designed to pass through a sewing machine rapidly. Sewing thread must be durable enough to withstand the abrasion, stretch and needle heat that occurs while sewing, garment finishing, stretching and recovery during wear (Coats Plc., 2014). Sewing threads are mainly produced in three-ply and occasionally two- or four-ply. Three S twisted single yarns are Z twisted to form a three-ply sewing thread. The finishing twist is in the Z direction to reach equilibrium and resist further tightening of twist. However, if the sewing thread is having an S twist as a finishing twist, the thread will get untwisted by the sewing action and may fray and break. As the lockstitch machine is the most severe in its handling of thread, the majority of sewing threads intended for use in a sewing machine are constructed with a finishing Z twist.

Sewing threads can be classified based on substrate or material, for example polyester, cotton, lyocell, meta-aramid, para-aramid, nylon, polypropylene and PTFE; based on construction, for example air-entangled, braided, poly-cotton core spun, poly-poly core spun, monofilament, spun, spun-stretch broken, textured, twisted multifilament and monochord; and based on finish, for example bonding, non-wick, anti-fungal, fire retardant, water repellent and anti-static finishes (Coats, 2003; American & Efird, 2014). Hundred percent polyester thread is used for sewing fabrics of all different fibre content; this is primarily due to the fact that cotton has less elongation property and is not suitable for high-speed sewing. In addition, the incompatibility of post-washing shrinkage between cotton sewing thread and cotton fabric may lead to shrinkage pucker. Hundred percent cotton spun thread is used only for garment dyeing. Poly-poly core spun thread is 40–50% stronger, has higher abrasion resistance, has a smaller diameter with the same breaking strength, shows optimised elongation behavior, has less shrinkage, exhibits fewer knots and thick and thin parts in between, runs with minimal tension, is suitable for high-speed and multi-directional sewing machines and has less fibre falling on the machine, so less maintenance compared to 100% cotton thread.

11.9.1 Sewing thread performance

Sewability is defined as a thread's ability to perform on the sewing machine. The thread must perform flawlessly at high machine speeds without breaking and without skipping stitches. The critical factors necessary for good sewability are thread strength, optimum twist levels, low fault levels, low and controlled elongation, tenacity and lastly even lubrication (Jana et al., 2008). While spun threads are the most common types, core-spun sewing threads are produced to achieve optimum strength-with-fineness of continuous-filament threads together with sewing performance and surface characteristics of spun-fibre threads. Continuous-filament threads are plied and corded thread produced from 100% synthetic filaments; they are twisted, set, lubricated and bonded with synthetic polymer. Multifilament single-ply threads are produced from a single ply of multifilament polyester yarn, suitably twisted and then treated with light bonding finish. Textured sewing threads have minimal 'S' twist, soft handle and are beneficial when used as underthreads where a particularly soft seam is required, such as the bottom-covering thread in 406/407/607 stitches in undergarments and serging overlock application in formal trousers.

Sewing threads are packaged and handled in different forms like cones, vicones and cops (Figure 11.28). Cones are more accurately described as frustrums of geometric cones onto which thread is cross-wound for stability and good off-winding performance. Vicones are parallel tubes or low-angle cones with an additional base in the form of a raised flange that may incorporate a small lip. Vicones are designed to contain any spillage, which may occur during unwinding of these smooth threads, with no snagging or trapping when the slag thread is taken up. Cops are small cylindrical, flangeless tubes onto which thread is cross-wound for stability. Apart from cones, vicones and cops there are spools, cocoons and pre-wound bobbins that are prepared for specific use but rarely used in the industry.

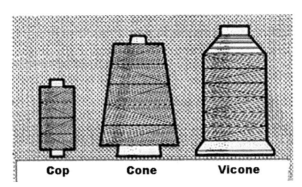

Figure 11.28 Different thread packages.
Reproduced with the permission of Coats Plc.

11.9.2 Sewing thread count and numbering system

Thread sizes are communicated through various numbers and numbering systems, which are derived by relating unit length and weight. All numbering systems used to indicate thread size are either fixed-weight or fixed-length systems. In fixed-weight systems, the number of hanks of 840 yards in one lb (Pound) is equal to one English Count (Ne). Similarly, the number of hanks of 1000 m/kg is equal to one Metric Count (Nm). In fixed-length systems, weight of 9000 m or 1000 m in grams is equal to Denier or Tex, respectively. As count number increases the yarn becomes finer in fixed-weight systems and the yarn becomes coarser in fixed-length systems.

Generally, metric count is used to describe synthetic, spun and core-spun thread, while filament threads are normally expressed with Decitex or Denier. As more than one ply of yarn is twisted into a thread, the resultant size of the thread can be derived by dividing (in the fixed-weight system) or multiplying (in the fixed-length system) individual yarn count by number of plies. Ticket numbering is a commercial numbering system, merely the manufacturer's reference numbers for the size of a given thread. Ticket number value can be calculated by dividing 1000 by the Tex number and multiplying by 3. Ticket numbers resemble the fixed-weight system; the higher the ticket number, the finer the thread, and vice versa. Finer sewing threads are preferred over coarser threads. Finer sewing threads can better fit into the fabric surface, hence, are less prone to abrasion. The seams that are subjected to higher stress, should be constructed with coarser sewing threads and vice versa. A ticket number in one type of thread will not be the same as in another. For example, ticket 40 cotton is not the same as ticket 40 core-spun.

11.10 Future trends and conclusions

The bed type, feed type and stitch type cannot be altered in a conventional sewing machine. However, currently convertible bed types (submerged to raised in eyelet buttonhole machines), convertible feed types (unison to top and bottom feed by Typical), convertible machine types (bar tack to button sewing by Typical), dual-feed types (drop feed and needle feed), and light- to medium-duty convertibility features are aiming to offer flexibility to users. Even automated workstations, which used to address a specific operation, are now offering flexibility (StitchWorld, 2009; Anon, 2012). Newer feed types like X-feed (Typical) and improvised differential feed (Megasew) are addressing the fabric feeding at an advanced level. The modular structure of machine heads by Typical Corp (Vetron) (Figure 11.29) and Duerkopp Adler (M-Type) will enable the use of exchangeable common parts between different machine types, resulting in easy repair, lower inventory and reduced cost of ownership. Fewer moving parts will reduce failure rate, increase ease of repair and lessen downtime (AMF Reece eyelet buttonhole machine).

In value-addition sequential and programmable sequin sewing, decorative effects achieved by multi-needle chain-stitch machines, multi-colour and multi-stitch overlock and flat-lock stitches (zebra stitch from Pegasus), jig-assisted quilting of large components, multi-head quilting machine and three-hole or six-hole button sewing are becoming product differential factors. In the area of information technology (IT)

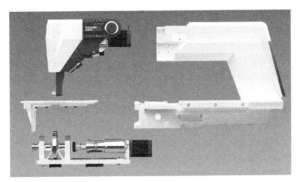

Figure 11.29 Modular sewing head.
From Typical Corporation product catalogue.

integration there are already USB-based operator log-ins (Vetron), networked sewing machines (Jana, 2003), touch-screen-based easy diagnostic modules for repair and mobile device-integrated sewing machines (Pfaff). With online connectivity becoming affordable, machine-to-machine (M2M) communication for computerized numerical control (CNC)-based machines and predictive maintenance will become a norm.

Automatic bobbin changers (ABCs) were developed way back in the 1990s. All the effort to popularise ABCs was unsuccessful because of their high cost and exclusivity. With current emphasis on removal of non-value-adding activity, ABCs are expected to be the norm in cycle-sewing machines and automats. Large-area pattern sewers with stationery sewing heads will give way to moving-head large-area pattern sewers for enhanced stitch quality.

Consumables like needle and thread are also keeping pace with developments, and authenticity of product is the key issue. Nano-particle-embedded markers will become the norm for product identification (Blue Sky technology from Amann) at any stage of use. Groz Beckert introduced titanium nitride-coated Gebedur needles in 1993; the same is now offered by a number of other needle manufacturers. Other high-performance needle coatings like ceramic and DLC (diamond-like coating) (The Trimph Needle Company) and diamond carboride (Schmetz) are aiming towards better insulation, cooling, chemical resistance, biocompatibility and wear resistance.

Sustainability and energy conservation will play a crucial role and use of recyclable and advanced lightweight high-performance material in machine parts will increase. Reduction of energy usage and alternate energy-driven sewing machines will become reality. Fabricating customised machine parts, work aids and attachments in a garment factory using 3D prototyping will be easier, quicker and more economical by the use of advanced polymers in place of metal. The most important and disruptive trend is the popularity of stitchless fabric joining using ultrasonic welding or adhesive-based bonding in apparel products (see Chapter 13).

11.11 Sources of further information and advice

The apparel manufacturing industry migrated from Europe, the United States and Japan to Asia during the 1980s and gradually to the least developed countries during the 2000s.

The legendary Pfaff and Duerkopp brands of Germany are now Chinese owned, and the epitome of industrial sewing-machine brands like Juki and Brother have lost leadership positions to the plethora of Chinese brands. Traditionally there were three major exhibitions — Bobbin in the United States, IMB in Germany and Japan International Apparel Machinery Trade Show (JIAM) in Japan — held once in three years showcasing the R&D of machinery manufacturers. Some of the most respected trade or popular journals were *Bobbin* from the United States, *World Clothing Manufacturer* from the UK and *JSN International* from Japan. Today Bobbin and IMB fairs have changed to Sewn Products Equipment & Suppliers of the Americas (SPESA) and Texprocess, respectively, with changed market focus, and magazines also have either changed their name and focus (*Bobbin* to *Apparel*) or just given up. Industry associations like the Textile Institute, AAMA (American Apparel Manufacturers Association), VDMA Textile Machinery Association and International Apparel Federation (IAF), and research institutes like [TC]2 and SATRA, used to compliment the academic institutes like The Nottingham Trent University (UK), Manchester Metropolitan University (UK), Fashion Institute of Technology (US), Royal Melborne Institute of Technology (Australia) and Institute of Textile and Clothing (HK) in nurturing research and manpower.

While the China International Sewing Machinery & Accessories (CISMA) exhibition in China and *StitchWorld* magazine from India are the new stars of the manufacturing industry today, there are no new books. *The Apparel Manufacturing Handbook* by Jacob Solinger, though out of print and circulation, is still considered the bible of apparel manufacturing. *The Technology of Clothing Manufacture* by Harold Carr and Barbara Latham; *Apparel Manufacturing Sewn Product Analysis* by Ruth E. Glock and Grace I Kunz and *Clothing Technology: From Fibre to Fashion* edited by Roland Kilgus are important resources. With online information significantly outpacing printed material, organisations are placing lots of interactive and engaging content on their websites; some of the important ones are:

www.apparelresources.com
www.stitchworld.net
www.garmento.com
www.tecexchange.com
www.strima.com
www.coatsindustrial.com
www.amefird.com
www.groz-beckert.com
www.schmetz.com
www.tc2.com

References

American & Efrid, 2014. American & Efrid (Online) Available at: http://www.amefird.com/products-brands/industrial-sewing-thread/ (accessed 15.08.14).
Anon, 2012. Sewing automats for multi product multi process manufacturing. StitchWorld (Issue December).

Brother Industries Limited, 2000. Industrial Sewing Machine Handbook. s.l.: Brother Industries Limited.

Carr, H., Latham, B., 1999. The Technology of Clothing Manufacture. Blackwell Science, Oxford.

Coats Plc, 2014 (Online) Available at: http://www.coatsindustrial.com/en/information-hub/apparel-expertise/thread-numbering. http://www.coatsindustrial.com/en/information-hub/apparel-expertise/thread-numbering (accessed 18.07.14).

Coats, 2003. The Technology of Thread and Seams. J&P Coats Limited, Glasgow.

Cooklin, G., Hayes, S.G., McLoughlin, J., 2006. Introduction to Clothing Manufacture, second ed. Blackwell Science, Oxford.

Garg, Y.P., 2010. Challenges of joining two plies. StitchWorld (Volume October).

Glock, R.E., Kunz, G.I., 2000. Apparel Manufacturing Sewn Product Analysis. Prentice Hall, New Jersey.

Groz Beckert, 2014. Groz Beckert (Online) Available at: http://www.groz-beckert.com/ (accessed 21.07.14).

Jana, P., April 2003. Automation in sewing: how does it work? and why it didn't? StitchWorld.

Jana, P., January 26, 2004. Trends in apparel manufacturing technology. Just-Style.

Jana, P., 2014a. Automation in sewing room: pocket attaching in shirt. StitchWorld (Issue May).

Jana, P., 2014b. Sewing Machine Resource Guide, first ed. Apparel Resources, New Delhi.

Jana, P., Heckner, R., Kamat, S., 2008. Sewability of thread: the intriguing factors. StitchWorld.

Jana, P., Khan, A.N., 2014. The sewability of lightweight fabrics using X-feed. Int. J. Fashion Des. Technol. Educ. 7(2), 133−142.

Juki, C., 1988. Basic Knowledge of Sewing. Juki Corporation, Japan.

Kilgus, R., 1996. Clothing Technology: From Fibre to Fashion. Verlag Europa-Lehrmittel, Hann-Gruiten.

Lower, J., May 28, 1987. Robotics advance softly. Bobbin 106−113.

Organ Needles, 2014. Organ-needles.com (Online) Available at: http://organ-needles.com/english/product/download.php (accessed 21.07.14).

Pfaff, 1983. Technical Bulletin. s.l.: Pfaff Industrial.

Rosenberg, J.M., 1983. Dictionary of Business and Management. Wiley, New York.

Solinger, J., 1988. Apparel Manufacturing Handbook, second ed. Bobbin Media Corp, Columbia.

StitchWorld, 2009. Flexible automation from Vibemac. StitchWorld (Issue June).

Stitchworld, 2012. X-feed from typical: the ultimate feed system. StitchWorld 23−24 (Volume February).

The Schmetz, 2001. The World of Sewing: Guide to Sewing Techniques. Ferdinand Schmetz GmbH, Herzogenrath.

Sewing-room problems and solutions

12

M. Carvalho, H. Carvalho, L.F. Silva, F. Ferreira
University of Minho, Guimarães, Portugal

12.1 Introduction

In this chapter some common defects in industrial sewing are analysed and their causes and solutions discussed. Although some of these defects can be considered mere aesthetic glitches — albeit not always negligible — in some situations these faults can compromise the functionality of the seams. Good examples are seams on load belts or other load-carrying products, and seams on airbags.

The problems that will be studied include seam pucker and seam undulation; faults caused by needle penetration; poor material feeding; ply-shift, folds and stitch distortion; inadequate stitch tension and tension balance; and irregular stitch length, skipped stitches and thread breakage. Many researchers have studied methods for automatic detection of some of these problems, which will be reviewed. Future trends will also be predicted, towards the design and development of sewing machines with further sensing and control techniques to overcome the problems herein analysed.

12.2 Seam pucker and other surface distortions

Generally, a seam is expected to lie flat in the fabric. The experienced sewing technician knows that in some types of fabrics, especially lightweight fabrics, this is rather difficult to achieve. Often the fabric gets rippled or undulated along the seam line. In most situations, this is a purely aesthetical defect of a seam, which may have several causes. Figures 12.1 and 12.2 show examples of seam pucker and seam undulation.

The deformation mechanisms of woven and knitted fabrics are usually very distinct. The term 'seam pucker' is most often used in relation to the deformation of woven fabrics. Dorkin and Chamberlain (1961) developed comprehensive work on seam pucker in early research and defined the main causes for seam pucker, which will be analysed in the following sections.

12.2.1 Differential fabric stretch/feed pucker

Differential fabric stretch, or feed pucker (Amann Group, n.d.-b), occurs when different fabric plies being sewn are fed in dissimilar conditions by the material feeding system of the sewing machine. This causes the individual plies to be stretched

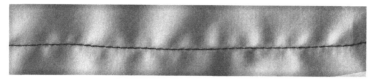

Figure 12.1 Seam pucker in fine woven fabric.

Figure 12.2 Seam undulation in stretch fabric.

differently, thus causing seam distortion when the materials recover from the suffered extension. Feed pucker can be identified easily by comparing the top and bottom plies sewed: the plies are puckered differently; usually the lower ply is more undulated than the upper one.

This kind of pucker is often observed when sewing two or more plies of fabric on machines with a drop-feed system. Drop-feed is the most common and simple feeding system, being composed of a presser-foot, a throat plate and a feed-dog (Figure 12.3).

The feed-dog has an approximately elliptical movement. After the withdrawal of the needle from the fabrics, the feed-dog rises above the throat plate and presses the fabrics against the underside of the presser-foot, whilst pushing it forward to a pre-determined distance, which defines the stitch length. This system has the drawback of not being able to provide the same feeding action on the upper and lower fabric plies. Only the lower ply has direct contact with the feed-dog, the mechanical element promoting fabric motion, while the upper ply has the tendency to get stretched because of the friction exerted on it by the presser-foot. Conversely, the problem of ply-shift, described later in this chapter, is also caused by the principle of the drop-feed system.

Feed pucker depends mostly on the fabric properties and can be managed in several ways. The reduction of presser-foot force and the use of specific presser-feet (e.g. Teflon coated) can reduce friction between presser-foot and fabric and thus minimise this problem.

More complex solutions involve using compound feeding systems, in which the feed-dog's feeding action is complemented by the action of the needle and/or the presser-foot itself. The proper choice of compound feeding system can solve the problem of ply-shift and minimise feed pucker, although the machines are more complex and expensive.

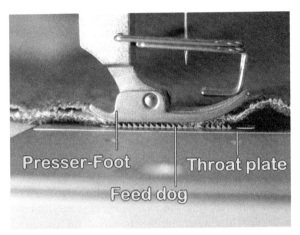

Figure 12.3 Drop-feed system.

The most immediate upgrade is the change to a needle-feed system, in which drop-feed is combined with the movement of the needle in the seam direction. The needle-feed system avoids ply-shift and thus reduces the effect of feed pucker, but it may be more prone to tension pucker. This is due to a major difference between the needle and the drop-feed systems: Feeding of the material occurs when the needle is in the material in needle-feed machines, whilst it occurs when the needle is outside the fabric in drop-feed systems. Stitch tightening occurs when the needle is outside of the fabric. This means that in a needle-feed machine, stitch tightening occurs when the feed-dog is under the stitch plate and thus not supporting the fabric from the lower side, leaving the fabric more loose and subject to tension pucker.

One of the more recent types of compound feeding system is the X-feed system by Xi'an Typical Industry Co., Ltd. (Xi'an, China). In this system, the drop-feed is combined with the needle-feed, but unlike a normal feed-dog, the feed-dog in this system is split into two parts, with each being active in about half of the sewing cycle. The fabric is thus supported between the presser-foot and feed-dog in a more even way throughout the stitch cycle. Jana and Khana (2014) studied this new feeding system and in their experiments found that this system reduced ply slippage and produced flatter seams than the traditional drop-feed. More details of X-feed is described in Chapter 11.

In a triple-feed machine, the drop-feed combines with the needle-feed and with a presser-foot, which also moves synchronised with the other two feeding elements. This system is able to virtually eliminate the problem of feed pucker, but its mechanical complexity results in a high price and a limitation in the machine's sewing speed. It is almost exclusively reserved for applications in heavy materials, such as leather goods.

12.2.2 Dimensional change of the sewing thread/tension pucker

Dimensional change of the sewing thread, known as tension pucker (Amann Group, n.d.-b), results from the high tensions suffered by the sewing threads during sewing, causing them to be stretched in the process. When the seam is finished and the joined textile parts relax, the threads tend to recover from the extension suffered. If this recovery is more pronounced than that of the fabric (which is usually the case in fine, non-elastic fabrics), the thread will exert tension on the fabric, contracting it and thus producing tension pucker. Again, this process may be immediate or take longer time spans and can also result from pressing or washing. To avoid this kind of problem, seams should be produced with as minimum tension as possible. Another common solution is to set up the feeding system of the machine to stretch the fabric slightly (using negative differential feed, to be analysed later, or tilting the feed-dog appropriately), in order to compensate for the thread's shrinkage. The problem is also minimised when the fabrics are sewn in a direction of higher elasticity, causing them to stretch a little, as seen in Figure 12.4, where the seam produced in the diagonal direction of the fabric structure (that presents higher elasticity) is less puckered. In this direction, there is also less susceptibility to structural jamming, which will be explained later.

12.2.3 Dimensional change of the fabric

Conversely, if the fabric shrinks more than the thread, the stitch may become loose and the seam may undulate, although in a different manner (as presented in Figure 12.2). This is very common when sewing stretch fabrics. In order to avoid this problem, sewing machines used for stretch fabrics are normally equipped with differential bottom feed (Figure 12.5). In this system, the presser-foot is split into two parts, whose movement amplitudes may be adjusted distinctly. If the rear feed-dog (called the main feed-dog) has a longer movement than the front feed-dog (called the differential feed-dog), then the material will be stretched. This is called negative differential bottom feed. On the contrary, if the differential feed-dog has a longer movement than the main feed-dog, a non-elastic material would be gathered near

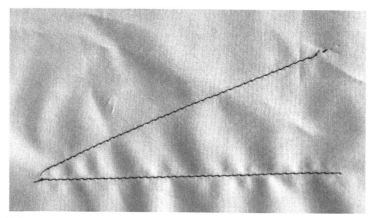

Figure 12.4 Comparison of seams produced in weft and in a diagonal direction on one ply of a fine fabric.

Figure 12.5 Differential feed system set to a positive difference at the beginning (upper image) and end of the feeding movement (lower image).

the needle (positive differential bottom feed). Stretch fabrics, however, tend to stretch when pushed along the presser-foot. If the differential feeding system is set to an appropriate positive difference in movement amplitude, one effect compensates the other. The fabric is sewn in a more relaxed state, and thus sewing thread recovery and fabric recovery are matched, avoiding pucker or undulation.

12.2.4 Seam pucker due to differential fabric dimensional instability

When two fabrics being sewn together, have different behaviour regarding the recovery from extension suffered during sewing, pucker may occur. Usually this happens immediately after sewing, but the recovery and puckering process can take longer. Dimensional changes may also be caused by pressing and washing. This type of seam pucker can be avoided by taking into account and matching, whenever possible, the dimensional behaviour of the fabrics being sewn.

12.2.5 Structural jamming

Structural jamming, also known as displacement pucker (Amann Group, n.d.-b), is produced by the presence of sewing thread in the fabric, taking up space and thus distorting its structure. The problem arises mainly in dense, lightweight fabrics, which have little space to accommodate the sewing thread and due to their thin yarns are easily deformed. The problem can be minimised using a greater stitch length and finer sewing thread and needle. Applying seams with an angle of 15° or more relative to the weft/warp yarns also reduces this kind of pucker, because in this diagonal direction there is more space between fabric yarns to accommodate the sewing threads. This effect adds to the compensation through higher elasticity, as can be observed in Figure 12.4.

12.2.6 Pattern mismatch and seam pucker

In this case, the machine operator tries to compensate variations in length of the pieces being sewn, by stretching one and easing in the other, so as to achieve a match on the seam ends. In this situation it is clear that the seam will be prone to a surface distortion, and the solution is also clear: solve the cutting inaccuracy that causes the mismatch in length.

12.3 Sewing defects caused by needles

Another main problem found in the sewing room is the defects caused by the interaction between the needle and the fabric. This can cause different types of damages in the fabric, namely:

- Rupturing of fabric yarns due to the collision with the needle tip.
- Friction between the fabric yarns and the needle.
- Thermal aggression of the needle on the fabric yarns.
- Friction between the fabric yarns and the sewing thread, although Blackwood and Chamberlain (1970) found that the damage produced on a fabric does not change significantly with thread in the needle, compared to a situation in which the fabric is stitched without thread.

The interaction with the needle is thus the main aspect to be analysed here, and several researchers have taken up the study on this subject.

Hurt and Tyler (1975) verified that the frictional properties of fabrics are one of the major factors affecting needle penetration forces, and can be improved with appropriate finishing processes. The other main variable is the needle size (Hurt and Tyler, 1976).

This result was confirmed by Leeming and Munden (1978), who found that the fabrics producing high penetration force values were generally those that exhibited more sewing damage using standard sewing tests. These two researchers developed the L&M Sewability Tester (U.S. Patent 3979951, 1976), a testing device in which an unthreaded needle penetrates a fabric sample at 100 penetrations per minute whilst measuring peak penetration force and counting the number of times the force exceeds a pre-defined threshold.

Rocha (1996) used a piezoelectric sensor inserted into the needle bar of an industrial overedge sewing machine, a measurement system that was further developed by Carvalho (2010). With this equipment, it is possible to measure needle penetration and withdrawal force at industrial speeds. Rocha found that, among other material properties, bending rigidity and drape factor were related to needle penetration forces. It was possible to demonstrate the relationship among needle size, sewing speed, fabric finishing and needle tip state with needle penetration forces (Carvalho et al., 2009).

Other variables with an effect on needle penetration force were found by Gurarda and Meric (2005) to be pre-setting temperature and the finishing process in cotton/elastane woven fabrics. These have significant effects on seam performance, needle penetration force and elastane fibre damage.

An important aspect to consider for fabric damage caused by needles is needle heating. Long seams at high speed tend to produce high needle temperature, which can cause damage to the fabrics, especially when sewing synthetic fabrics. Hot needles not only cause damage to the fabric but they may also deteriorate the sewing thread, causing thread breakage or a reduction in the strength of the seams, a problem that is again worse when using synthetic sewing threads. Temperatures above 180 °C have been measured by Liasi et al. (1999) using infrared radiometry. At 140 °C, the threads suffered a reduction of about 40% in breaking strength. The researchers determined that the sewing speed, the type of material being sewn and thread tension were the main factors influencing needle heating.

This problem can be reduced by using lubricants or air cooling of the needle and sewing threads. Some sewing machine manufacturers provide devices for this purpose. A more straightforward solution is to reduce sewing speed and/or thread tension.

Needle manufacturers offer many options to minimise the problems related to needle penetration. The main factor for good performance is the needle point, which can have different shapes according to the materials sewn. The needle points can be divided into two main classes: cloth points and cutting points. The latter are used for leather and other continuous materials that have to be cut to achieve penetration (Figure 12.6). Cloth points are designed to penetrate between yarns of structured fabrics. These can coarsely be divided into round points and ball points. Round points are generally used for woven fabrics, whilst ball points are designed for

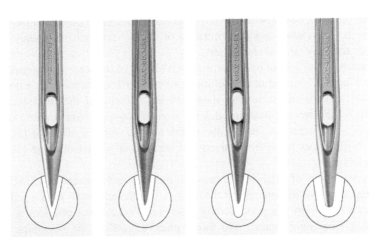

Figure 12.6 Examples of cloth points, from left to right: sharp round point, round point with rounded tip, light ball point and heavy ball point.
From Groz-Beckert.

knitted fabrics. However, many exceptions exist, and often only experience and case-to-case analysis allows an adequate choice of the needle tip. A proper choice of the needle type and size (along with the use of an adequate-quality thread size and type, with good frictional characteristics) is important to avoid a certain type of sewing defect promoted by the needle, while penetrating the fabrics to form the stitch, damaging the fabrics by rupturing the yarns due to contact with the needle tip.

Other characteristics of needles that manufactures try to optimise are the design of the needle's cross-section and the needle's surface finishing. Both have a direct effect on the friction between the fabric and the needle, thus contributing to manage the problem of needle heating. Surface finishing of the needle is also very important to avoid thread breakage during high-speed sewing.

12.4 Material feeding and associated problems

Over the years, sewing machine manufacturers have developed and refined different types of feeding systems adapted to the materials and shapes of the seams to be produced. These allow situations of even feeding, in which the fabric plies are to be fed by the same length, or various types of differential feeding, in which materials are gathered, stretched or otherwise fed differently one relatively to the other(s). To achieve the desired effect or quality of the seam, the machines have to be set by the technicians with seamstresses based on empirical judgement and the quality of the seam is largely dependent on these skills, regarding both machine set-up as well as correct handling of the fabrics being sewn. Still, several defects originated by the feeding system occur, especially in high-speed sewing. Some of these have already been highlighted previously with respect to seam pucker.

Another class of problems is related to the impact on the feed-dog on the presser-foot, with the fabrics in between, which often results in lack of control of the fabrics at high speeds. The causes and consequences of this behaviour have already been the subject of study by several researchers.

The analysis carried out by Eileen Johnson (1973) of the feeding system of a lockstitch sewing machine led to the definition of presser-foot 'bouncing'. Eileen Johnson tested two lockstitch sewing machines and concluded that the stitch length produced by the sewing machine did not remain constant over a range of speeds and presser-foot pressures, partly due to the aforementioned effect of presser-foot bounce. Differences between obtained and nominal stitch densities up to 35–40% were observed.

These variations of stitch length result in irregular and uneven seams, influencing seam properties like the strength of the seam. Other defects, such as skipped stitches, may also arise due to this poor feeding behaviour. Yet another problem is ply-shift, originating *misaligned* seams: two fabric plies, aligned at the beginning and end before sewing, appear misaligned at the end of the seam after sewing. This defect occurs mainly in drop-feed machines, due to the same reasons previously laid out to explain differential fabric stretch, that is, better contact of the feed-dogs with the lower fabric ply. Jana and Khana (2014) found the new X-feed drop-feed system to be more efficient in this regard, too, substantially reducing ply-shift.

Presser-foot *bouncing* is a problem in most of the existing feeding systems, even the more complex ones. Several researchers have investigated this subject, such as Frank and Mo (1974), Matthews and Little (1988), Clapp et al. (1992), Araújo et al. (1992), Rocha et al. (1992a,b, 1996a,b), Bühler and Hennrich (1993, 1994), Chmielowiec and Lloyd (1995), Alagha et al. (1996), Silva (2002) and Carvalho (2010).

Sewing test-rigs have been developed, and some devices to avoid this adverse behaviour have been proposed and tested. Barrett (1992), Barrett and Clapp (1995), Stylios and Sotomi (1995) and Carvalho et al. (2012) have presented concepts for active actuation on the presser-foot, as well as monitoring systems and control strategies for fabric feeding, based on the use of electromagnetic and pneumatic linear actuators. The approach has been to improve feeding dynamics through innovative mechanisms or to control presser-foot force dynamically, on the basis of actual sewing behaviour or on sewing speed. Some of these approaches have been implemented in commercial machines, as will be described later in the chapter.

12.5 Problems in stitch formation

The previous topics have been addressed to the review of the occurrence of seam pucker and other surface distortions, and defects caused by needles and fabric feeding systems. The present section will review the problems related to stitch formation: stitch imbalance, stitch distortion and skipped stitches.

12.5.1 Stitch imbalance

When a machine is set with the right ratio of thread pre-tensions on the threads being processed into the seam and with an absolute value of tension adequate to the materials and objectives of the desired seam type, a stitch is balanced. Stitch balance is fundamental for a quality seam, both from the aesthetic as well as from the functional point of view. Requirements on thread tension may vary according to the seam's final use. A joining seam, for instance, requires higher tension than a seam for edge finishing ('serging' seam), because it has to hold the joined fabric plies firmly together. A seam on a stretch fabric uses lower tension than on a non-stretch fabric, because it has to provide some elasticity. It is up to the operator or technician to set up the machine for each type of sewing condition in order to assure optimal seam balance and tension setting. To achieve optimal balance, the interlacing points of the different threads intervening in a particular stitch type should occur according to the ideal stitch geometry that can be found, for instance, as described in the ISO 4915 standard. For example, a balanced 504 stitch type has the looper threads crossing at the centre of the edge of the material being sewn and the needle thread holding the seam efficiently, assuring the necessary resistance and elasticity (Figure 12.7). The stitch balance is directly related with the pre-tension values set during machine set-up. An indicator for correct stitch balance and tension are the resulting thread

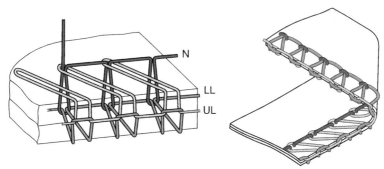

Figure 12.7 Correctly balanced overedge stitch type 504. Left: Schematic representation. (UL, upper looper thread; LL, lower looper thread; N, needle thread). Right: Realistic representation.

consumption values for each thread. Expected values of thread consumption can be computed from the geometrical parameters of the seam, that is, material thickness, stitch length and width (Amann Group, n.d.-a). Incorrect adjustment results in deviation from the predicted consumption values.

When the needle thread loop is too loose, the joined fabric plies are not adequately held together, which may be acceptable in some situations (serging seams) and unacceptable in others (joining seams). If any of the interlacing spots of the threads do not match the pre-defined positions, there is an imbalance between the thread tensions. This will normally influence the seam's performance negatively, and should thus be avoided.

12.5.2 Stitch distortion

Stitch distortion is a problem especially in knitted garments, which affects the size stability of the garments. In this case, the thread tension adjustment is not the cause of the problem, because the stitch shows balance before and after the defect. The distortion may have several causes, such as thread irregularity, improper handling by the operator, sharp speed variations, etc. Being an isolated occurrence, this kind of defect is often very difficult to detect. An isolated stitch distortion is shown in Figure 12.8.

12.5.3 Skipped stitches

One of the most obvious, serious and sometimes difficult-to-detect defects are skipped stitches. Besides being aesthetically unpleasant, they represent a weak point in the seam. A skipped stitch occurs when one or more of the interlacements between sewing threads fail to happen (Figure 12.9).

The causes for skipped stitches include (Carvalho, 2003):

- Sewing threads with inadequate twist amount and/or direction, affecting the formation of loops necessary for the interlacement of the sewing threads (in general sewing machines use threads with a 'Z' twist).
- Inadequate needle diameter, thread linear density or a combination of both.

- Machine incorrectly threaded.
- Damaged or wrongly positioned needle.
- Needle and loopers at inadequate distance, damaged or not synchronised.

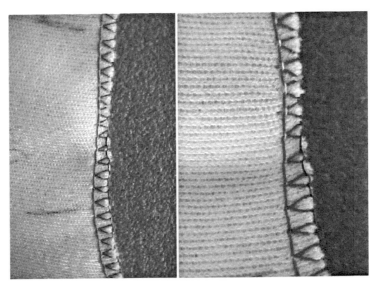

Figure 12.8 Distorted stitch.

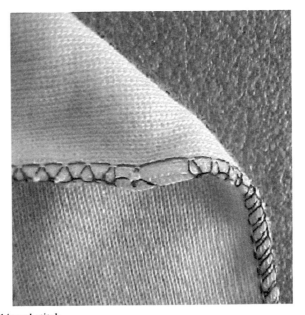

Figure 12.9 Skipped stitch.

12.6 Thread breakage

Thread breakage can be an annoying problem in the sewing room, although it is often quite easy to avoid. Constant thread rupture not only causes productivity loss due to the time needed to re-thread the machine: in some situations, the whole seam has to be remade, and in other, the interruption of the seam is an unrepairable defect (e.g. top-stitching a shirt collar).

As seen before, thread breakage may result from overheated needles due to thermal aggression. Also, bad-quality or worn-out needles may cause problems due to excessive friction. Very often the problem lies in the machine. Incorrect threading or defects in the machine thread guides cause the thread to weaken and fail. The correct setting of the synchronisation and distance between needles and loopers, or other stitch formation elements, are also important.

Thread breakage is more likely to occur at high speeds and when using higher thread tensions. Reducing these two factors certainly minimises the problem, but due to productivity or functional requirements it cannot always be done.

Most often, the problem resides on the quality of the sewing threads. The appeal to save on thread costs does not normally pay off. The cost on thread in a garment typically does not exceed a few percent of the total price. Productivity loss due to thread breakage, besides non-repairable defects, is potentially more costly. Moreover, a good-quality thread is an asset for strong, stable and durable seams.

A summary of the general sewing defects mentioned throughout the chapter, their causes and possible recommended solutions can be found in Table 12.1.

12.7 Future trends

Many of the problems described in the previous sections result from adverse process conditions that cause systematic or random defects. Poor properties of the materials being processed – fabrics and threads; wearing-out of the sewing machine's components or needles; incorrect settings or bad choice of components such as feed-dogs, presser-feet or needles; all of these can be the source for the problems described. The machine operators and quality control team have to be alert to detect these situations as early as possible, because often these defects go unnoticed for whole batches.

Traditional sewing machines are blind regarding the process, unable to detect fault situations. The set-up of the machines is also carried out based on the technician's experience and a trial-and-error process. Until now, no quantitative feedback method of the machine during its operation exists. Hence, the operators and the quality control team can resolve the problems from their experience.

Machines with more sensing and control capabilities would certainly ease the work of operators and technicians and assure higher levels of quality and control over the produced seams. This is especially important for seams in the context of technical textiles, in which specific functional requirements may be imposed. It should be emphasised that monitoring and controlling of the sewing process itself is meant,

Sewing-room problems and solutions 329

Table 12.1 Summary of sewing defects, causes and possible remedies

Sewing defects	Causes	Remedies
Seam pucker	Feeding system (differential fabric stretch or feed pucker)	Proper adjustment of the stitch length, sewing threads and needles, according to the type of fabric (structural jamming appears mainly in dense, lightweight fabrics)
Other fabric undulations	Dimensional change of sewing threads or tension pucker	Correct adjustment of thread tensions
	Dimensional change of fabrics	Proper selection of the feeding system (on knits and stretch wovens, use differential feed and minimum of pressure-foot pressure)
	Differential fabric dimensional instability (fabrics recovery after sewing, or after pressing and washing)	Reduction of sewing speed
	Structural jamming (insertion of sewing threads in fabrics)	Better fabrics handling
	Pattern mismatch	Improve the finishing of the fabrics to be sewn
		Solve the cutting inaccuracy that causes the mismatch in length of patterns
Rupture of fabric yarns	Needle choice (tip and cross-section)	Proper selection of quality and appropriate needles (type and size); needle's cross-section and surface finishing also play an important role
Thermal aggression on the fabrics	Needle penetration (large penetration forces tend to produce more sewing damage)	Proper selection of quality threads (with good frictional characteristics)
	Friction between the needle and fabric yarns	Use of adequate lubricants or air cooling on the needles and sewing threads
	Friction between the sewing threads and fabric yarns	Reduction of sewing speed and thread tension
Stitch length variations (irregular and uneven seams)	Inadequate use of feeding systems (producing non-uniform fabric feeding action)	Correct selection of the sewing machine feeding system
Ply shift	Empirical setting of the sewing machine feeding system	Reduction of the presser-foot force and use of appropriate presser-feet and feed-dogs

Continued

Table 12.1 Continued

Sewing defects	Causes	Remedies
Folds Feed pucker	Presser-foot force; sewing speed Fabric properties Poor fabric handling	Correct setting of the machine for the sewing operation Correct handling of the fabrics being sewn Reduction of sewing speed
Stitch formation defects: Stitch imbalance Stitch distortion Skipped stitches	Fabric properties Stitch type and use of improper threads Use of inadequate (or damaged) needles Sewing thread tension Machine incorrectly threaded or with the needle and loopers at inadequate distance or not synchronised Poor fabric feeding Improper fabric handling Use of rapid sewing speed variations	Quality (size and type) of the sewing threads for the fabrics Correct adjustment of thread pre-tensions and use of adequate thread tensions to the fabrics being sewn and seam type Selection of adequate needles (type and size) Machine correctly threaded and adjusted (loopers, feeding system and speeds) Correct and careful handling of the fabrics being sewn Reduction of sewing speed
Thread breakage	Incorrect threading of the sewing machine Bad-quality or worn-out needles Improper sewing machine thread guides (missing thread guides or with sharp edges) Synchronisation and distance between needles and loopers High sewing thread tension Friction between the needles and the sewing threads with the fabric yarns (causing thermal aggression to the fabrics due to overheated needles) Bad-quality sewing threads or incorrect lubrication	Proper selection of quality threads (with good frictional characteristics, such as poly cotton core-spun sewing threads) Better lubrication of sewing threads Proper selection of needles Proper machine adjustments (regarding all the machine's stitch formation components) Reduction of sewing speed and thread tensions

not material handling, although some relation might exist. Material handling is an entirely different, complex subject.

The objective evaluation of the textile materials to be sewn, e.g. woven fabrics could improve performance in the sewing process.

The use of Kawabata evaluation system (KES) or fabric assurance by simple testing (FAST) evaluation systems allows objective measurement of factors such as extensibility, bending and shear properties, compression, friction, surface irregularity and abrasion — factors that, together with fabric construction, have a direct influence on the behaviour of materials as they are handled and processed. More details of this are described in Chapter 3.

Studies can be designed in order to establish the relationship between these parameters and the performance of the sewing process. The output of these studies including this knowledge previously acquired, can lead to more adequate setting-up of the sewing machines, by proper adjustments or selection of needles and sewing threads. In addition, information on how to improve the characteristics of the textile fabric, by adjusting the manufacturing process, with different finishing treatments, for example, could be used.

The principles of complete intelligent manufacturing systems were defined by Stylios (1996) by pointing out four main tools:

- A fabric measurement system to determine the properties that completely define each material.
- A sewability/tailorability prediction system to predict the interaction between the machines and the material, guiding the set-up of the machines and supporting the design of the materials for optimum sewability/tailorabilty.
- Sewing machines that can set up themselves and dynamically control their mechanisms as a function of the material properties, able to detect and correct fault situations automatically.
- Autonomous or centrally integrated systems with self-learning capabilities able to adapt the control models dynamically.

In the last decades, much work at the academic level has been pointing the way to intelligent textile and garment manufacture. The market has also seen some successful presentations in this context, although only a small step towards fully integrated process control has been taken. The "Docu-Seam System" by German sewing manufacturer Pfaff (Pfaff, 2014) is a sewing system that measures thread force, stitch length setting and the backtack lever of the machine. It was originally developed for production of the seams where the side airbags are deployed. A backtack during this seam, incorrect stitch length setting, excessive tension, skipped stitches or other factors may affect the correct operation of this safety equipment. Another example of a system offered commercially where a little more control over the process is exerted, in this case regarding the feeding system, is Pfaff's "SRP-system" (Speed Responsive Presser-Foot). This system adjusts force on the presser-foot according to sewing speed, producing more regular stitch length and reducing feeding pucker (Pfaff, 2009).

More comprehensive and integrating approaches have been followed by several academic researchers, aiming to develop both offline sewability testing/process planning tools, as well as inbuilt intelligent and self-learning sewing machines and control systems. Some examples include "neuro-fuzzy control systems" that are

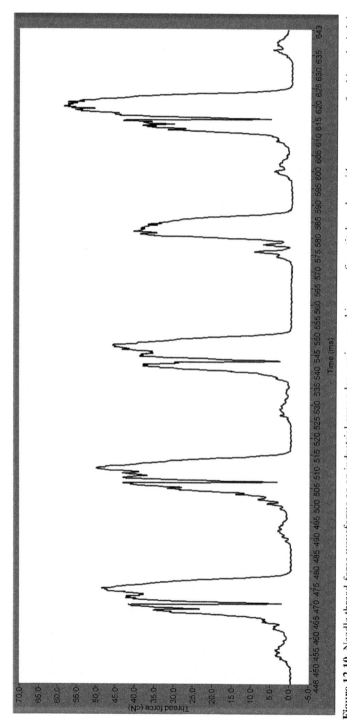

Figure 12.10 Needle thread-force waveforms on an industrial over-edge sewing machine over five stitch cycles, with occurrence of a skipped stitch in the fourth cycle.

able to set the machines automatically according to material properties (Stylios and Sotomi, 1995); the control of the force on the machine's presser-foot adapting to the material being sewn, using information from sensors (Carvalho et al., 2012); optimum thread tension control, thread tension balancing and stitch formation defect monitoring (Carvalho and Ferreira, 2006); and integrated software for testing and development of machine control/monitoring algorithms and material-testing tools. It has been possible to demonstrate skipped stitch and stitch deformation detection, thread balance and tension prediction and monitoring, presser-foot bouncing detection, needle penetration and withdrawal force measurement, among other functionalities (Carvalho et al., 2008). Examples are given in Figures 12.10 and 12.11.

Figure 12.10 presents a needle thread-force signal measured on an industrial overedge sewing machine during the occurrence of a skipped stitch. Observing the peaks of tension in the sewing threads involved in stitch formation, it is quite straightforward to detect the skipped stitch. It has been shown that even a minor sporadic stitch distortion, like the one presented in Figure 12.8, is easily detected by processing the peak values and relations between the individual peaks in a stitch.

Figure 12.11 shows presser-foot vertical displacement during two stitch cycles on the same machine at two different sewing speeds, measured with an linear variable differential transformer (LVDT) attached to the machine.

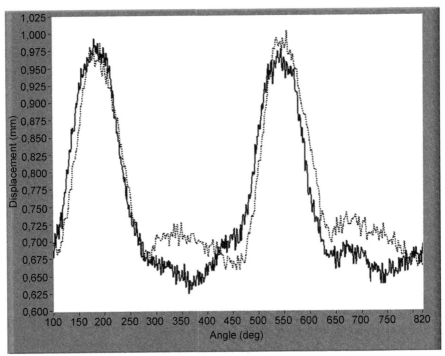

Figure 12.11 Presser-foot displacement waveforms at low-speed (2000 spm (stitches per minute), solid line) and high-speed (4700 spm, dashed line) situations.

It can be observed that the presser-foot describes different trajectories depending on sewing speed. At the higher speed, when the feed-dog lowers below the throat plate (approximately between angles 250° and 300° of the stitch cycle), the presser-foot bounces up again and eventually loses contact with the fabric during needle penetration.

12.8 Conclusions

The introduction of advanced technologies in industrial practice is certainly a progressive process. Whilst in some cases they involve costly equipment, demanding careful cost-benefit analysis, in other cases they have still to be further developed and tested. Nevertheless, many of them will see common use in the future. The need to assure reliability and predictability in the behaviour of technical textiles, providing functionality and safety for users, is raising the demand for these new tools.

12.9 Sources of further information and advice

Reading on practical and theoretical aspects of industrial sewing can be found in many textbooks and scientific publications. The references given are good sources for the attentive reader.

Another interesting source of good advice on sewing problems is the technical documentation offered by sewing machine, needle and thread manufacturers. A selection of some interesting websites follows:

Pfaff-Industrial, http://www.pfaff-industrial.com (section 'Service and Support')
Pegasus, http://www.pegasus.co.jp/ (section 'Customer Information')
Duerkopp-Adler, http://www.duerkopp-adler.com (section 'Support')
American & Efird thread supplier, http://www.amefird.com/ (section 'Technical Tools')
Amann sewing threads, http://www.amann.com/ (section 'Service')
Schmetz needles, http://www.schmetz.com/ (section 'ServiceHouse')
Groz-Beckert needles, http://www.groz-beckert.com (section 'Products and Services')

References

Amann Group, n.d.-a. Determining Your Sewing Thread Requirements. Retrieved in October 4, 2010 from: http://www.amann.com/en/download/industrial-sewing-threads.html.
Amann Group, n.d.-b. Preventing Seam Pucker, in Service & Technik -Information for the Sewing Industry. Retrieved in April 18, 2014 from: http://www.amann.com/fileadmin/download/naehfaden/b_nahtkraeuseln_EN.pdf.
Araújo, M., Little, T.J., Rocha, A.M., Vass, D., Ferreira, F.N., 1992. Sewing Dynamics: Towards Intelligent Sewing Machines, NATO ASI on Advancements and Applications of Mechatronics Design in Textile Engineering, Side, Turkey.

Alagha, M.J., Amirbayat, J., Porat, I., 1996. A study of positive needle thread feed during chainstitch sewing. J. Textile Inst. 87 (Part 1, No. 2), 389–395.

Blackwood, W.J., Chamberlain, N.H., 1970. The Strength of Seams in Knitted Fabric. Technical Report, No.22. Clothing Institute.

Bühler, G., Hennrich, L., 1993. Control 'On-Line' de la Formación de las Costuras. Punto Técnica y Moda 11 (4), 239–242, e244.

Bühler, G., Hennrich, L., 1994. Control 'On-Line' de la Formación de las Costuras. Punto Técnica y Moda 12 (1), 42–46 and 51.

Barrett, G.R., 1992. Sewing Dynamics: The Autodamp, Technology for the Elimination of Presser Foot Bounce. North Carolina State University Leaflet, 7/92.

Barrett, G.R., Clapp, T.G., 1995. Coprime factorization design of a novel Maglev presser foot controller. Mechatronics J. 5 (2/3), 279–294.

Carvalho, H., 2010. Optimisation and Control of Processes in Apparel Manufacturing: New Tools for Efficiency and Quality Assurance in Industrial Sewing Technology. VDM Verlag Dr Müller, Saarbruecken.

Carvalho, H., Rocha, A.M., Monteiro, J.L., 2009. Measurement and Analysis of Needle Penetration Forces in Industrial High-speed Sewing Machine, vol. 100. Taylor and Francis, London. No.4, pp. 319–329.

Clapp, T.G., Little, T.J., Thiel, T.M., Vass, D.J., 1992. Sewing dynamics: objective measurement of fabric/machine interaction. Int. J. Clothing Sci. Technol. 4 (2/3), 45–53.

Chmielowiec, R., Lloyd, D.W., 1995. The measurement of dynamic effects in commercial sewing machines. In: Proceedings of The Third Asian Textile Conference, vol. II, pp. 814–828.

Carvalho, H., Silva, L.F., Rocha, A., Monteiro, J., 2012. Automatic presser-foot force control for industrial sewing machines. Int. J. Clothing Sci. Technol. 24 (1), 20.

Carvalho, M.A.F., 2003. Estudo das Relações entre os Parâmetros de Controlo, Propriedades dos Materiais e Condições de Regulação numa Máquina de Costura Corta-e-Cose' (Study of the Relationship between Control Parameters, Material Properties and Set-up Conditions in an Overlock Sewing Machine) (Ph.D. thesis), University of Minho, Portugal.

Carvalho, M.A.F., Ferreira, F.B.N., 2006. High performance. In: Sewing – Guaranteeing Seam Quality through Control of Sewing Dynamics, Textiles for Sustainable Development. Nova Science Publishers, Inc, ISBN 1-60021-559-9, pp. 161–171 (Chapter 14).

Carvalho, H., Silva, L.F., Monteiro, J., Rocha, A., 2008. Parameter monitoring and control in industrial sewing machines – an integrated approach. In: Proceedings of the IEEE International Conference on Industrial Technology, CD-Rom. IEEE, Chengdu, China, ISBN 978-1-4244-1706-3.

Dorkin, C., Chamberlain, N., 1961. Seam Pucker, its Cause and Prevention. Technological Report No.10.

Frank, T.F.P., Mo, E.S.M., 1974. A simple test for assessing presser foot bounce. Clothing Res. J. 2 (2), 81–84.

Gurarda, A., Meric, B., 2005. Sewing needle penetration forces and elastane fiber damage during the sewing of cotton/elastane woven fabrics. Text. Res. J. 75 (8), 628–633.

Hurt, F.N., Tyler, D.J., 1975. Seam Damage in the Sewing of Knitted Fabrics. II – Material Variables. HATRA. Research Report, No.36.

Hurt, F.N., Tyler, D.J., 1976. Seam Damage in the Sewing of Knitted Fabrics. III – the Mechanism of Damage. HATRA. Research Report, No.39.

ISO 4915, 1991. Textiles – Stitch Types – Classification and Terminology. ISO Standard.

Jana, P., Khana, N.A., 2014. The sewability of lightweight fabrics using X-feed mechanism. Int. J. Fashion Des. Technol. Educ. Published online: April 17, 2014.

Johnson, E.M., 1973. Some factors affecting the performance of high speed sewing machines. Clothing Res. J. 1 (1), 3–35.

Leeming, C.A., Munden, D.L., 1978. Investigations into the factors affecting the penetration force of a sewing needle in a knitted fabric and its relationship with fabric sewability. Clothing Res. J. 6 (3), 91–118.

Leeming, C.A., Munden, D.L., 1976. Testing Fabric Sewing Properties. US Patent 3979951.

Liasi, E., Du, R., Simon, D., Bujas-Dimitrejevic, J., Liburdi, F., 1999. An experimental study of needle heating in sewing heavy materials using infrared radiometry. Int. J. Cloth Sci. Tech. 11 (5), 300–314.

Matthews, B.A., Little, T.J., July 1988. Sewing dynamics, Part I: measuring sewing machine forces at high speeds. Text. Res. J. 58, 383–391.

Pfaff Industrial, 2009. What are the causes of seam pucker retrieved May 13, 2014 from: http://www.pfaff-industrial.com/pfaff/en/service/faqs/generalsewing/seampucker/view?searchterm=Speed responsive.

Pfaff Industrial, 2014. Pfaff Docu-Seam System retrieved May 13, 2014 from: http://www.pfaff-industrial.com/pfaff/en/product-range/Engineered%20workplace/3745%20Premium.

Rocha, A.M., 1996. Contribuição para o Controlo Automático dos Parâmetros de Costura: Estudo da Dinâmica da Penetração da Agulha e da Alimentação do Tecido (Contribution to automatic control of sewing parameters: Study of needle penetration and fabric feeding dynamics) (Ph.D. diss.). University of Minho, Portugal.

Rocha, A.M., Sousa, E.J., Lima, M., Araújo, M., Outubro 1992a. Estudo da Dinâmica do Processo de Costura numa Máquina de Corta-e-Cose. Revista de Robótica e Automatização 9, 12–17.

Rocha, A.M., Lima, M., Sousa, E., Araújo, M., 1992b. Evaluation of sewing performance and control of sewing operation. In: Second International Clothing Conference, University of Bradford, UK.

Rocha, A.M., Lima, M., Ferreira, F.N., Araújo, M., 1996a. Developments in automatic control of sewing parameters. Text. Res. J. 66 (4), 251–256.

Rocha, A.M., Ferreira, F.N., Araújo, M., Monteiro, J., Couto, C., Lima, M., 1996b. Mechatronics in apparel: control, management and innovation on the sewing process. In: Proceedings of the Mechatronics'96 Conference, vol. II. Universidade do Minho, Guimarães, Portugal, pp. 109–114.

Silva, L.F., 2002. Estudo de Mecanismos Alternativos de Controlo do Sistema de Alimentação de Máquinas de Costura Industriais (Study of Alternative Control Systems for the Feeding Systems of Industrial Sewing Machines) (Ph.D. thesis). School of Engineering, University of Minho, Guimarães, Portugal.

Stylios, G., Sotomi, J., 1995. A neuro-fuzzy control system for intelligent overlock sewing machines. Int. J. Cloth Sci. Tech. 7 (2/3), 49–55.

Stylios, G., 1996. Principles of intelligent textile and garment manufacturing systems. Assembly Autom. 16 (3), 40–44.

Alternative fabric-joining technologies

E.M. Petrie
Independent Consultant, Cary, NC, USA

13.1 Alternatives to sewing

13.1.1 Competitive methods of joining garments

Garment manufacturers are developing innovative clothing designs that are placing significant and critical demands on the way garments are joined. Manufacturers are also facing increased economic and environmental pressures to increase processing speeds, lower costs, add value to finished products, and amortize capital equipment over more than one use. As a result, these manufacturers are seeking new and alternative methods for joining garments.

Garments are made by joining several patterned substrates together. These substrates can be various types of woven or nonwoven fabric that are joined together or to accessories that could include linings, buttons, zippers, tapes, and decorative pieces. The quality of the joint or seam in terms of strength, durability, permeability, flexibility, and comfort can have a significant effect on the value of the final product.

Today, the most dominant method of joining these materials is by sewing or stitching. In addition to sewing, numerous other joining methods are available for garments. They can be categorized by the basic mechanisms that bring about the joining, as shown in Figure 13.1. Of these alternative processes, adhesive bonding and thermal welding are gaining the most prominence.

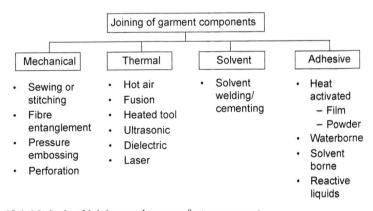

Figure 13.1 Methods of joining used to manufacture garments.

The traditional sewn seam produces high strength and low stiffness, which makes it ideal for most common clothing manufacturers. Disadvantages with traditional sewing include discontinuous joints producing perforated seams, sewing thread deteriorating over time, thicker material at the point of joining, and production speed limitations. As a result, innovative alternative processes have been and are continuing to be developed. Such processes have a significant commercial impact on specific applications, and they are rapidly gaining broader market share on realization of their added value potential.

The alternative methods next to sewing include adhesive bonding, conventional thermal (e.g., hot air and heated tool) welding, and advanced thermal (e.g., ultrasonic or laser) welding. These joining methods are the focus of this chapter. A general description of these processes, along with applicable substrates and applications, is provided in Table 13.1.

Adhesives can set or cure by carrier (solvent or water) evaporation, chemical reaction, or thermal activation. Chemically reactive adhesives solidify primarily by a chemical reaction of one or more components in the adhesive formulation. It should be noted that solvent welding cementing processes or solvent-borne adhesives are in disfavor due to environmental, safety, and health concerns and regulations. As a result, waterborne adhesives and heat-activated adhesives are replacing solvent-based adhesives in many applications.

In thermal-welding processes, adhesion occurs by melting the substrate surfaces so that they flow into one another and then set on subsequent cooling. Heating can be achieved using a range of different methods such as conventional external heating sources, dielectric or high-frequency welding, ultrasonic welding, and laser-assisted welding.

Arguably, thermal-welding processes can be considered a subset of adhesive bonding. With thermal-joining processes, the substrate itself (in molten form) acts as the adhesive. Once the substrate is liquefied, the principles of adhesion determine the strength and durability of the bond. In addition, thermal methods can be used to activate a hot-melt adhesive rather than the textile substrate.

13.1.2 Selecting a joining process

Usually, the decision of which joining process to use involves several trade-offs. An analysis of the ultimate requirements is crucial in identifying probable joining process. When this is performed, the potential for using adhesives or welding over other methods of joining becomes apparent. The selection of a joining process is a critical factor that will influence the entire manufacturing process as well as the final performance of the garment. The selection will be dependent on various production, performance, and consumer requirements including:

- The type and nature of substrates to be bonded.
- Availability of manufacturing equipment.
- Cost (material and process).
- Production speed necessary for producing a practical and economic joint.

Table 13.1 General description of adhesive-bonding and welding processes for joining garments

Process	Description	Substrates/applications
Adhesive bonding	Adhesive bonding uses a separate material at the joint interface, which binds either chemically or mechanically to the substrate. The adhesive may be chemically or thermally reactive or may bond on evaporation of a carrier (water or solvent).	Most synthetic or natural fabrics (woven and nonwoven) independent of synthetic fibre content. It can also be used to join garments to nontextile accessories (glitter, foil, etc.).
Conventional thermal welding	Welding is a thermal process requiring melting of fabric materials. A separate heat-activated adhesive material can also be used. Heating is achieved by direct contact of the fabric with a heated-tool surface or hot air.	Fully or partially synthetic fabrics (woven and nonwoven) with thermoplastic components that are chemically and physically compatible when fused together.
Advanced thermal welding	Advanced welding is similar to conventional thermal welding except that the heating is achieved by indirect contact of the fabric with a source such as ultrasonic horn, electromagnetic field, or laser. A separate heat-activated adhesive material can also be used.	Fully or partially synthetic fabrics (usually woven) with thermoplastic components that are chemically and physically compatible when fused together.

- Strength and the expected stresses that the joint will face in service.
- Durability and the expected environments that the joint will face in service.
- Aesthetic appeal.
- Wearing comfort.

In most cases, the garment joint must be highly resistant to exposure to sunlight. The joint must be sufficiently soft to provide good hand and drape, yet must be tough and strong enough to have good life and be resistant to scuffing and abrasion. The joint must adhere even after periods of flexing in cold or heat, and high levels of moisture resistance are necessary in outdoor clothing.

Consumers purchase apparel on the basis of hand, drape, colour, texture, and fit. They expect long wear without scuffing or picking and no delaminating or opening of seams. In many cases, the apparel will be dry-cleaned, and consumers will expect the garments to return from the cleaners unchanged. These consumer requirements can be translated into requirements for the adhesive or welded garment joint.

It is usually necessary to compromise on some of these requirements when selecting a practical joining system. Some properties and characteristics that are desired will be more important than others, and a thoughtful prioritization of these criteria will be necessary in selecting a process.

Once sewing, adhesive bonding, or thermal welding is selected based on the requirements noted above, the garment manufacturer must define the specific process to be used (e.g., adhesive bonding with thermally activated adhesive films, ultrasonic welding, heated-tool welding, and so forth). Adhesive-bonding and thermal-welding processes are difficult to define and control because many factors must be considered, and there is no universal material or process that will fulfill every application.

To achieve optimum economics, reliability, and performance; one must carefully plan every stage of the joining process. One needs to optimize the entire bonding process and not merely a single part of the process. Considerations need to be given concurrently to the substrates, joint design, surface pretreatment, quality control, application and curing methods, as well as other subprocesses (Figure 13.2).

When the joining process is first considered by the designer or manufacturer, seldom is anything firmly fixed except possibly the substrates that will be used. Alternative substrates, processes, etc. should always be considered. For example, a slight

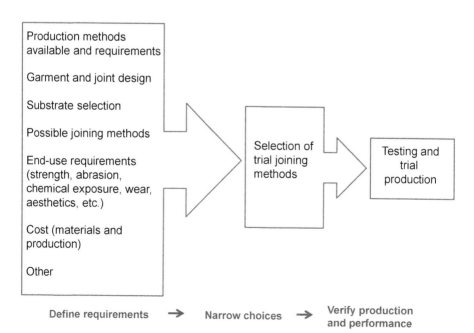

Figure 13.2 Adhesive selection requires a number of early considerations.

change in the application methods or the type of substrate could open the door for consideration of an entirely new joining technology.

13.2 Adhesive bonding

13.2.1 Adhesion

Adhesive bonding uses a separate material between the two layers of materials, and bonding occurs through the effects of heat, pressure, chemical reaction, or their combination. The type of adhesive that is used will primarily depend on the type of textile to be joined and the end-use application. The specific process and form of adhesive (liquid, film, powder) to be employed will generally be determined by the type of adhesive and the production speed that is required.

Adhesive bonding is often used to manufacture seamless garments. However, the technology is also used to seal the insertion of holes made by needles, provide extra strength to seams, create stiffness, and to waterproof seams. Applications are primarily in sports, medical, and protective clothing as well as shoemaking and the manufacturing of other apparel accessories. Adhesives can be used on synthetic or natural fabrics, and they can be used to make joints between dissimilar materials, such as fabrics to metals or fabrics to foams.

The advantages and disadvantages of garment adhesive bonding are summarized in Table 13.2. While adhesives have several important advantages, the material cost, production requirements, and additional weight of the adhesive can be disadvantageous in certain applications.

There are several theories that explain the phenomenon of adhesion. The most common theories are described in Table 13.3. No single theory explains adhesion in a general or universal way. Some theories are more applicable for certain substrates and applications, and others are more appropriate for different circumstances. The theories that primarily relate to garment joining are absorption, chemical bonding, and mechanical joining.

13.2.2 Adhesive materials

A variety of different chemical-base polymers are available for adhesive formulation, each with their own set of properties. For example, polyurethanes (PUs) are flexible and have good adhesion, but some may discolour. Polyvinyl acetate emulsion adhesives are relatively inexpensive, but they are fairly brittle and have limited water resistance.

Base polymers are classified as being either thermoplastic or thermosetting. Thermoplastic polymers are polymers that melt when heated and then resolidify when cooled. Thermoset polymers are those that do not melt when heated, but at sufficiently high temperatures they exhibit creep under load and even decompose. Polymers gain their thermosetting nature by a process called cross-linking, where the adhesive molecules chemically react with one another to form a three-dimensional network. The

Table 13.2 Advantages and disadvantages of adhesive bonding

Advantages	Disadvantages
• Provides large stress bearing area. • Provides excellent toughness and abrasion resistance. • Can be used to provide a moisture-resistant barrier. • Can be made to be thermally and/or electrically conductive. • Joins all shapes and thicknesses. • Joins any combination of similar or dissimilar materials. • Provides smooth contours. • Provides continuous, nonperforated seams. • Heat, if required, is too low to affect heat-sensitive substrates. • Provides attractive strength-to-weight ratio. • Often less expensive and faster than other methods of joining.	• Surface must be carefully cleaned or chemically treated (if a low-surface-energy textile). • Long cure times may be needed and fixturing could be necessary. • Limitation on upper continuous operating temperature (generally 175–200 °C). • Heat and pressure may be required for cure. • Rigid process control usually necessary. • Inspection of finished joint is difficult. • Useful life depends on the service environment. • Environmental, health, and safety considerations are necessary. • Special training may be required; selection of the optimal adhesive and bonding process may be difficult and relates only to the intended application.

base polymers that are more commonly used in textile adhesives and their properties are summarized in Table 13.4.

Textile adhesives, like other adhesives, are generally highly compounded formulations, and different base polymers may be blended to produce specific properties. For example, a hot-melt adhesive may consist of a blend of co-polyesters and modified polyethylene.

Modern-day adhesives are often fairly complex formulations of components that perform specialty functions. Very few polymers are used without the addition of some modifying substance such as a plasticizer, tackifier, or inert filler. The selection of the actual ingredients will depend on the end-properties required, the application and processing requirements, and the overall cost target of the adhesive.

The industry has settled on several common methods of classifying adhesives, such as method of reaction, chemical composition (as indicated by the base polymer in the formulation), and end-use application. Adhesives can also be classified by form and reaction mechanism, as shown in Table 13.5.

Some adhesives solidify simply by the evaporation of a carrier liquid (water or solvent). Others harden as a result of going from a molten liquid to a cooled solid. Still others solidify by means of a chemical reaction. Some adhesive systems may require several mechanisms to harden. For example, thermosetting waterborne adhesives require the water carrier to first evaporate and then chemical cross-linking occurs before a final bond is achieved.

Table 13.3 Common adhesion theories

Theory	Operating mechanism
Adsorption	Adhesion results from molecular contact and the surface forces that develop. For these forces to develop, the adhesive must make molecular contact with the substrate surface—a process known as "wetting."
Chemical bonding	Adhesion occurs due to the formation of chemical bonds across the interface. For this to occur, the substrates and adhesive have to be chemically reactive or have "functional" groups in their molecular make-up.
Mechanical	The adhesive must penetrate the porosity or roughness on the surface, displace the trapped air at the interface, and lock-on mechanically to the substrate.
Electrostatic	The electrostatic theory states that electrostatic forces are formed at the adhesive–adherend interface. These forces account for resistance to separation.
Diffusion	Adhesion arises through the interdiffusion of molecules in the adhesive and adherend. Solvent or heat welding of thermoplastic substrates is considered to be due to diffusion of molecules.
Weak boundary layer	Adhesion can be prevented if there is a weak boundary layer between the adhesive and the substrate. Weak boundary layers can be corrosion, contamination, moisture, etc. In these cases, failure occurs within the boundary layer and not the adhesive.

The most common textile adhesives are available as solutions in water or solvent, as dispersions in water, or as solids that melt under the application of heat. However, 100% reactive liquid adhesives such as epoxies and moisture-cured urethanes are also used.

The mechanism of reaction is an important determinant in the adhesive-selection process. For example, if the processing requirements dictate that only a room-temperature-curing adhesive system can be used, then temperature resistance may be sacrificed. There are very few room-temperature-curing adhesives that exhibit good elevated-temperature resistance because of the lack of extensive cross-linking and the weaker nature of the polymers formed with room-temperature-curing formulations.

Most textile adhesive formulations are available in a waterborne liquid or solid (film or powder) form. When supplied as a film, the adhesive is unsupported, reinforced with fabric or nonwoven, or coated on both sides of a carrier material. These film adhesives are made flowable by heating to a melt and then gain strength on cooling back to a solid form. Film adhesives offer convenience, since no coating operation is required by the user, energy (drying costs) is minimized, and waste is eliminated;

Table 13.4 Properties of textile adhesives formulated from various base polymers

Base polymer	Types available	Properties	Applications
Acrylics	• Solutions and aqueous emulsions • Both thermoplastic and thermoset formulations available	• Very wide adhesion range • Excellent resistance to discolouration, light, and oxidation • Curing types are available that have wash and dry-cleaning resistance	• Pressure-sensitive adhesives • Laminating adhesives
Vinyl acetate	• Hot-melt and aqueous forms • Thermoplastic	• Good adhesion range • Flexibility • Limited heat and chemical resistance • Relatively low cost	• Nonwovens • Temporary adhesives
Styrene-butadiene	• Solvent solutions; aqueous dispersions (latex), hot melt (block copolymers) • Primarily thermoplastic but some thermoset formulations available	• Properties vary with type • Some heat-cure formulations are available • Good heat and chemical resistance • Some types are prone to oxidation	• Personal hygiene products • Textile, nonwoven lamination
Polyamide	• Hot melt • Thermoplastic	• Good heat and chemical resistance	• Fusible interlining applications • Powders for outerwear

Polyester	• Hot melts, solvent solutions • Thermoplastic; solvent solutions can be either thermoplastic or thermoset	• Chemically related to the polyester fibre • Good water and solvent resistance • Reactive types have good heat and chemical resistance • Good adhesion, especially to polyester fabrics and films	• Interlinings • Laminating adhesives
Polyurethane (PU)	• One- and two-component liquid (100% solids, solvent-based, waterborne), reactive hot melts, films • Both thermoplastic and thermoset formulations available	• High adhesive strength to many materials • Good elongation, toughness, and abrasion resistance • Good water and solvent resistance • Reactive types have good heat resistance • Some types yellow on exposure to ultraviolet	• Polyether-PU films for Spandex and hydrophobic textiles • Reactive PU hot melt for breathable membranes like Sympatex, nonwovens, and foils • Films for textile seam sealing
Polyvinyl chloride	• Solvent solutions, 100% solids plastisols • Thermoplastic	• Limited adhesion • Good water resistance	• Paste dot adhesive for nonwovens
Polyolefins	• Films, reactive hot melts • Thermoplastic	• Good adhesion • Good resistance to moisture • Lower density (less adhesive and weight)	• Personal hygiene products • Textile, nonwoven lamination

Table 13.5 Classification of adhesives by form and reaction mechanisms

Adhesives that harden by chemical reaction	Adhesives that harden by solvent or water loss	Adhesives that harden by cooling from the melt
• Two-part systems • Single-part, cured via catalyst or hardener • Moisture-curing adhesives • Radiation (light, ultraviolet, electron beam, etc.)-curing adhesives • Adhesives catalyzed by the substrate • Adhesives in solid form (tape, film, powder, etc.)	• Contact adhesives • Pressure-sensitive adhesives • Reactivatable adhesives • Resinous adhesives	• Hot-melt thermoplastic adhesives • Hot-melt applied pressure-sensitive adhesives • Powder and film reactive adhesives

but film adhesives have a higher material cost. Because film adhesives are available as continuous stock, they are ideal for laminating fabrics. The advantages and disadvantages of the various adhesive forms are summarized in Table 13.6.

Several new approaches have been introduced into the garment industry primarily in response to the need for faster-setting adhesives. These include reactive hot-melt systems that cross-link on exposure to relative humidity in the ambient environment and solventless liquid adhesives that cure on exposure to ultraviolet (UV) light.

Collano, for example, has developed a series of moisture-curing PU hot-melt adhesives for textiles (Collano, 2013). These provide high initial strength due to their hot-melt mechanism, but they cross-link to a heat- and moisture-resistant adhesive on exposure to ambient moisture once the joint is made. As a result, these adhesives have excellent resistance to dry-cleaning, washing detergents, and even sterilization processes. In addition, PU adhesives have low processing temperatures (90–140 °C) and are ideal for heat-sensitive substrates.

Light-curing adhesives based on acrylic polymers have become prime candidates for fast adhesive-bonding processes given their ability to cure within milliseconds. Light-curing adhesives contain photoinitiators, which absorb light to start their deterioration and also kick off the curing reaction. Depending on the intensity and range of the light source, full-adhesive cure can easily be achieved within seconds as long as the optimal light wavelength is present for the specific photoinitiator used.

13.2.3 The bonding process

It must be realized that the adhesive or sealant itself is only one factor in developing a successful adhesively bonded joint. The manufacturing methods used in producing an assembly will also determine the initial degree of adhesion and service characteristics

Table 13.6 Advantages and disadvantages of various types of textile adhesives

Type	Advantages	Disadvantages
Solvent-based	• Oldest technology • Widest range of formulations • Faster drying than waterborne • High bond strength and durability	• Regulatory restrictions • Waste disposal • Safety (flammable), toxicity • High cost due primarily to petrochemical solvent
Waterborne	• Inexpensive to moderate cost • Generally not regulated • Fewer health and safety problems • Easy clean-up and storage • Increasingly wide range of formulations available	• High energy required to evaporate water • Corrosion of equipment may be an issue • Slow drying process • Generally low solids content (low thickness build) • Wetting of surfaces sometimes difficult • Poorer heat and moisture resistance than solvent-based types
Thermoplastic hot melts	• Simplicity • No curing mechanism needed • Low energy demand • Low footprint • Instant bond in many cases • Easy storage • No waste	• Limited heat and chemical resistance • Specialized application equipment required • Heat may damage substrates • Short open time and loss of tack on cooling
Reactive liquids	• No volatiles emission • Low energy demand • Maximum heat and chemical resistance • Maximum strength and durability	• Expensive • May require long cure times • Heat may be necessary to initiate or accelerate cure • Multiple-component systems must be carefully metered and mixed • Storage life or working life (once mixed) could be too short

of the joint. A typical flowchart for the adhesive-bonding or sealing process is shown in Figure 13.3.

Next to adhesive selection, prebond surface preparation is the most important process that will affect joint strength and reliability. Aqueous processing or wet treatments carried out to finish the fabric or garment (e.g., bleaching, dyeing, printing, and chemical finishing) can modify the nature of the fibre surface and affect adhesion. The removal of fibre impurities and sizing materials may be necessary to achieve optimal adhesion.

In order to improve adhesion, fibres are usually subjected to controlled surface treatments. The main purpose of surface modification is to modify the chemical and physical structure of the substrate surface to make it more amenable to adhesive bonding, but without influencing the fibre's bulk mechanical properties.

Thermoplastic fibres often are more difficult to "wet" (see adsorption theory, Table 13.3). This is especially true for the thermoplastics such as polyolefins and linear polyesters. Methods used to increase the wettability and improve adhesion include the surface treatments shown in Table 13.7. The effects of surface treatments generally decrease with time, so it is important to carry out adhesive bonding as soon as possible after surface preparation.

The literature shows that it is possible to improve the properties of the adhesive joint by surface treatment of the fibres to increase interfacial strength (Luo and Van Ooij, 2002; Holme, 1999). Oxidation prebond treatments are currently the most widely used. Plasma treatments offer some advantages and are becoming more popular. They have the potential to increase the surface functional groups and thereby increase the number of chemical bonds between the treated fibre and the adhesives. Surface modification of textile fibre is still an area under active study and more techniques may emerge in the near future.

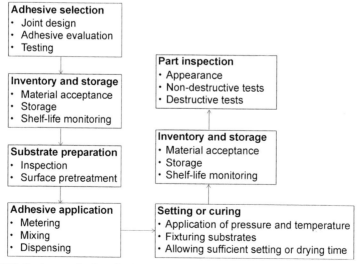

Figure 13.3 Basic steps in the adhesive-bonding process.

Table 13.7 Surface treatments for improving adhesion to textile fibres

Surface treatment	Applicable fibres	Characteristics
Washing and cleaning	All fibres	• Washing and cleaning removes weak boundary layers that may be associated with sizing, contamination, etc.
Flame	Most synthetic fibres	• Oxidizes the surface introducing polar groups
Corona or electrical discharge	Polyolefins, nylon, polyesters, other low-surface-energy fibres	• Oxidation and introduction of active groups
Ultraviolet radiation	Polyolefins, nylon, polyesters, other low-surface-energy fibres	• Chain scission of surface molecules followed by cross-linking • Surface oxidation
Plasma (atmospheric and low-pressure processes)	Nearly all low-surface-energy fibres	• Cross-linking of the surface • Surface oxidation with the formation of polar groups • Grafting of active chemical species to the surface

Atmospheric plasma treatment has been developed that negates many of the disadvantages of older, low-pressure plasma technology.[1,2] With atmospheric plasma treatment, the plasma is sustained at atmospheric pressure. It can be configured to work with a variety of feeding systems including continuous web. With atmospheric plasma, a high concentration of atoms and radicals are delivered to a surface to clean, remove, modify, or deposit materials of choice. The reaction chemistry depends on the gases fed to the plasma. Atmospheric plasma has been used effectively on films, papers, foams, nonwoven materials, woven materials, fibres metals, and powders. Operational speeds vary by substrate and can be up to 1000 ft/min.

13.2.4 Heat-activated adhesives

Thermoplastic adhesives can be applied as a molten liquid to encapsulate the fabric and then harden by cooling. These heat-activated adhesives generate joint strength almost instantly after cooling. The most common sources of heating include hot air or heated

---

[1] Atomoflo, Surfix Technologies, LLC.
[2] Plasma3, Enercon Industries Corporation.

tools, but indirect heating methods such as those generated by ultrasonic, dielectric, or laser may also be utilized.

Heat-activated or hot-melt adhesives are formulated so that they melt at a lower temperature than the melt temperature of the fabric to be bonded. The fabric and its outer surfaces, therefore, are unaffected by the process. Production rates of up to a few m/min are possible.

The major uses of heat-activated adhesives are in applications where fast production speed and reduced environmental hazards from solvent or liquid waste have significant importance. The primary strengths and weaknesses of heat-activated adhesives compared to other forms of adhesives are listed in Table 13.8.

Seam-sealing tape is the most recognized form of heat-activated adhesive. This is applied in film form, and heat is used to activate or liquefy the polymer. On cooling of the seam under pressure, the adhesive goes from a molten liquid to a cohesive solid, and a strong bond is formed.

Chemical types of heat-activated adhesives include polyolefin, polyamide, polyester, PU, and styrene butadiene copolymers. These are described in Table 13.9. There are copolymer variants of each type, allowing a wide range of applications and performance properties.

There are also a number of applications in the garment industry that require the use of heat-activated powdered adhesives. They can be applied to all sorts of textile products, although nonwovens are the most popular substrate. Powder adhesives are especially important when the manufacturer wants to provide a strong but discontinuous bond such as in the case of breathable fabrics.

A thermoplastic adhesive powder can be applied onto a web by gravimetric feed and then bonded with heat and pressure. The main advantages of using powder

Table 13.8 Advantages and disadvantages for using heat-activated adhesives

Advantages	Disadvantages
• Form bonds rapidly (high assembly speeds and short fixturing time) • Clean, easy handling • Little waste • Easy disassembly and repair of joints • No problems with solvent or solvent vapors • Good storage life and simpler storage requirements • Precise bond-line control can be achieved through temperature and pressure • Equipment available for automated assembly • Easily maintained equipment • Minimal floor space required	• Bonds lose strength at elevated temperatures • Some bonds may exhibit creep under stress and moderate temperatures • Adhesive may be sensitive to moisture and chemicals • Some substrates may be sensitive to heat of application • Hot melt used in bulk form (heated tanks) may be subject to oxidation and require a nitrogen blanket

Table 13.9 Properties of hot-melt textile adhesives

Adhesive	Special characteristics	Usual adherends	Price range
Low-density polyethylene (LDPE)	Low cost, inert.	Used for fusible interlinings.	Low
High-density polyethylene (HDPE)	Low cost, inert, higher temperature resistance than LDPE.	Used in fusible interlining, shirt collars, and cuffs.	Low
Ethylene vinyl acetate (EVA)	Good tack and adhesion, relatively low cost.	Used in footwear and to bond leather, good flexibility.	Low
Polyamide (PA) and copolymers	Wide range of properties and melting points, better heat resistance than polyethylene and EVA but more expensive, some reactive types available.	Used in garments where solvent resistance is required (e.g., dry-cleaning).	Medium to high
Polyester (PE) and copolymers	Wide range of properties, temperature resistance, good durability, high cost, some reactive types available.	Specialty applications where heat and chemical resistance is important.	High
Polyurethane (PU)	Good adhesion to many substrates, good durability, good flexibility and toughness, expensive, reactive types are available.	Used in laminates for protective clothing, shoe manufacture; reactive hot-melt PU used in applications requiring heat and moisture resistance.	High
Styrene butadiene copolymers (SBCs)	Available as both a pressure-sensitive and nonpressure-sensitive adhesive, good flexibility and toughness, resistant to both hot and cold temperatures.	Used to join difficult-to-bond substrates (e.g., polyolefin fibre).	Medium

adhesives are their cleanliness and minimal waste and maintenance costs. Spills and overspray may be collected and reused if not contaminated. The major disadvantage is controlling distribution of the particles. If coating uniformity is poor, large variations of joint strength may result with powder adhesives.

The use of thermoplastic powders for bonding nonwovens is common. These powders consist of polyamides or polyesters with melting points of 100–140 °C. They are generally incorporated into the nonwoven web and melted by heating with infrared radiation or circulating air. The viscosity of the molten powder is low enough to ensure that web bonding occurs mainly at the fibre cross-over points. This process is often termed "fusing," and the equipment used consists of various heated presses and hand irons. Continuous fusing systems are also possible where the garment part with the adhesive or interlining is passed over a heat source and pressure is applied during or after heating.

13.2.5 Liquid adhesives

Liquids are the most common form of adhesive, and they can be applied by a variety of methods. Liquids have an advantage in that they are relatively easy to transfer, meter, and mix; however, these processes require time and cost. Liquid adhesives also wet the substrate easily and provide uniform bond-line thickness. However, they have the disadvantages of sometimes being messy, requiring clean-up, and having a relatively high degree of waste.

Spray, dipping, and mechanical-roll coaters are generally used on large production runs of textile parts. Spraying usually is evaluated against other methods on the basis of cost. Spraying applies the liquid resin fast, reduces the required drying time through better evaporation of solvent or water carrier, and is capable of reaching areas that are inaccessible to other manual application tools.

Spray bonding is used for applications that require high loft or bulk. Several spraying methods can be used to apply adhesives, including conventional air spray, hydraulic cold airless spray, hot spray, and hot airless spray. The main types of spray equipment are described in Table 13.10.

Table 13.10 **Main types of spray equipment**

Type	Characteristics
Air spray	Utilizes air pressure to spray a fine mist of atomized adhesive on the substrate. Generally unsuitable for small parts. Maintenance is high due to nozzle clogging and overspray. Several passes necessary for thick coating build-up.
Hot spray	Adhesive is first heated, permitting the use of high-viscosity materials. Heavier coatings are possible than with air spray. Drying is accelerated due to evaporation of solvent in the spray.
Airless spray	Uses hydraulic pressure rather than air pressure. Saves energy and over spray. Can be applied both hot and cold.

When large flat surfaces and webs of materials are to be coated, mechanical-roller methods are commonly used to apply a uniform layer of adhesive via a continuous roll. Such systems are used with adhesives that have a long working life and low viscosity.

Roll-coating via gravure rolls is used for print bonding. Print bonding is used for applications that require only a part of the area of the web to be bonded. Printing patterns are designed to enhance strength, softness, hand, absorbency, and drape. In addition to roll-coating methods, saturation bonding is used with liquid adhesives. Saturation bonding is used in conjunction with processes that require rapid and heavy adhesive addition. It is used, for example, on fabric applications that require strength, stiffness, and maximum fibre encapsulation. This is achieved by totally immersing the web in an adhesive bath and removing excess adhesive by vacuum or roll pressure.

Bulk adhesives such as pastes or mastics produce heavy coatings that fill voids, bridge gaps, or seal joints. These systems can be extruded through a caulking gun or trowel. Since the thixotropic nature of the paste prevents it from flowing excessively, application is usually clean, and not much waste is generated. High-viscosity liquids, pastes, and mastics are ideal adhesives for application by robots.

13.2.5.1 Solvent-based systems

Solvent-based systems may be considered the origin for any trend line in adhesive development. They represent the oldest technology and a wide range of chemistries. Solvent-based adhesives have been valued through the years for providing fast drying conditions and good adhesive properties, especially durability.

The major disadvantages of solvent-based adhesives, and the issues triggering their replacement, are regulatory restrictions on solvent emissions, waste disposal, and factory vapor levels. Many solvent systems are also flammable and require safety and health-management processes.

Another disadvantage of solvent-based adhesives is their cost. The solvents are only in the formulation to reduce viscosity and allow good coating characteristics. Once the adhesive is applied, the evaporated solvent must be either recovered or treated. Solvents are generally of petrochemical origin, and as a result have a high price and price volatility associated with crude oil.

13.2.5.2 Waterborne systems

Liquid, waterborne adhesives are commonly used for garment manufacture. Waterborne textile adhesives consist of a dispersion of very small particles of polymer in an aqueous medium—known as emulsions. Waterborne adhesives have become increasingly popular because of regulatory compliance and economic reasons.

There are many types of waterborne laminating adhesives with widely differing applications and performance properties. Common waterborne textile adhesives include natural rubber latex, co-polymers of acrylic esters and acids, vinyl and vinylidene chloride, vinyl acetate, and chloroprene. Other, less common adhesives include starches, dextrin, and other polymers of natural origin. Characteristics of common waterborne adhesive used by the textile industry are given in Table 13.11.

Table 13.11 Characteristics of common waterborne adhesives used in the textile industry

Adhesive	Special characteristics	Usual adherends	Price range
Natural rubber latex	Excellent tack, good strength. Surface can be tack free and yet bond to similarly coated surface. Poor aging and ultraviolet (UV) resistance. Good moisture resistance.	Felt, fabric, paper, metal, elastomers	Medium
Polyvinyl acetate	Versatile in terms of formulation and applications, but rather rigid bonds. Odourless, good resistance to oil, grease, acid. Fair weather resistance.	Paper, wood, fabric, and most porous substrates	Low
Vinyl copolymers	Odourless and fungus resistant. Excellent resistance to grease and oils. Good bond strength.	Paper, wood, fabric, and most porous substrates	Medium
Acrylic	Good low-temperature properties, poor heat resistance, excellent resistance to UV. Clear and colourless.	Textiles, paper, metallic foils, plastics	Medium
Polyurethane dispersions	Excellent cohesive and adhesive properties. Good toughness and abrasion resistance.	Fabric, elastomers, metal, plastic	Medium to high
Neoprene latex	Superior to other rubber adhesives in most respects. Fast strength development. High temperature and weathering resistance. Resistant to UV, mild acids, and oils.	Fabric, leather, elastomers, metal	Medium to high

Although waterborne adhesives generally have poorer moisture and thermal resistance than their solvent-based counterparts, the introduction of cross-linkers into the formulation has enabled waterborne adhesive to meet many of the performance criteria required narrowing the performance gap between solvent-based and waterborne adhesives. The addition of cross-linkers, however, will reduce the working life of the adhesive, and this may be a barrier to its use in certain applications.

13.2.5.3 Reactive liquids

Reactive liquid adhesives can be chemically cross-linked to form durable, three-dimensional molecular structures. These bind chemically and/or mechanically with the fabric surface to generate high joint strength. They are generally one-part elevated-temperature-setting or two-part room-temperature-setting polymer systems. However, one-part systems that cure at room temperature when in contact with the substrates are also available (e.g., cyanoacrylates).

Reactive liquid adhesives can consist of waterborne emulsion- or solvent-based adhesives. These are adhesive systems similar to those described above but here a cross-linking agent is added to the formulation. They most notably are acrylics and PUs where a cross-linking agent is added to the adhesive immediately before application to provide improved heat and chemical resistance.

13.2.6 Advantages and disadvantages of adhesive bonding and welding

Sewing is considered as the best way of achieving both strength and flexibility in a seam. However, several factors are driving manufacturers away from traditional sewing and toward adhesive or welding technologies. These alternative processes are generally first considered for niche applications where their advantages are most apparent, then they make advancements into more conventional garment manufacturing once their full value is realized.

Adhesive bonding and welding are the most commonly used where a sealed or reinforced seam is required. They also offer value in that they can be used in high-speed, automated production processes. Examples of various applications for adhesive bonding or thermal welding are shown in Table 13.12.

Adhesive bonding and thermal welding collectively have several distinct advantages over conventional sewing processes. Adhesive bonding and welding are often used to join garments requiring high seam strength (e.g. sports clothing, shoes, protective clothing), since stresses are transmitted more effectively than by other joining methods. Welded seams are usually stronger than conventionally sewn seams because the welding process does not use a needle to create holes in the fabric and weaken the fabric structure. In conventional sewn seams, if one stitch is broken, the whole seam is compromised. Because of the elimination of holes, adhesive bonded or welded seams are also less permeable to water, and many waterproof garments are constructed today using these methods. Adhesive bonding and welding

Table 13.12 Examples of applications where adhesive bonding and thermal welding is usually first considered

Optimum application areas for adhesive bonding and/or thermal welding	Examples
Joining of dissimilar materials	Combinations of textiles with metals, rubbers, plastics, foamed materials, ceramics, glass, etc.
Laminated structures	Sandwich construction: Shoes, interlinings, belts, etc.
Structural applications	Heavy load-bearing structures (e.g., footwear, sports clothing) or where abrasion is prevalent
Bonded inserts	Studs, rivets, tube construction
Sealed seams	Rainwear, sports clothing, protective clothing
Components of particular dimensions	Where joining areas are large or there is a need for shape conformity between bonded parts
Temporary fastening	Where the intention is to dismantle the joint—sustainable clothing that might require reuse or recycling or temporary fixturing to apply printing, accessories, etc.
Thin or aerodynamic shapes	Sportswear, lightweight clothing
"Smart clothing"	Combination of clothing with electronics (for heating, sensing, etc.)

technologies are also considered to be more wide-ranging method of attaching accessories, since many plastics, elastomers, and certain metals can be joined to textiles with adhesives or by thermal welding. Adhesives are usually the first choice in applications where they can perform a secondary function such as providing water impermeability, electrical or thermal conductivity in "smart garments," increased abrasion resistance, and so forth.

However, adhesive bonding and thermal welding do have certain limitations. Seams made with these processes tend to be noticeably stiffer than seams that have been joined using traditional stitching methods. Seams cannot be thermally welded when the fabric is predominantly made from natural fibres although adhesive bonding is possible on all types of fabric.

13.3 Conventional thermal welding

13.3.1 Direct thermal welding

Thermal welding is the joining and sealing of thermoplastic garment materials without the use of adhesives or other means (sewing, staples, other fasteners). There is ambiguity in the term "welding," as some may refer to using heat-activated adhesives as a welding process. However, the conventional use here is that thermal welding occurs without an additional adhesive and essentially uses the substrate itself as the adhesive.

The main growth area for adhesives is in heat-activated tapes or films, where there is substantial overlap with thermal-welding processes. Thermal bonding is increasingly used at the expense of adhesive bonding for the following reasons:

- Thermal bonding can be run at high speed, whereas the speed of adhesive bonding is often limited by the drying or curing stage.
- Thermal bonding requires less energy than the energy needed for drying and curing adhesives.
- Thermal bonding requires relatively little factory space compared with the drying and curing ovens that may be required by adhesive-bonding processes.

The main advantage of thermal-welding processes is the speed at which a bond can be made. Other advantages include increased durability, lighter weight, reduced bulk of seams resulting in reduced abrasion, and good stretch and recovery performance in elastic products. As a result, thermal welding is being increasingly used for manufacturing of underwear and clothing articles that directly touch the body. The primary disadvantages are stiffness and possible fibre degradation caused by excessive heat conducted through the fibres.

With thermal welding, surface preparation is not as critical as with adhesive bonding. However, some form of surface cleaning may still be necessary although difficult chemical or physical treatments to increase the surface energy are not usually required.

Thermal welding is confined to fabrics that have high thermoplastic synthetic fibre content such as nylon, polyester, polypropylene, acrylic, and certain fibre blends. As a general rule, the textiles to be welded must be made of the same thermoplastic or be physically and chemically compatible with one another. Fabrics that have a thermoplastic coating (e.g., polyvinyl chloride, PU, and polypropylene) are also amenable to thermal-welding processes, since in these cases the coating acts as the adhesive. Weldability will vary with the type of material, its thickness, and its form (coating, film, fibre).

The major parameters in all thermal-welding processes are temperature, pressure, and time (Table 13.13). The optimum conditions will depend on the type of welding method employed and the type and thickness of the fabrics being joined. Thermal-welding processes are usually equipped with a feeding system and electronic sensors to allow for monitoring and control of critical parameters. Thermal welding can be categorized as either point welding, where the bond is discontinuous; or rotary welding. With rotary welding, the fabric moves continuously through the machines while the welding is accomplished.

Several methods of generating heat can be used in welding processes. In conventional heat-welding processes, the heat is applied to a fabric substrate through conduction, and the surface of the fabric is melted in the process. Advanced welding processes are those where the heating source is not in direct contact with the substrate.

Table 13.13 Parameters affecting quality of the thermally welded joint

Parameter	Considerations
Temperature	• Temperature should be measured at the joint line, but for quality control it is usually measured at the tool surface (temperatures about 20 °C higher than the joint). • Temperature at the joint must be above the melt temperature for the substrate (can cause fibre degradation).
Pressure	• Typically pressure at the joint should be in the range of 0.1–1.0 N/mm^2.
Dwell time	• Generally in the range of 2–15 s, which is enough for heat to conduct through to the joint. • Dwell time is generally considered to be controlled by the production speed and the time that the substrate is in contact with the heated tool.
Cooling time	• Cooling time is the time that pressure should remain on the joint after the heat source is removed. • Sufficient time must be allowed for the substrate to go from the viscous, molten state to a gel state with sufficient cohesive strength (generally 2–10 s).

13.3.2 Hot-air welding

In hot-air welding processes, hot air is directed between the plies of the materials to be joined to melt the substrate surface. A cold roller or other tool is then used to facilitate flow of the melt and wetting of the fibres. Finally, cooling of the melt occurs via contact with the cold surface and a cohesive joint is formed. If the fabrics are thermoplastic, one can use the fabric material itself as the hot-melt adhesive. However, if the fabric has a high melting temperature, then another thermoplastic material (e.g., film, coating, powder) with a lower melting temperature can be introduced into the joint region.

A common method of hot-air welding is to use a hot-air nozzle to deliver heat, as shown in Figure 13.4. This process is sometimes referred to as hot-air wedge welding. The main components in the system are a resistance heater, blower, temperature controller, and feed and cooling rollers. The hot-air wedges and nozzles are typically 10–40 mm wide. Since there is no contact with the substrate with a hot tool, impressions and soiling of the fabric are eliminated. The same equipment can be used for applying and heat activating hot-melt adhesive film between the substrates as described in the previous section.

The temperature is measured using a sensor in the hot-air stream. The temperature needs to be significantly higher (100–250 °C) than the adhesive material's melting point. Hot-air systems typically allow operation up to 600–700 °C. Typically production speeds are on the order of 0.7–4.0 m/min for thick materials and faster for thinner materials. Pressures are generally in the range of 0–7 bar. Higher pressures are used for tightly structured fabrics, and very low pressures are used for foam and nonwoven materials.

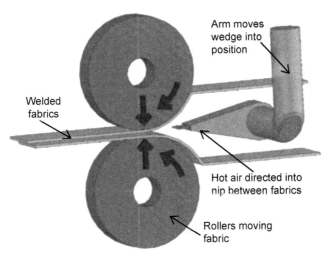

Figure 13.4 Hot-air wedge-welding equipment.
From Jones (2013).

It is possible to set the roller gap independently of the roller pressure by using an adjustable shim or depth stop on the roller wheels. This should be set to less than the combined thickness of the materials and can be set to zero to allow the joint thickness to be fully controlled by the applied pressure.

Flame lamination is another hot-air bonding process that is used mainly to continuously join fabric to foam. A foam sheet is passed over an open flame to create a thin layer of molten polymer on the foam surface, which is then mated with a fabric by passing the assembly through a set of rollers (Figure 13.5). The molten foam surface acts as an adhesive when it cools. PU foam is the most frequently used material in flame lamination.

13.3.3 Heated-tool welding

Heated-tool welding is an excellent method of joining many thermoplastic garment materials. In this method, the surfaces to be joined are heated by holding them directly against a hot surface. Electric-strip heaters, soldering irons, hot plates, and resistance blades are common methods of providing heat. When the substrate surface becomes molten, the parts are removed from the hot surface. They are then immediately joined under pressure and allowed to cool and harden. The molten substrate acts as a hot-melt adhesive providing a bond between the substrates.

As with hot-air welding, the process settings for temperature, duration of heating and cooling times, and pressures will depend on the fabric. Adjustments will be required until the desired bond quality is achieved. The thickness of the molten layer is an important determinant of weld strength. Dimensions are usually controlled through the incorporation of displacement stops at both the heating and mating steps in the process.

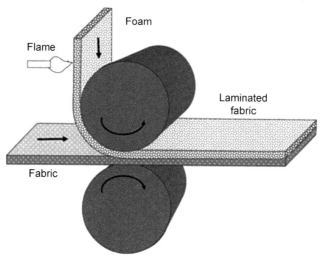

Figure 13.5 Flame laminating equipment.
From Shim (2013).

13.3.3.1 Hot-wedge welding

A common method of heated-tool welding is hot-wedge welding, which is similar to the hot-air wedge-welding technique mentioned earlier. In hot-wedge welding, a metal wedge is used to deliver heat to the substrate by direct contact immediately before it passes between the drive wheels where the pressure is applied. The wedge is generally heated by internal cartridge heaters. In some applications with heavy-duty fabric, another set of rollers is used to hold the fabric against the heating wedge. The heated wedge can be 7–75 mm wide.

Heated-wedge welding consumes less power and produces less noise than hot-air welders. However, hot-wedge welding can result in surface irregularities and contamination on the joint surface since the fabric is in direct contact with the tooling. Fabric surface irregularities can also affect the uniformity of heating, which hampers the quality of the weld. As a result, wedge welding is considered more suitable for relatively simple products that are made from less technically advanced fabrics having regular and smooth surfaces.

PFAFF (www.pfaff.com) has developed a very sophisticated hot-wedge welding system called the PFAFF 8320. The equipment is programmable with a touch screen that allows accurate control of all parameters. A revolutionary new two-axis engaging system allows the wedge to be adjusted accurately.

Hot-wedge welding is generally used for joining of heavy fabrics and films and in making impermeable seams in healthcare and protective clothing. The various fabric types that can be joined using hot-wedge welding are shown in Table 13.14.

13.3.3.2 Other thermal-welding processes

Electric-strip heaters, hot plates or irons, and resistance blades are common methods of providing heat locally. Usually, the heating surface that will be in contact with the

Table 13.14 Fabrics that can be joined using hot-wedge welding

Fabric group	Fabric subgroup	Suitability of hot-wedge welding			Comments
		Yes	Some cases	No	
Natural					
Cotton	Woven, knitted, fleece, nonwoven		x		Requires adhesive film
Cellulose	Woven, knitted, fleece, nonwoven		x		Requires adhesive film
Wool	Woven, knitted, fleece, nonwoven		x		Requires adhesive film
Silk	Woven, knitted, fleece, nonwoven		x		Requires adhesive film
Glass	Woven, knitted		x		Requires adhesive film
Carbon	Woven, knitted		x		Requires adhesive film
Synthetic					
Acrylic	Woven, knitted		x		Requires adhesive film
Regenerated Cellulose					
• Acetate	Woven, knitted, fleece, nonwoven		x		Requires adhesive film
• Viscose	Woven, knitted, fleece, nonwoven		x		Requires adhesive film

Continued

Table 13.14 Continued

Fabric group	Fabric subgroup	Suitability of hot-wedge welding			Comments
		Yes	Some cases	No	
Nylon	Woven, knitted, fleece, nonwoven	x			With or without adhesive film
Polyester	Woven, knitted, fleece, nonwoven	x			With or without adhesive film
Polypropylene	Woven, knitted, fleece, nonwoven	x			With or without adhesive film
Polyethylene	Woven, knitted, fleece, nonwoven	x			With or without adhesive film
Coated					
• Polyurethane		x			Without adhesive film
• Polyvinyl chloride		x			Without adhesive film
• Polytetrafluoroethylene				x	
Laminated		x			Preferably with a film interlayer
Elastic fabrics		x			With an elastic film interlayer
Natural-synthetic blends		x			Synthetic melts into natural fibres or use adhesive film layer

From Jones (2013).

fabric is coated with a fluorocarbon such as Teflon for nonsticking. In all these processes, the parts are held on the heated tool until sufficient fusible material has been developed.

The direct heat-welding operation can be completely manual or semiautomatic or fully automatic for fast, high-volume production. Heated wheels or continuously moving heated bands are common tools used to join substrates together. With heated bands, sections of the same belt can be used to provide heating or cooling. This enables consolidation and completion of the joining process while the parts are pressed together for a longer time than that can be achieved by a heated roller alone.

A variation of direct thermal welding is impulse welding. In this process, welding is carried out using a hot bar where the heating element is a resistive nichrome wire. By using electrical resistance heating, the tool can be heated and cooled rapidly, providing a well-controlled welding cycle.

Thermal methods are also used to manufacture molded materials for the garment industry. Molding is a method of forming a three-dimensional article having flexibility and durable shape retention. In order to mold a textile or fabric, it must have certain basic properties. It should have the ability to stretch when hot. It also should have a degree of heat strength to resist the molding operation. With molded textiles, it is generally necessary to treat or impregnate a textile with a heat-activated polymer and then form the product under certain conditions of time, temperature, and pressure. Molded textiles are commonly used as structural materials in the automotive industry. In the garment industry, a typical application is a molded outer layer (e.g., foam-backed knitted outer layer).

13.4 Advanced thermal-welding processes

A number of advanced thermal processes other than hot-air or heated-tool processes have been developed for the joining of garment materials. The most prominent are ultrasonic, dielectric, and laser welding. Because of equipment cost, required skill level, and relative inflexibility, these processes are used mainly in specialty applications. However, most garment manufacturers see broader application for these technologies in the future as the cost can be justified by the added-value propositions.

Similar to conventional thermal welding, these advanced heating processes can be used to activate thermoplastic adhesives. However, the focus here will be on joining substrates without the use of an added adhesive material.

In the conventional thermal-bonding process, surfaces to be joined are melted individually by direct contact with hot air or heated elements and then brought together. The main disadvantage of this technique is that excessive heat conducted through fibres can cause fibres degradation. Advanced thermal welding processes offer a more controlled and faster method of joining as the heat is generated primarily at the joint interface where the bonding is to occur.

13.4.1 Ultrasonic welding

Ultrasonic welding has become a well-accepted method for joining high-volume, relatively small plastic parts. In this process, an ultrasonic generator is used to produce oscillations of one substrate against a stationary second substrate. This, in turn, causes intense frictional heating between the two substrates, which is sufficient to rapidly generate a molten weld zone. With pressure and subsequent cooling, a strong bond can be obtained. The basic parts of a standard ultrasonic-welding device are shown in Figure 13.6.

The ultrasonic-welding process consists of four phases. In phase 1, the horn is placed in contact with the substrate, pressure is applied, and vibratory motion is started. Heat generation due to friction melts points of direct contact, and the molten material flows into the joint interface. In phase 2, the melting rate increases, resulting in increased weld displacement, and the part surfaces fully meet. Steady-state melting occurs in phase 3, as a constant melt layer thickness is maintained in the weld. In phase 4, the holding phase, the vibration ceases, maximum displacement is reached, and a high joint strength occurs as the weld cools and solidifies.

In the ultrasonic-bonding process, the material is fed between a vibrating ultrasonic horn and a stationary anvil (plunge mode) or a moving wheel (rotary mode). High-frequency mechanical vibrations (20–40 kHz) are transmitted through the substrate to generate a frictional heat built-up at the joint interface and to achieve a sufficient temperature to melt and bond the materials. Because the heat necessary for bonding is produced at the interface and not from the outside of materials and carried to the joint

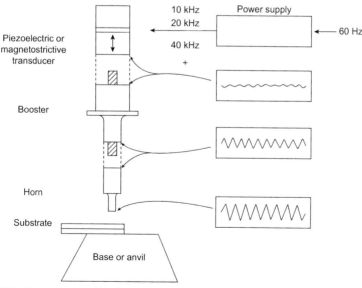

Figure 13.6 Equipment used in the standard ultrasonic-welding process. From Grimm (1995).

via conduction, the ultrasonic-welding process starts exactly at the interface and heat degradation of the fibres is minimized.

Ultrasonic horns can vary in size up to a maximum of about 24 cm by 4 cm if rectangular, or about 9 cm in diameter if circular. The actual shape of the horn relates to the particular welding operation to be achieved. One of the advantages of ultrasonic welding is that the same process can incorporate an ultrasonic cutting edge to enable "cut and seam" processing.

Ultrasonic-bonding technology has entered garment manufacturing as an advanced technique for joining synthetic materials and blends to produce continuous impermeable seams or localized points of joining. Typical uses of ultrasonic welding are the application of motifs to garments, the curing and sealing to length of ribbons and straps, and the shaping of small garment parts that would otherwise be costly and time-consuming to sew.

Fabrics may be 100% synthetic or blends with up to 40% natural fibres content. For nonsynthetic fabrics or blends with more than 40% natural fibre content, heat-activated materials (thermoplastic adhesive films or fabric coatings) are placed between two pieces of fabric. The material must have uniform thickness, yarn density, tightness of weave, elasticity of the substrate, and style of knit which are the factors that can influence the welding ability. The various fabrics that can be joined by ultrasonic welding are shown in Table 13.15.

Several companies produce automated ultrasonic-welding equipment for the garment industry. For example, Swiss manufacturer Schips (www.schips.com) has introduced several ultrasonic machines for hemming, continuous assembly of two fabrics with bonding tape, and welding of folded elastic to garment fabric. PFAFF (www.pfaff.com) has developed ultrasonic "cut and seam" equipment.

13.4.2 Laser-assisted welding

The technology of laser-welded plastic parts has been available for the last 30 years. However, only recently has the technology and cost allowed such joining techniques to be considered broadly as a high-speed garment-welding process (Nayak and Khandual, 2010).

Laser welders produce small beams of photons and electrons that are focused onto a workpiece. Power density varies from a few to several 1000 W/mm^2, but generally low-power lasers (less than 50 W/mm^2) are used for welding polymeric substrates.

In the laser-welding process, heating occurs by passing the beam through one of the fabric substrates so that the energy can be absorbed by the lower fabric or by a thermoplastic coating or adhesive film within the joint interface. Once the polymer is molten, it is then held under pressure and cooled by a sliding clamp or a roller. Melting occurs only at the joint interface, so that outer surfaces of the fabrics are unaffected. Welding rates of at least 30 m/min are possible depending on the fabrics being processed.

The methods of laser welding of fabrics fall into two categories: direct welding and transmission welding. These are primarily characterized by the laser type and wavelength, as described in Table 13.16. In transmission laser welding, an absorber material

Table 13.15 Fabrics that can be joined using ultrasonic welding

Fabric group	Fabric subgroup	Suitability of ultrasonic welding			Comments
		Yes	Some cases	No	
Natural					
Cotton	Woven, knitted, fleece, nonwoven		x		Requires adhesive film
Cellulose	Woven, knitted, fleece, nonwoven		x		Requires adhesive film
Wool	Woven, knitted, fleece, nonwoven		x		Requires adhesive film
Silk	Woven, knitted, fleece, nonwoven		x		Requires adhesive film
Glass	Woven, knitted		x		Requires adhesive film
Carbon	Woven, knitted		x		Requires adhesive film
Synthetic					
Acrylic	Woven, knitted		x		Can be cut via ultrasonic welding, but bonding requires an adhesive film
Regenerated Cellulose					
• Acetate	Woven, knitted, fleece, nonwoven		x		Requires adhesive film
• Viscose	Woven, knitted, fleece, nonwoven		x		Requires adhesive film

Nylon	Woven, knitted, fleece, nonwoven	x	With or without adhesive film
Polyester	Woven, knitted, fleece, nonwoven	x	With or without adhesive film
Polypropylene	Woven, knitted, fleece, nonwoven	x	With or without adhesive film
Polyethylene	Woven, knitted, fleece, nonwoven	x	With or without adhesive film
Coated		x	Without adhesive film
• Polyurethane		x	Without adhesive film
• Polyvinyl chloride			Welding is inhibited by plasticizers
• Polytetrafluoroethylene			Spray coats can be welded, thick coats need a film interlayer
Laminated			Preferably with a film interlayer
Elastic fabrics		x	With an elastic film interlayer
Natural-to-synthetic blends			Synthetic melts into natural fibre or use adhesive film layer

From Jones (2013).

Table 13.16 Two types of laser welding

Laser-welding method	Type of laser	Characteristics
Direct	Carbondioxide (CO_2) lasers (10,600 nm wavelength)	Welding occurs when the laser beam is aimed directly at the interface by the top surface of the fabric. The laser can be directed at the nip between two rollers so that heating and cooling can be achieved almost simultaneously.
Transmission	Nd:YAG (neodymium-doped yttrium aluminium garnet) or diode laser (400–1500 nm wavelength)	The location of the weld is controlled by placing an absorber material at the joint interface. The absorber heats to generate melting locally at the interface only.

(e.g., a carbon black coating) is generally used to create localized heating at the joint interface. Direct laser welding does not require an absorber material.

There are several advantages associated with the laser welding of garments. The external texture and appearance of the fabric is retained, strong air and waterproof seams can be made efficiently, production is fast and almost fume free, and the colour of the parent materials is unaffected by welding. The process is also amenable to automation, and multiple layers can be welded in one pass. On the downside, some fabrics need to be precoated with absorber, production usually requires investment in new equipment, and repair and rework can be difficult.

TWI (www.twi.co.uk) pioneered the laser welding of fabrics in the late 1990s. They also developed ClearWeld™, a clear transmission-type laser-joining method marketed by Gentex. ClearWeld™ garment construction includes making seams in products like gloves, fleeces, jackets, and shirts. Figure 13.7 illustrates a typical seam made with the ClearWeld™ process. TWI has also collaborated with Coleg Sir Gar to develop textile-welding machines that include three thermal techniques: transmission laser welding, direct laser welding, and ultrasonic welding (TWI, 2007).

13.4.3 Dielectric welding

Some polymeric materials will be heated in an alternating high-frequency electric field. This process is similar to the heating of foods or other materials in a conventional kitchen microwave oven. The alternating electric field forces the polymer molecules to oscillate in order to stay aligned with the field. This causes the molecules to rub against one another and generate heat by friction. With highly polar molecules, this heating is sufficient to melt the polymer. When the electric field is removed, pressure is applied and held until the weld cools.

Because of the cost of the equipment and the nature of the process, applications for dielectric welding are generally high-volume, commodity-type articles of

Figure 13.7 Waterproof laminated fabrics welded with a Nd:YAG laser using the ClearWeld™ technique. Note the lack of any surface melting—only the interface between the two fabric layers has been melted. *(Image illustrating the use of ClearWeld™ reproduced courtesy of TWI Ltd.)* From TWI (2007).

clothing. The advantages and disadvantages of dielectric welding are summarized in Table 13.17.

A typical dielectric welding tool consists of two platens that are electrically connected to a generator that imposes an alternating electric field. The platens are used both as electrodes and apply pressure to the joint area. Various parameters will affect the quality of the dielectric weld. These include the type of fibre in the fabric, its propensity to heat in a dielectric field (called the material's dielectric loss), the thickness of the fabric, and the melt temperature of the fabric. Time, pressure, and field frequency (directly related to temperature) will also affect the quality of the joint. The frequency of the field being generated can be from the radiofrequency range (13–100 MHz) up to microwave frequency (2–20 GHz).

Generally, heating occurs in the entire volume of the polymer that is exposed to the electric field, and the main difficulty in using dielectric heating is in directing the heat to the joint interface. However, high-frequency susceptors (similar to absorbers in laser welding) can be used to make the substrate surface more conductive so as to generate localized heating in the interface region. Since the field intensity decreases with distance from the source, dielectric heating processes are normally used only with thin substrates.

Dielectric heating can also be used to generate the heat necessary for curing polar, thermosetting adhesives such as epoxies. It can be used to quickly evaporate water

Table 13.17 **Advantages and disadvantages of dielectric welding**

Advantages	Disadvantages
• Clean and quick. • Easily controlled and automated. • Heating is localized, minimizing fabric distortion. • Material is heated through the bulk rather than by conduction. • Possible to heat contact surfaces without damage to outer surfaces. • Produces a hermetic seal.	• Restricted to the range of polymeric materials that are highly polar. • Not a continuous process, but limited to platens or other electrode configurations. • Potential risk to operators from high-frequency radiation although dielectric welding equipment is usually shielded.

from a waterborne adhesive formulation. Dielectric processing of waterborne adhesives is commonly used in the furniture industry for very fast drying of wood joints in furniture. Common white glues, such as polyvinyl acetate emulsions, can be dried in seconds using dielectric heating processes.

Dielectric welding is only applicable to materials that are polarizable or have highly polarized molecules with short molecular chains. Typical fibres include polyamide (nylon) and polyester. Fabrics based on polyolefin copolymer fibres can be dielectrically welded if the copolymer is polar or blended with a polar fibre. Dielectric welding is used for both woven and nonwoven fabrics. Various coatings or adhesives can be dielectrically welded including those based on polyvinyl chloride, PU, polyamide, and polyester. As a result, dielectric welding is often used to join fabric to polymeric film. Dielectric welding is employed in the manufacture of protective clothing such as rainwear, waterproof PU clothing, protective clothing, etc.

13.5 Conclusions

In addition to the move toward water-based adhesive products and away from solvent types, several trends have been noticeable in the textile industry over the past decade. These include:

- The use of advanced thermal joining processes such as ultrasonic welding.
- The use of adhesives (nonblocking hot melts and reactive "B"-staged adhesives) for offline coating and combining later in the process or at another facility.
- Bonded multilayer laminates.
- Flame-retardant adhesive systems.
- High-strength contact cements (for shoe manufacture and installing liners to outerwear).

The most significant volume trends, however, are the increasing use of hot-melt adhesives and moisture-curing PUs as replacements for either solvent- or water-based systems.

Higher standards of performance in modern applications are pushing the boundaries of adhesives and welding processes. This, of course, requires the development of newer and improved technologies. But perhaps the leading driver for joining is "performance on demand." Significant demands have been placed on joining processes to be fast, easy, and low cost.

References

Collano, 2013. Moisture Curing PUR Hotmelt Adhesive for Textiles and Technical Textiles. Collano Adhesive AG, Switzerland. www.collano.com.

Grimm, R.A., March 1995. Welding Process for Plastics. Advanced Materials and Processes, ASM International.

Holme, I., 1999. Adhesion to textile fibers and fabrics. Int. J. Adhes. Adhes. 19, 455–463.

Jones, I., 2013. The use of heat sealing, hot air, and hot wedge to join textile materials (Chapter 11). In: Joining Textiles: Principles and Applications. Woodhead Publishing, Limited, Cambridge, pp. 353–373.

Luo, S., Van Ooij, W.J., 2002. Surface modification of textile fibers for improvement of adhesion to polymeric matrices: a review. J. Adhes. Sci. Technol. 16 (13), 1715–1735.

Nayak, R., Khandual, A., 2010. Application of laser in apparel industry. Colourage 57 (2), 85–96.

Shim, E., 2013. Bonding requirements in coating and laminating of textiles (Chapter 10). In: Joining Textiles: Principles and Applications. Woodhead Publishing, Limited, Cambridge, pp. 3099–3352.

TWI, 2007. Laser textile joining puts stitching in the shade—a New Wales project. Connect 146 (1/2), 1.

Seamless garments

N. Nawaz, R. Nayak
School of Fashion and Textiles, RMIT University, Melbourne, VIC, Australia

14.1 Introduction

The seamless technique dates back to the fourth and fifth centuries AD. It was then used in the production of socks in Egypt during the Coptic era. The socks were constructed without seams, of closed loops by some complex mechanism. Medieval caps, socks, gloves, and hoses were all knitted without seams to the shape of the human body, giving correct fit and stretch (Black, 2005). At a later date, the fishermen and their womenfolk on the northern coastlines also produced upper-body garments without any apparent seam. Hand knitting also falls under seamless technique, as there is no seam but the different parts are joined by loops. In the early1990s, seamless technology was limited to the production of socks and hand gloves, but not anymore. The seamless technique has revolutionized the garment sector, and it is now the new trend in fashion worldwide. This new concept is becoming very popular in western countries, as there are advantages to wearing a seamless garment. Seamless garments are like a second skin and are comfortable and make very elegant outer and underwear.

In 1940, the manufacture of shaped knitted skirts was patented in the United States, using a "Fléchage" technique to improve drape and fit along with reducing production cost. In 1955, automatic knitting of traditional berets through shaping of components was reported. In the 1960s, Shima Seiki further used the tubular-type knitting principle to produce gloves. By 1995, the same company fully developed shaped seamless knitting machines. In 1997, seamless garments comprised a mere one percent of knitted production. Today's figure is more than 11% and it is still growing. Swimsuits, bras, and underwears have all embraced the concept, but now new body-size machinery is helping to swell the seamless market into new sectors. The technology to develop seamless garments now can compete with the traditional manufacturing processes (Black, 2005; Hunter, 2004a). Table 14.1 introduces the sequential historical developments of seamless technique (Choi, 2005).

14.2 Seamless technique

The seamless technique is relatively simple. The process evolved from cut-and-sew production to fully fashioned garments to seamless garments. The cut-and-sew production is created by the use of one entire panel of fabric, the garment production requiring several post-knitting processes including cutting and sewing. With this process up to 40% of the original fabric can be wasted. Fully fashioned garments are

Table 14.1 Historical developments of seamless technique (Choi, 2005)

Year	Historical developments
1589	William Lee in England invented the first flat-bed frame to create hosiery.
1800	The flat knitting machine was fitted with sinkers, which controlled stitches in order to knit single-jersey tubular articles such as gloves, socks, and berets.
1863	Issac W. Lamb invented the first operational V-bed flat knitting machine including the latch needles.
1864	William cotton of Loughborough patented his rotary-driven machine that uses a flat bed to produce fully fashioned garments.
1940	The manufacture of shaped knitted skirts using the "flechage" technique was patented in the United States.
1955	The *Hoisery Trade Journal* reported on the automatic knitting of traditional berets through shaped sections.
1955	Shima seiki introduced seamless entire-garment knitting at ITMA.
1960	Shima seiki company further explored tubular-type knitting to produce gloves commercially.
1960	Courtaulds established British patents on the idea of producing garments by joining tube knitting.

produced by widening or narrowing a piece of fabric by loop transference in order to increase or decrease the number of wales (done mainly on a V-bed flat knitting machine). The cutting process is eliminated in this process, but it requires a post-sewing or linking process (Nayak et al., 2006). Seamless garment knitting creates a complete garment by several different feeders with minimal or no cutting and sewing processes. The garment is knitted into shape, rather than knitting the cloth and then cutting and reassembling the pattern pieces into a garment. All the machine needs to do is keep openings for head, arms, and legs.

Knitting the products in one piece has the benefit that they are seamless at the sides and they have a knitted-in waistband that does not pinch or roll (Maison, 1979; Raz, 1991). This technology has been in use for many years in the hosiery industry, but new innovations based on variations of the hosiery knitting machine, in conjunction with the advanced yarn technology, have expanded the range of possibilities. It is now possible to make seamless, pre-sized double and single-jersey garments with knitted-in waistbands and trims. It is also possible to create a variety of designs such as stripes, dimensional textures, and intricate jacquards.

Unlike conventionally knitted garments the seams of which tend to break up the continuity of the garment design, the seamless process allows patterns and designs to remain uninterrupted across the entire garment—front to back, over the shoulder,

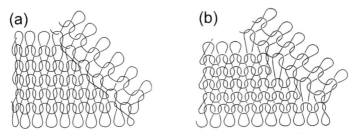

Figure 14.1 Formation of courses in seamless knitting: (a) diminishing of loops in each row and (b) diminishing of loops in every two rows.

and down the sleeves. In addition, without seams there is the opportunity to create single-knit garments that feature truly functional reversibility without the added weight and bulk of double knits. They are the result of a positive fusion of fashion and technology and, on the consumer side, provide a very welcome answer to the search for something new that should incorporate innovation, wear comfort, and easy care properties at reasonable prices. Although, these products are mainly to be found in the underwear sector, particularly for ladies underwear (78%), they have also spread to other sectors including men's underwear (13%), outerwear (2%), swimwear (2%), sportswear (4%), and sanitary products (1%). The various techniques involved in the production of seamless garments are: (1) course shaping, and (2) wale shaping

1. **Course shaping**: The principle involves diminishing or extending successively the length of the courses being knitted alternately. It can be described as knitting in which wales contain a differing number of loops. In most of the cases the number of wales throughout the knitting is unaltered. In course shaping there are two alternate methods:
   a. The number of loops diminishes in every row. If the diminution is by more than one loop, small floats occur (see Figure 14.1(a)).
   b. The number of loops diminishes in every two rows. There are no floats, but small holes can result when knitting on all wales is recommended (see Figure 14.1(b)).
2. **Wale shaping**: The principle involves knitting by increasing or reducing the number of wales internally within a flat piece of fabric or a tube of knitting by keeping the number of courses same (see Figure 14.2). There are 3 types of wale shaping such as: (a) tubular knitting, (b) running-on (picking up), and (c) casting-off (knitting off).
   a. **Tubular knitting**: The human body can be easily covered by garments of tubular shape. Tubular knitting is done by knitting the constituent thread(s) of the fabric separately.

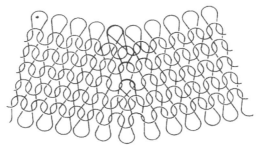

Figure 14.2 Formation of wales in seamless knitting.

b. Running-on (picking up): This is a knitting process of placing course loops or selvedge loops onto the needles of a knitting machine. Knitting is done perpendicular to previously formed portions or with a different number of wales from one another. In this process knitting is commenced on the edges of previously formed knitted fabric.

c. Casting-off (knitting off): This technique is limited to hand knitting with pins or hand-operated knitting machines. This is the process of structure sealing the last knitted course of a piece of fabric.

All or any of these processes are utilized by the knitting machines for manufacturing seamless garments.

14.3 Common seamless products

1. **Berets**: Beret, a seamless hat made of wool, had originated in France. The beret is knitted on a single-needle flat knitting machine. The three-dimensional shape of a beret is formed by knitting a series of interlinked triangular sections. The full width of a beret is knitted at the beginning and is completed by course shaping. After knitting is completed, it is milled, dyed, dried, and blocked. Sometimes brushing may be applied to the finished hat (Choi, 2005; Evans-Mikellis, 2012).

2. **Half hose or socks**: The sock is the first seamless garment used by both sexes. The shape of sock is created by stitch shaping and course shaping. The knitting is commenced at the leg opening with a welt-and-rib construction having an elastomeric thread in the rib to aid the grip. When it reaches the heel, the knitted tube is held and knitting is continued to complete up to the toe in a reciprocal manner. Then, the two halves are secured together by a closing seam. The decoration of the sock can be done by jacquard, semi-intarsia, and wrap-stripe embroidery or by structural design (Choi, 2005; Evans-Mikellis, 2012).

3. **Protective gloves**: After an inventive effort of two to three decades, these seamless garments can be produced in automatic knitting machines. The problem arises while shaping the gloves with cuff, palm, correctly angled thumb, and the four fingers. These problems are solved by complex knitting motions, which produce complete gloves with closed fingers and thumb and elasticated wrist. In modern machines, the speed and efficiency are combined with knitting of alternate left and right gloves in tuck stitch and colour jacquard. The gloves are extensively used for precision parts manufacturing and medical wear. These gloves are lightweight, flexible, and comfortable for workers in the electronics, food handling, paint, plastics, and other high-precision business sectors that require high levels of safety in addition to contaminant-free cleanliness. The same principle has been applied for manufacturing of five-toe socks in which each separate toe is individually knitted. These socks are used with thronged sandals but give rise to splaying and discomfort due to layers of fabrics between the toes (Choi, 2005; Evans-Mikellis, 2012; Spencer, 2001).

4. **Pantyhose**: The "banana"-type pantyhose, introduced by Pretty Polly in the 1960s, was knitted toe to toe as a single tube and then splitting subsequently in the widened panty sections. Single-cylinder, double-cylinder, and flat knitting machines can be used for production of pantyhose.

    Recently, Italian machine manufacturer, Saveo-Matec, used a knitting machine with two cylinders in which the upper cylinder is inverted over the lower one. Knitting commences from the waistband, each cylinder knitting an elastomeric turned welt. Both cylinders knit simultaneously, each producing one leg of pantyhose, which are joined at the crotch with a small intersecting flat machine. Now new pick-and-place machines are entering

the market, which can pick up the two legs of the garment, orient them, cut and seam them together, and sew the toes (Choi, 2005; Evans-Mikellis, 2012).
5. **Upper-body garments/apparel**: Normally these (as sweaters) are knitted on V-bed flat machines. Tubes are knitted for the body and sleeves and spaced on a needle bed having a precise number of needles between each sleeve and body. As knitting progresses, sleeves are merged to the body (Choi, 2005; Evans-Mikellis, 2012).
6. **Upholstery**: Three-dimensional seamless seat upholstery with inherent stretch characteristics is now produced by several companies (Teknit, Courtaulds, General Motors Corporation, Lear Corporation, etc.), which provides better fit for a seat, providing better seat trimming and enabling users to feel more comfortable by eliminating ridges caused by a sewing or a linking process (Choi, 2005; Evans-Mikellis, 2012).
7. **Medical textiles**: Recently, tubular-type knitted structure such as bandages, orthopedic supports, and medical compression stockings have been developed by three-dimensional flat bed knitting machines (Choi, 2005; Evans-Mikellis, 2012).

14.4   Raw materials

Seamless garments are primarily made from a combination of microfibre yarns, lycra, and cotton. Microfibre yarns are made up of fine filaments of nylon that produce the softest and most sensual fabrics. Unlike regular nylon, microfibre is similar to regular cotton garments. The technology was derived from the hosiery market, where nylon has clearly been the product of choice, but the use of DuPont's **CoolMax**® to successfully produce seamless garments proved that polyester is also highly compatible with this technology.

Natural fibres give a smooth look and soft touch, have reduced pilling, and are used in bodysuits, lace, or jacquard knits. A new arrival among seamless wear fibres is Sensil® by Nilit Ltd. Sensil® Arafelle is a 100% nylon 6,6 false twisted yarn from the same company. This textured yarn has a more natural and cotton-like handle than flat or combination yarns, and its improved quality ensures knitting efficiency and less pilling, as well as its assurance of an easy one-step dye process.

A garment that conforms smoothly to an individual's body shape and moves effortlessly with every move one makes is possible using seamless technology with Tactel® and Lycra® yarns. The alternate needle selection knitting makes fabric open and less elastic, particularly in the welt and the cuff areas. In this context, elastic, flexible, and durable yarns such as wool are recommended. In the current global scenario wool such as cashmere, and angora, and manufactured fibre such as acrylic, are used to create seamless outerwear. Viscose and polyamide with Lycra or other elastomerics are also among the other choices.

14.5   Seamless knitting machines

Seamless garments can be knit either on a circular knitting machine or on a flat (V-bed) knitting machine (warp-knit double-layer machines that can create more open work and lace effects, a greater variety of structures). The garments knitted on a circular machine may need minimal cutting and minimal seam joining on one body tube and two sleeve tubes as well as the finished edges. Consequently, seamless knitting on circular machines is not true seamless knitting. On the V-bed flat knitting machine, loop transference for performing shaping and designing structures can be performed by selecting alternate needles. However, this makes the garment more open and less elastic, which requires the use of more elastic yarns on the seamless knitting machines. There are many companies offering various types of machines producing garments by seamless technology. Some of these companies are **Shima Seiki, Stoll, Santoni, Matec, Sangiacomo, Orizio,** etc. (Choi, 2005).

Santoni, part of the **Lonati** group, the biggest supplier of the circular knitting machinery, has 14 models. Some of them include **SM8** Top Plus (two-colour designs in the welt), **SM8** Top2 (three-way technique at all feeders), **SM8VE** (high-speed eight-feeder model with four selection points and a single yarn-cutting line), and **SM9** (Twin Section). SM8 operates with a diameter between 10″ and 16″ and gauges of 16−32″, whereas SM9 has a diameter of 14−22″ and gauges between 12″ and 15″. The **SM9-ST** is a special version with diameters of 16″ and a 24″ gauge. The models of the series SM8 are fitted with a 16-stage needle selection with piezoelectric drive. The machines of the SM9 range have a monomagnetic selection system, both in the cylinder and in the dial. The yarn selector utilizes threading in groups with seven colours or $6 + 2$ on the models of the SM8 range and threading in groups with four colours on the models of the SM9 and SM9-S range.

HF models, that is, **HF 50** (diameter 5″) and **HF 90** (diameter 9″), are available from **Matec** in Scandicci, Italian region of Tuscany (Lonati Group). The gauge is 16−34″. The needle−needle selection takes place by means of four or eight selector points with a monomagnet, both for weft patterns and for float stitch patterns and with cut threads. The patterns permit up to five colours plus ground colour. The step-motor-controlled stitch cams are pneumatically controlled in groups for up-and-down movement, for multi-colour patterning. The machines are fitted with threading systems that control up to nine yarn carriers or guides per system.

SRA of the Lonati group has a specific machine for the treatment of body-size products. It consists of special boards that enable processing in autoclaves of garments such as bodies, panties, etc. in different sizes and from different fibre compositions. With the aid of a special loading station, the operative can carry out a 360° control of the product.

The **Jumbo** (single cylinder with eight systems, each one of which has piezoelectric actuators with 16 selection steps and seven yarn carriers per delivery station) machine of **Sangiacomo** is available with a diameter of 10−16″ and gauges of 16−32″ (Choi, 2005).

Orizio, the circular knitting machine manufacturer, has two versions: **MTM/BE Bodysize** and **MTM/CE Bodysize**. The diameter is 21″ and the gauge is 20−28″ (Choi, 2005).

The **Shima Seiki** company invented the Wholegarment® machine (**SWG-V**) having 5−18 gauge (needles per inch) and knitting width ranging from 50 to 80 inches, which uses a latch needle for loop transference. A newer version of this machine uses a special twin-gauge needle configuration (a pair of needles working together in each needle slot). The **SES-S.WG** uses the standard latch needle and spring-type sinkers at the same pitch, making it possible to knit fine-gauge shaping as well as integral knitting and multiple-gauge knitting. The **SWG-X** for fine-gauge knitwear has either 12 or 15 gauge and uses a slide needle and a pull-down device. This is the only machine that can knit a complete garment without an alternate needling technique. The **SES-C. WG** can knit coarser-gauge complete garments using compound needles (the hook and hook closing portions are separately controlled) and a take-down system using a pull-down device. This compound needle provides higher operational stability. The **FIRST** (another model) uses a slide needle and a unique two-piece slide mechanism. This eliminates the transfer spring and allows stable knitting, better quality, and higher productivity (Choi, 2005).

Shima Seiki knitting machines with CAD system are also available today. Using CAD system, the knit patterns can be created and all data can be saved to a diskette. The saved data can be transferred to the Shima Seiki knitting machine, which can be operated in the required design. Shima Sieki SDS ONE® CAD is a totally integrated knit production system that allows all phases including planning, design, evaluation, and production (Spencer, 2001). The most important feature in this system is a loop simulation program that permits quick estimation of knit structures without any kind of actual sample-making (Hunter, 2004b). In this program, there is an option to view the knit problems and to try out diverse knit structures on the computer system before beginning actual knitting. However, the CAD patterning of the seamless garment knitting is comparatively complicated in comparison to fully fashioned knitting due to the alternate needle selection process during knitting.

STOLL is also a major machinery producer for seamless knitting. Knitting machines made by STOLL are similar to the machines made by Shima Sieki. The STOLL SIRIX® (M1) CAD system is a complete design, patterning, and programming system utilizing two windows to graphically develop knitting programs for STOLL machines in a similar manner to the Shima Sieki CAD system (Spencer, 2001). The STOLL markets complete garment-knitting machines known as **Knit and wear**®. **Knit and wear**® flat-bed machines from **STOLL** use a gauge range of E2.5—E9.2 and a knitting width of 72—84 inches. The five different variations of the machine are **CMS 330 TC, CMS 340 TC, CMS 330 TC-T, CMS 330 TC-C**, and **CMS 340 TC-M**. All the machines use latch needles (Choi, 2005).

14.6 Advantages of seamless garments

Seamless garments have the following advantages compared to the garments having seams:

1. Improved aesthetic value and comfort: As there are no seams in the garment, the appearance of a garment is enhanced. Also, there are no chances of seam puckering, mismatching of patterns, and fitting-related problems. The discomfort felt due to the seam is overcome in seamless garments. Seamless garments consist of limited buttons, hooks, zippers, or clasps to

contend with and have superior comfort and smooth and sleek fit, which increases the ease of wearing for the wearer.
2. Cost saving: As many processes like fabric inspection, storing, spreading, cutting, stitching, etc. are omitted in the seamless technique, there is high savings in labor cost, floor space, cost of machines, power, etc. Seamless garments take 30–40% less time to make than a cut-and-sew version, minimizing the traditional labor-intensive step of cutting and sewing and thus saving cost and time (Bhosale et al., 2013).
3. Waste reduction: The fabric wastage that occurs during spreading and cutting and subsequent stitching operation is not there. So there is saving of yarn and fabrics. Additionally, yarn consumption can be minimized by complete garment knitting as well as by effectively analyzing yarn feed through the computerized system on the machine. The DSCS (digital stitch control system) on the Shima Seiki machine determines how much yarn is required for each stitch. The size of each can be controlled, and there is much less stress on the yarn at the sinker (Bhosale et al., 2013; Hunter, 2004c).
4. Lower lead-time: As many processes are eliminated, the time required for the particular process is saved. In addition, there is no dependence on the fabric supplier or supplier of other stitching accessories, which is one of the causes of production delays in garment manufacturing units. As these processes are eliminated in seamless garment manufacturing, this leads to a lower lead-time.
5. Flexibility: It provides flexibility to knit special features like incorporating moisture-management properties in the garments using different types of yarns in their structure. It provides more creative possibilities for knitwear designers. One can get a better design in the seamless technique, as there is simultaneously the designing of the fabric and the shape of the garment, so the two are better integrated. Multi-gauge knitting is also possible through this technique. Multi-gauge application capability provides the opportunity of conversion of gauges in the same machine. As a result, time and cost is saved that would be required to invest in different machines for every gauge (Legner, 2003). Moreover, seamless technology is versatile in nature. Due to its versatility there are infinite opportunities in the market, both locally and internationally. The seamless concept can be applied to make garments intended for use in a wide range of applications including underwear, swimwear, control-wear, leisure-wear, sleepwear, ready to wear, and active wear.
6. Quality and durability: Using seamless technology to make a garment eliminates the process of cutting and sewing steps. Due to the decreased number of steps involved in making the garment, the risk of defects and damages is also minimized. A single complete piece production method is claimed to provide more consistent quality. Thus seamless garments are usually long-lasting due to seamless knit—no fraying and stitch defects (Bhosale et al., 2013).
7. The other advantages are more constant product quality, better trimability for finished edge lines, just-in time production, and mass customization (Evans-Mikellis, 2012; Bhosale et al., 2013).

14.7 Disadvantages of seamless garments

Despite the advantages described above, seamless garments have the following disadvantages (Choi, 2005; Evans-Mikellis, 2012):

1. There are technical limitations in the seamless technique to knit every garment type/shape currently produced by cutting and sewing. The main problem has been the fabric take-down in keeping equal tension of each loop (i.e., stitch) (Hunter, 2004a).

2. Another problem is caused during alternate needle selection, which makes fabrics more open and less elastic than conventional fully fashioned garments. This problem occurs mainly in the welt or the cuff areas (Mowbray, 2004).
3. The machines used for manufacturing seamless garments are costlier, and more skilled operators are required (Bhosale et al., 2013).
4. A fault during knitting (particularly a hole or a barré) damages the whole garment (Knit Americans, 2001).
5. While lightweight, next-to-skin garments score advantage points for comfort, this is slightly less important for outerwear garments, where the seam can become a style/fashion feature (Bhosale et al., 2013).
6. Seamless garments are costlier as compared to seamed garments (Bhosale et al., 2013).

14.8 Applications of seamless garments

Seamless garments are preferred by the wearer for their comfort, snug-fitting, durability, and aesthetic characteristics. In addition, seamless garments are cheaper and do not easily fail on waistband and side seams. Seamless technology has wider application in various areas such as upholstery, automotive and industrial, sports textiles, medical textiles, and intimate apparel apart from general apparel. Some of the major applications are discussed in the following section.

14.8.1 Upholstery

The recent computerized knitting machines for seamless garment construction provide technically as well as aesthetically advanced design possibilities. Three-dimensional seamless seats in office chairs can be achieved easily using sophisticated computerized system. The fabrics possess stretch characteristics to follow contours of seat. Engineered design can help in increased knitwear performance involving various knit structures. This recent development in upholstery manufacturing using knits enhances appearance, better seat trimming while eliminating ridges.

14.8.2 Automotive and industrial textiles

The design of automotive seat covers by the usage of predictive computer models not only provide comfort and durability, but also assist in the quick change in the design and knitted tube size. It adds on quality, provides ergonomic seat design and time saving.

In industrial applications, fibres such as Kevlar® offers seamless filament knit gloves and apparels which are lightweight, flexible, and comfortable for workers in electronics, food-handling, paint, plastics, and other high-precision industrial sectors, which require high levels of safety in addition to contaminant-free cleanliness.

14.8.3 Sports textiles

Sports apparel demand high-performing garments to enhance a consumer's performance in addition to comfort. Seamless apparel construction focuses on supporting muscles and areas where it is needed the most. An engineered fit, micro massaging features and performance innovation, results with blending of various technical fibres and yarns, to achieve the requirements. The development of advanced second-skin textiles has led to renewed interest in seamless garment construction for sports applications. A diverse range of products such as hand gloves, hats, socks, sports underwear, and tees are some of the obvious applications.

14.8.4 Medical textiles

There is a high demand for seamless products in the medical sector in various applications such as bandages, orthopedic supports, medical compression stockings, gloves, and many more. With the incorporation of high-performance fibre combined with technical developments, it is possible to design these products with unique features which are comfortable in addition to their other desired features. Specialized aesthetics, feeling of wellness, additional functionality, and other desired properties have helped in quick recovery. The use of seamless products can lead to certain medical procedures delayed or even avoided.

14.8.5 Intimate apparel

Intimate apparel produced by seamless technique gives seam-free, easy-care, comfort and fit, and gives a feel similar to one's second skin. Due to these, they are preferred for today's lightweight intimate apparel. In order to meet the increased demand for seamless garment, companies such as Santoni, Shima Seiki have introduced various new machines for the production of under wear, swim wear, and sanitary garments.

14.9 Future developments

The three-dimensional seamless knitting has diverse capabilities, and numerous products such as fashion garments, upholstery, medical garments, etc. that can be manufactured using seamless knitting technology. Based on the use of seamless-knitting technology for a diverse range of products, seamless knitting is forecasted to continue growing and could be one of the largest next-generation knitting technologies. Seamless-knitting technology is also becoming popular for functional clothing such as seamless body armor for females due to its fitting characteristics for female body geometry. Recently, seamless knitting technology has been used to produce seamless female body armor to improve fit and comfort suitable for female body geometry using 100% Kevlar and Kevlar/wool fabrics (Mahbub et al., 2014). Seamless technology is widely used in therapeutic clothing and gloves, and in compression garments, allowing ultimate compression, comfort, fit, and performance (Supacore; Skinniesuk; Absolutemedical).

14.10 Conclusions

The seamless technique is pioneering in apparel markets because of its smooth fit, comfort, invisibility, and easy care properties. Seamless knitting machines have the capability not only to create shaped knitting but also to make various knit structures in the complete garment by utilizing alternate needle selection. This process, however, creates fabrics more open and less elastic than conventional fully fashioned garments. Nevertheless, complete garment knitting provides major benefits for the market as well as for technical production, as discussed earlier. In the case of seamless knitting, manufacturers do not have to rely on cutting and sewing processes. As a result, seamless knitting saves production time and cost and also minimizes production waste. Moreover, product quality is more consistent during seamless knitting, offering greater wear comfort to the wearer. The trends worldwide suggest that seamless garments are becoming popular among the masses, especially with the youth. There is a potential for seamless garments to hold 50% of the industry's sales within next 10 years by providing educational and training facilities and adding versatile design features.

References

Black, S., 2005. Knitwear in Fashion. Thames & Hudson, Inc, New York.

Bhosale, N., Jadhav, B., Pareek, V., Eklahare, S., 2013. Seamless Garment Technology, Application and Benefits. Published on http://www.fibre2fashion.com/industry-article/49/4854/seamless-garment-technology1.asp.

Choi, W., 2005. Three dimensional seamless garment knitting on V-bed flat knitting machines. J. Text. Appar. Technol. Manag. 4 (3), 1–33.

Evans-Mikellis, S., 2012. New product development in knitted textiles in new product development. In: Horne, L. (Ed.), Textiles - Innovation and Production. Woodhead Publishing Ltd., Cambridge, UK.

Hunter, B., 2004a. Complete garments-evolution or revolution? (Part 1). Knit. Inter. 111 (1319), 18–21.

Hunter, B., 2004b. Is knitwear software too rigid? Knit. Inter. 111 (1312), 40.

Hunter, B., 2004c. Loop tension and fabric quality. Knit. Inter. 111 (1312), 40.

Knit Americans, 2001. Seamless Sweaters Here to Stay, Winter, pp. 22–23.

Legner, M., 2003. 3D products for fasion and technical applications from flat knitting machines. Melliand Inter. 9 (3), 238–241.

Maison, L., 1979. Flat Knitting Machines. ITF Maille, France, p. B-4.

Mahbub, R., Wang, L., Arnold, L., 2014. Design of knitted three-dimensional seamless female body armour vests. Inter. J. Fashion Des. Technol. Educ. 7 (3), 198–207.

Mowbray, J., 2004. A new spin on knitwear solutions. Knit. Inter. 111 (1312), 34–36.

Nayak, R., Mahish, S.S., 2006. Seamless garments: An overview. Asian Textile J. 15 (4), 77–80.

Raz, S., 1991. Flat Knitting: The New Generation. Meisenbach Bamberg, Meisenbach pp. 34–37, p. 14, pp. 62–63.

Spencer, D., 2001. Knitting Technology, a Comprehensive Handbook and Practical Guide. Woodhead Publishing Limited, Cambridge, England.

Viewed from http://supacore.com/pages/supacore-compression-garments (on 10.10.14).

Viewed from http://www.skinniesuk.com/ (on 10.10.14).

Viewed from http://compression-therapy.absolutemedical.com/viewitems/gloves-and-gauntlets-20-30mmhg/jobst-ready-to-wear-gloves-20-30mmhg (on 10.10.14).

Part Three

Garment finishing, quality control, care labelling and costing

Garment-finishing techniques

15

S. MacA. Fergusson
RMIT University, Melbourne, VIC, Australia

15.1 Introduction

Typically, casualwear and sportswear are the major products being produced by several garment manufacturers. Today, the garment manufacturers are facing harsh competition due to the global economic crisis, stricter regulations in international trade and the stiff rise of the prices of raw materials and chemicals (Nadvi et al., 2004; Ramaswamy and Gereffi, 2000). Apparel manufacturers have to produce a diverse product mix as consumers are difficult to understand and predict (Deaton, 1980). Consumer's choice is shifting from traditional designs towards luxury high-fashion items. Almost all consumers demand both moderate pricing and frequent style changes (Ulrich et al., 2003). The frequent style changes provide new challenges for the manufacturers in their efforts to fight for the dwindling consumer dollar. The garment-finishing processes can assist to a certain extent to fulfil the requirement of fast-moving fashion and to add functionality.

Garment finishing consists of a series of finishing operations performed in the garment to improve its aesthetics, handle and functional properties. The processing operations can be either or both mechanical and chemical in nature, which are performed in stitched garments as single or batch. Several finishing techniques applied to the fabric can be applied to manufactured garments. However, specialty machines are needed for garment finishing operations. In addition, many of the finishing operations may not be economical to perform in garment form. Hence, it is imperative to complete the finishing operations in fabric form unless there are unique features that can only be added in the garment form or incorporation of any functionality to the garment.

The term 'garment finishing' was a buzzword for the process in the denim industry; now the term has been extended to a range of ready-made garments such as shirts, T-shirts, trousers and jackets and even to all other types of clothing. Various chemicals are used for value addition to garments through effects including various feels such as soft, supple, dry feel, bouncy feel; and to adding functionalities such as water/oil repellency, wrinkle free, moisture management, stain protection and durability to the garment.

Garment dyeing, one of the finishing operations, allows the manufacturer to produce special colour effects that may not be feasible from continuous processed fabric. The demand from retailers for rapid response to fashion and colour changes has resulted in some speciality garment manufacturers producing products that can meet these requirement using fabric that has been previously prepared for dyeing when the garment is made. The made-up garments are then processed to their respective colours by specialised garment dyers. Thus, short runs of a specific product are therefore possible with the advantage of more economical garment production when only an uncoloured fabric is being used. This reduces wastage and lowers the cost of stock, when only a single fabric type is required.

In addition to the garments prepared from woven fabrics, knits (especially single jersey and interlock) are also the major products suitable for garment finishing. The garments prepared from traditional natural fibres or their blends can be suitably processed in garment form. Some specialty fabric types such as blends of wool/nylon and wool/cotton have recently been processed in garment form in order to meet specific market demands of comfort and performance.

The traditional process of pressing is still the dominant operation in almost all the garment manufacturing units, which adds aesthetic value to the garment and improves the attractiveness of a garment at the point of sale. However, the recent technical advancements have assisted in the garment finishing techniques to achieve improved functionality and/or to create customised garments. This chapter covers the garment-finishing techniques that are recently being used by the manufacturers. The finishing techniques performed by mechanical and chemical means, which are used for garments made from various fibres, are discussed. An overview of denim product finishing and the pressing method are also given.

15.2 Garment finishing for functionality

The sequence of operations and various types of garment-finishing techniques are discussed in the following section.

15.2.1 Sequence of operations

There must be a logical sequence in the finishing of garments so that they are presented to the consumer in the best possible condition. The following operations should be followed to ensure that all possible mistakes are being removed prior to garment finishing:

1. Careful visual inspection of the garment for sewing faults, loose threads, untidy seams, uniform positioning of buttons, etc. Loose threads may become entangled in machinery during wet processing or pressing; these should be attended to prior to further operations.
2. Performing the necessary garment-finishing operations followed by the subsequent steps (if any) needed for adjusting the chemical formulations.
3. Wet processing of the garments for structural relaxation or to remove any residual chemicals or any visible stains acquired during the making-up process. During this processing step, it is possible to carry out some chemical treatments to enhance the garment feel or performance.
4. Tumble-drying of the garments, which assists in further relaxation.
5. Final pressing and packaging.

15.2.2 Permanent crease and wrinkle-free treatments

Garments made from 100% cotton — particularly trousers, slacks or pleated skirts — may require the addition of a suitable permanent pleat treatment (Kang et al., 1998). If the fabric has already been treated during manufacture with the necessary chemical, that is DMDHEU (di-methylol dihydroxy ethylene urea), the processing route is relatively simple; however, it must be remembered that the resin treatment has not been cured and therefore no wet treatment can be applied to the garments prior to the fixation stage (Wei and Yang, 1999). The necessary creases are applied to the garment by hot-head pressing at a temperature of at least 150 °C. The pressed garments are then given a heat treatment for 5 min at 150 °C in a suitable oven. Following this treatment, if necessary, the garments can be washed to remove any stains and given a final pressing.

If the treatment is required on garments made from cotton, linen or viscose, the appropriate chemical, usually DMDHEU, can be applied by using a robotic spray system that utilises a spray machine with two robotic arms and two mannequins, allowing higher production to be achieved (e.g. Picasso SS330, VAV Technology GmbH). This system ensures that the chemicals are applied uniformly to all parts of the garment. Garments can then be dried, hot pressed and the resin treatment cured as previously outlined. Finally, the garments may be rinsed in warm water to remove any unreacted resin, hydro-extracted, dried and finish-pressed. This system is ideally suited for the application of easy-care finishes that may be required for specific end-uses where the chemical has not been previously applied at the fabric-manufacturing stage. Surface finishes such as stain release and softeners may also be applied using this technology.

Similar processes can be adopted for wrinkle-free finishing, where the chemical treatment helps in the better recovery from creasing. The finishing chemicals can help to achieve crease recovery, dimensional stability, reduced pilling and particularly with knit goods and improved appearance after several washes. To achieve a good result in finishing, it is absolutely essential that the garments are well prepared, and that the recipes and processes are strictly followed and exactly monitored.

15.2.3 Water/oil repellent treatment

Hydrophobic properties are achieved by the application of the water/oil repellent treatment to the substrates (Bahners et al., 2008). The main product groups for this treatment are: (1) metal salt paraffin dispersion, (2) polysiloxane and (3) fluorocarbon polymers. The surface of the substrates must be covered with molecules in such a way that their hydrophobic radicals are ideally positioned as parallel as possible, facing outwards, during the chemical finishing with these products.

Metal salt paraffin dispersions (such as aluminium) are products positively charged due to the trivalent aluminium salt, which produces a counter-polar charge on the fibre surface. Polysiloxanes form a fibre-encircling silicone film with methyl groups perpendicular to the surface. The hydrophobicity of the finish is affected by the film formation and direction of the methyl groups. Fluorocarbon polymers also form a hydrophobic film where the fluorocarbon radicals are perpendicular to the fibre axis, which prevents wetting of the fibre surface. The extremely low interfacial tension of the fluorocarbon chain towards all chemical compounds is responsible for its high hydrophobic and oleophobic properties.

15.2.4 Antimicrobial treatment

Different types of antimicrobial finishes used in other areas such as food preservatives, disinfectants, swimming pool sanitisers or wound dressings, can also be used for textiles (Gao and Cranston, 2008; Joshi et al., 2009). The antimicrobial finishes are potent in their bactericidal activity, which is indicated by the minimal inhibitory concentration values (Nayak et al., 2008). However, repeated laundering of the textiles leads to the gradual loss of the biocides. In addition, their attachment to the surface of a textile or incorporation into the fibre substantially reduces their activity and limits their availability. Due to these reasons, the finishes need to be applied in large amounts to the textiles to sustain durability, for effective control of the bacterial growth. Different chemicals such as organic compounds (amines or quaternary ammonium compounds, biguanide, alcohols, phenols and aldehydes), mineral compounds (metal ions, oxides and photocatalysts), organometallic compounds and natural compounds are used for antimicrobial finishes (Nayak and Padhye, 2014a; Simoncic and Tomsic, 2010).

15.2.5 Flame-retardant (FR) finish

Flame-retardant (FR) finishes are essential to reduce flame propagation, hence to achieve FR properties (Horrocks, 1986; Tesoro et al., 1972). The FR finishing of fabrics can be divided into wash-resistant or non-wash-resistant finishing, depending on the end-use application. In the case of garments, FR finishes that are non-durable can be applied to avoid the constraint in the application techniques involved. Although these non-durable finishes are fast to dry-cleaning, they are not fast to repeated laundering.

15.2.6 Enzyme washing or bio-polishing

The application of enzyme treatments on cotton and regenerated cellulose materials such as lyocell (Tencel®) has become widely accepted (Harnden et al., 2001; Nostro et al., 2001). The process referred to as bio-polishing has the advantage of preventing pilling, as the enzyme "cellulase" hydrolyses the loose surface fibres on the yarns, causing them to break off and thus leaving a smoother, more uniform fabric. A softer fabric with improved colour brightness is also achieved by this technique. Enzyme treatments are important for the finishing of lyocell, which was invented in 1991 and is sold under the trade name Tencel®, manufactured by Lenzing. This fibre has a tendency to fibrillate when the fibre is wet. These fine fibrils on the surface tend to peel up and if not removed, show as pills on the fabric surface. Cellulase treatments also enhance the surface features of the fabric, giving it a smooth, silky appearance.

Treatment temperatures range from 50 to 60 °C and pH ranges from 4.5 to 6.5 depending on the severity of effect required.

The application of a protease treatment to silk was introduced as an alternative to the degumming process using alkaline soap solutions (Freddi et al., 2003; Gulrajani et al., 2000). Alkaline soap has a deleterious effect on the silk, resulting in a harsh feel to the material. Enzyme degumming with protease removes the sericin without damaging the fibre. Results of enzyme treatment have shown that the fibre is stronger than that obtained by traditional alkaline soap treatments.

Cheng et al. (1998) reported on the enzyme washing of silk "crepe de Chine" fabric using commercially available protease. This work showed that the degree of surface change to the fabric surface was dependent on the enzyme-dosage level. The higher the dosage, the greater the damage to the fabric, and thus a lowering of fabric strength and an increase in surface fuzziness. Protease treatments for wool garments were introduced during the early 1990s. This treatment modifies the surface protein of the fibre and as a result reduces the surface fibre and thus reduces the tendency for pilling to occur.

15.2.7 Garment dyeing

In garment dyeing, fully fashioned garments such as pants, sweaters, shirts and skirts are dyed after manufacturing is completed (Partridge, 1975). Most garments are made of cotton or a cotton-rich blends which may contain other fibres such as wool, nylon, silk, acrylic, or polyester as a minor component in the blend. Traditionally, garments are manufactured from pre-dyed fabrics before the cutting and sewing. Garment dyeing has been gaining importance and popularity due to cost savings and fashion trends in recent years, and will continue to grow in the future (Bone et al., 1988). A major drawback of garment dyeing is the risk of maintaining a large inventory of a particular style or colour in today's dynamic marketplace.

Two types of equipment, namely paddle machines and rotary drums, are generally used for garment dyeing. Paddle machines gently move the garments using paddles similar to a paddle-wheel on a boat. A high liquor ratio is required for paddle machines, and they may have limitations in shade reproducibility. On the other hand, rotary drums work on the principle of stationary liquor and movement of the material. These machines are sometimes preferred for garments such as sweaters, which require gentler handling.

For dark shades it is normal to use reactive dyes in the dyeing of 100% cotton garments. This can pose significant problems in relation to fastness of the finished garment, particularly performance to wet treatments such as washing and perspiration if the hydrolysed reactive dye is not completely removed. Traditionally, garment dyers used after-treated direct dyes for both pale and dark shades. Pale shades were after-treated with a cationic resin compound such as Tinofix ECO® (Huntsman Chemicals). Dark shades were often dyed using after-coppered direct dyes that would give adequate wet-fastness properties.

Customers require their garments to be washable under particularly variable conditions, that is from ambient (25–30 °C) to hot (40–60 °C) temperatures, depending on the domestic washing machine employed. Reactive dye will in general meet the demands of high wet fastness. Liquor ratios in garment dyeing vary from about 20:1 to 30:1, as large amounts of electrolyte are required to achieve adequate exhaustion; which can create some environmental issues. The need to remove the hydrolysed dye (dye attached to the fibre surface but not chemically bound to the fibre) can create some problems in the soaping operation. As a general rule, the most appropriate product for soaping reactive dyes is a protective colloid based on sodium polyacrylate. This chemical will remove the hydrolysed unfixed dye from the fabric and hold it in solution ready for discharge to effluent.

Some fabric blends can be a problem; for example, the dyeing of a cotton wool blended fabric where the wool portion is only about 10%–15%. In such a case, the pH of the dyeing must not exceed 8.0. In reactive dyeing, the pH of the fixation step is usually about 10–10.5 and is achieved using sodium carbonate and a temperature of 60–70 °C. A lower pH of 8.0–8.5 can be used, but the dye bath exhaustion would be lower. This lower pH will not damage the wool to the same extent. After-coppered direct dyes may in fact be more suitable for such a blend, giving good light and wet fastness.

15.2.8 Other functional finishes for garments

Often, garment finishing includes softeners, soil-release finishes and finishes for ultraviolet (UV) protection. Softeners can alter the handle of the garment, and the degree of softness depends not only on the chemical character but also on their position in the textile. Soil-release finishes facilitate the removal of stains from various fabrics that usually show some resistance to stain removal by normal cleaning processes. The UV finishes ensure that the clothes reflect the harmful rays of the sun, reducing a person's UV exposure and protecting the skin from potential damage.

Not all garments may require wet processing. Some customers request that garments are washed prior to pressing in order to develop the specific handle required. During wet processing, additional chemical finishes can be added to enhance garment properties such as handle modifiers, antimicrobial materials and non-stain or comfort-enhancing products. In subsequent processes following washing, the garments can be dried by commercial dryers.

Drying of the garments is a two-step process. The first step is removal of excess moisture that is not chemically bound to the fibre. Several methods are available; these include squeeze rollers or mangle and hydro-extraction (spin-driers). The squeeze-roller or mangle method is no longer popular due to excessive creasing that may be introduced into the garment. Hydro-extraction, while not a continuous process, is very efficient at removing the water trapped between the fibres within the yarns. Following the removal of excess moisture, the second stage is to tumble-dry the garments. Temperature of tumble-drying should not be excessive so as to cause yellowing of the fibre, particularly in pale shades. Some fabrics may be treated with handle modifiers at the manufacturing stage; some of these handle modifiers are sensitive to temperature and may cause shade changes to occur. Temperatures of 60–75 °C should be considered as a maximum.

15.3 Knitwear finishing

The finishing of knitwear garments requires special attention to processing, as this will ultimately result in improved quality. The knitwear sector consists of a range of fibre and fibre blends. Wool is still possibly the most important section of knitwear production, followed by 100% acrylic and wool acrylic blends. Mohair, alpaca and cashmere have all figured into the production of knitwear, but essentially their production is relatively small compared to both wool and wool acrylic blends.

Prior to the introduction of the Shima Seiki whole-garment knitting machine, the majority of finishing operations took place prior to final garment making up. That is, the garment bodies, sleeves, etc. were finished separately to overcome differences in relaxation due to varying tensions that might have applied during the knitting stage.

The finishing process can be divided into two sections, wet finishing and dry finishing. Dry finishing can also be described as steam finishing. In more recent times, solvent finishing has been used for the treatment of wool knitwear materials. In relation to the different types of yarns used in manufacture each of these processing methods will be discussed.

15.3.1 Dry finishing

The final steaming operation is the key to attractive product presentation to the customer. During final pressing, creases should be removed and the garment checked for size (see Section 15.5 for detailed information on pressing). During crease removal, excessive stretching should be avoided, as this could result in shrinkage at the first wash. Whilst hand-held irons may be used, the most appropriate system is a flat steaming table that incorporates a vacuum system that will cool the garment quickly after steam application. The Hoffman-New Yorker SBT knitwear steam table provides the knitwear finisher with a versatile manual steaming and vacuum table; or alternatively the Hoffman-New Yorker SBTA automatic model can be used.

The finisher can be assured of producing a quality garment from the use of this equipment. The Woolmark Company (Australia) can supply extensive practical information on the processing of all types of wool garments.

15.3.2 Wet finishing

Where woollen spun yarns are used in the knitting process, it is necessary to remove the lubricants used during the spinning operation by scouring or washing. Other contaminants from manufacture may also be present; these include floor dirt and machine lubricants. The scouring or washing process also brings about relaxation of the knitted loops so that a stable garment is produced. Woollen yarns may not be manufactured from non-shrink or washable wool, and therefore careful attention must be observed in the wet processing so as not to cause excessive shrinkage.

Smith drums (Figure 15.1) are ideal for washing due to low mechanical action on the garments. For milling shrinkage, the most appropriate machine is a barrel washer (Figure 15.2), which will give increased mechanical action to promote consolidation and surface cover. The typical conditions discussed below can be used for the wet finishing of knits prepared from wool.

Typical washing conditions at 40 °C
Liquor to material ratio 30:1
Suitable detergent 0.5 g/L
Wash gently for 5−10 min
Rinse two cycles at 40 °C.

If further consolidation, that is milling shrinkage, is required, this can be accomplished in a separate bath at 40 °C as follows:

Liquor ratio 30:1
0.3 g/L non-ionic detergent
Wash 5−30 min depending on the degree of surface finish required
Drain the bath and rinse well at 30 °C
Hydroextract and tumble-dry.

Figure 15.1 Smith drums used for wet finishing of wool.

Garment-finishing techniques

Figure 15.2 Barrel Washer used for increased mechanical action.

Unlike woollen yarns, worsted spun yarns contain only minimal quantities of lubricants, and therefore the finishing treatment can be less vigorous. Relaxation of the tensions produced in knitting is important to prevent relaxation shrinkage of the garment on initial washing by the consumer. Relaxation can be accomplished either by wet finishing or by simply steaming the garments. If wet finishing is to be undertaken, a similar process to that described for woollen knitwear can be adopted. The mechanical action during the washing stage should be minimal: a short run for 1–2 min followed by soaking would suffice. A Smith drum or overhead paddle would be suitable machinery. The washing procedure is described below:

Water at 40 °C, liquor ratio 30:1
0.1–0.2 g/L suitable detergent
Run 2 min, stand for 10 min
Rinse well in warm water
Tumble-dry.

15.3.3 Solvent finishing

Solvent processing was introduced by manufacturers to reduce water and effluent charges. The basic process is similar to dry-cleaning and now uses perchloroethylene (Perc) as the solvent medium rather than the traditional Stoddard solvent. During the process a small amount of water is added with the detergent system to assist in the removal of water-soluble contaminants. The system is totally enclosed and all the solvent is recovered by distillation.

15.4 Denim garment finishing

In denim manufacturing, it is an amazing fact that one style of jeans is converted into different colours, which is achieved by garment washing. In today's denim manufacturing, many denim finishes are done to give the jeans a broken-in look, much like what unwashed (dry denim) denim looks like after it has been worn many times. There are two main types of garment washes for denim: chemical and mechanical.

Various chemical finishes such as bleaching, enzyme washing and acid washing are the major types of chemical finishing used for denim. The bleaching is performed to discolour the denim in specific locations. The degree of bleaching action depends on the strength of the bleach, temperature and duration of the treatment. The environmental friendly finishing process 'enzyme washing' is used to create prominent effects at seams, pockets and hems (Koo et al., 1994; Heikinheimo et al., 2000). More recently, a range of enzyme treatments have been developed by AB Enzyme GmbH specifically for the treatment of denim garments. ECOSTONE® is the registered trademark of this bio-polishing process developed by AB Enzyme GmbH. The advantages of the enzyme process are:

- Improved abrasion, colour removal without damaging the fabric strength
- Simple and easy treatment
- Environmentally friendly
- Greater variety of effects are achievable
- Shorter washing times.

BIOTOUCH® is another range of products developed by the same company for the treatment of materials made from cotton, linen, viscose and their blends with other fibres such as polyester or nylon. Garments treated by this process have smoother surface properties, are pill free and as a general rule have better drape and a softer handle.

Acid washing is also known as stone washing, which is accomplished with pumice stones. Due to severe abrasive action, surface fibre damage occurs. In some instances the pumice is impregnated with potassium permanganate (an oxidative bleaching agent), this chemical being less damaging to the cotton fibre than sodium or calcium hypochlorite. Following this bleaching and stone washing, the garments are washed with sodium bisulphite (reducing agent) to complete the removal of the brown colour generated by the permanganate wash. This treatment has the effect of surface bleaching the indigo as well as giving the garments a softer handle and creating a distressed look.

Mechanical denim finishing includes the process of stone washing and microsanding. Mechanical stone washing is similar to the chemical washing process of the acid wash. Here also stones are tumbled with freshly dyed denim, but it does not require any chemicals. The final look of the product depends on the size, shape and hardness of the stones used. In several instances the metal buttons and rivets get damaged due to the mechanical action. There are three types of micro-sanding: sandblasting, machine sanding and hand sanding. The sandblasting process is the most common, which is accomplished by passing a very abrasive substance through a nozzle at the denim at high speed and pressure that creates many different patterns

Garment-finishing techniques 397

Figure 15.3 Example of X-Burner technology.

in the denim. Machine sanding is performed by using a machine (similar to the machine used to sand down wood furniture), whereas hand sanding is done with a fine-grain sandpaper by hand.

The most recent development in the processing of denim has been the use of laser technology to imitate the human hand. VAV Technology GmbH produces the X-Burner system, a dry method of producing patterned effects on denim. Figure 15.3 illustrates the effects available using the X-Burner technique. This system allows a wide range of effects to be produced efficiently with minimum damage to the fabric. The VAV Technology for automated denim processing increases productivity, uniformity and quality of the finished product.

15.5 Pressing (factors and equipment)

Garments are pressed to remove any creases, and present the garment to the customer in attractive condition suitable for sale. Garment presentation to the consumer is a vital step in the finishing of a product. The opinion of the customer is an integral step in brand recognition. Poorly presented product will have a detrimental effect on the brand's quality and therefore product saleability. A badly creased garment will lower its retail value and thus the manufacturer's sale margin. Pressing therefore is an important step in the production process. Pressing should accomplish the following:

- Removal of all manufacturing creases and wrinkles
- Clarity of pleats if there are pleats present (such as in skirts and trousers)
- Uniformity of collars and cuffs if present
- Stabilising the garment, particularly in the case of wool knitwear to retain the desired shape
- Relaxation of any stresses induced during garment manufacture.

15.5.1 Factors affecting pressing

In order to achieve good pressing quality, there are four basic parameters that need to be controlled to meet optimum performance: heat, moisture, pressure and cooling with vacuum. The importance of each parameter is discussed in the following section.

Heat is required in most pressing operations to enable the fibres to soften and thus stabilise the garment shape. Temperature selection is of utmost importance, as an incorrect temperature setting can cause damage to fibres and yarns.

Moisture is introduced by the use of steam. Steam at different pressures has different moisture contents; the higher the steam pressure, the lower the moisture in the steam. The presence of moisture is required to aid in fibre swelling and thus shape stabilisation. Different fibres require different amounts of moisture. For example, natural fibres such as cotton and wool and regenerated cellulose fibres such as bamboo viscose and viscose rayon require the presence of moisture in the steam, and therefore steaming tables are usually preferred. On the other hand, synthetic fibres require heat to promote swelling and therefore relaxation of the structure. Excessive moisture may cause fabric shrinkage and colour bleeding.

Pressure is applied to the garment during pressing to give good crease retention and permanency. Excessive pressure may result in garment or crease distortion.

Vacuum is applied at the completion of the pressing operation. This draws cool air through the garment, reducing the garment temperature, lowering the moisture content and increasing shape retention. Particularly important for garments made from wool and wool blends, this also applies to cotton and viscose blends with synthetic fibres such as polyester and nylon.

15.5.2 Pressing equipment

There is a large range of equipment available for the pressing of garments, from the simple hand steam irons to the sophisticated vertical front (BRI-1400/101) and back pressing robot manufactured by the Viet Group (Figure 15.4) for the pressing of jackets. The development of this machine has the advantage of making the pressing operation more pleasant for the operator as well as reducing the skill level required.

In the operation of this sophisticated equipment, the jackets are loaded onto the pressing former automatically and remain on the hanger at all times during the operation. As the human body is three dimensional, both the front and the back of the garment are pressed while the jacket is hanging on the body former. In this position it hangs as if on the human body, thus making the alignment of seams easier. Since

Garment-finishing techniques

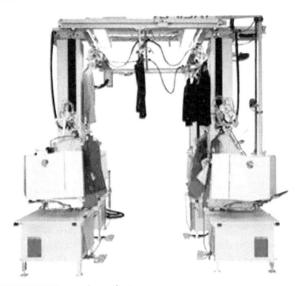

Figure 15.4 BRI-1400/101 pressing robot.

the jacket is in the hanging position during the pressing operation, the lining is in the most appropriate position for even pressing.

As the garment is always in the hanging position there is even pressure applied during pressing. The design of the body form is such that there is little chance of the garment moving during pressing. This improves quality and consistency of the final product. The operator is always in the vertical position rather than being bent, so there is little danger of injuries occurring. Figure 15.5 shows the BRI 1200/101 jacket front

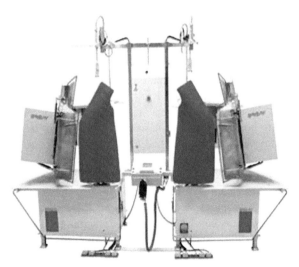

Figure 15.5 BRI-1200/101 jacket front finisher.

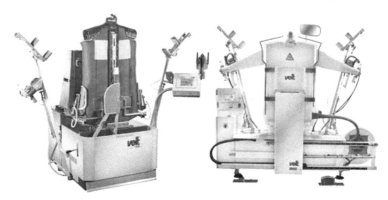

Figure 15.6 Veit 8326 shirt finisher and Veit 8900 shirt press.

finish pressing. Similar to the BRI-1400/101 (Figure 15.4) machine, the jacket hangs on the former so that the alignment of the edges and seams is made easier for the operator. For shirt finishing, two high-performance machines have been developed by the Veit Group: the Veit 8326 shirt finisher and the Veit 8900 shirt press (Figure 15.6).

Special attention is given to the position of the hem and the sleeves, to ensure that the shirt is finished in optimum condition. Excessive stretching of the hem is avoided by the specially developed hem-tensioning device. The advantages of this machine are:

- Shorter cycle times
- Increased energy savings
- Lower radiant heat to the surroundings
- Automatic unloading.

The UADD B58 automatic left and right front jacket press from Hoffman Machinery Co. is fitted with a unique 40-program micro-processor control. The unique feature is automatic fine-pressure control with 2-psi increments. It is possible to connect the micro-processor to a PC network in order to monitor production.

In addition to the above automated machines, hand irons are the most common type of pressing equipment used by households. These irons are heated electrically with the provision of steam supply and temperature control. There are various shapes of the irons and the weight ranges from about 1 to 15 kg. There are several types of pressing tables available for these irons, which may include a simple table or a table with vacuum arrangement to hold the garment or section of a garment in place and dry after pressing. Additional parts can be attached to the table to support various parts of a garment so that a suitable shape is available for each part.

Steam presses are used to assist in better shape retention and improve the efficiency of pressing. The steam presses can be of various shapes with automatic operations. There are provisions of stem supply to all the parts, vacuum and altering the pressure. Some designs can be fitted with a programmed logic circuit to work in varying cycles depending on the type of garment. In some designs, additional extensions such as bucks or matching heads can be attached when the shape of the garments changes. Many other types of pressing equipment are available that will enhance the final quality of the garments produced. Depending on the type of product being produced, different equipment will be required. Some examples include carousel press, specially designed press for trousers and skirts, a steam air finisher and a steam tunnel.

Specially designed machines are available for creasing and pleating. Creasing equipment is used to press the edges of clothing components so that they are easily sewn. For example, the cuffs and patch pockets are formed into shape by the working aid, and are pressed to retain the shape, which makes the sewing operation easier. Pleating machines create a series of creases following a specific pattern or randomly, depending on the type of the cloth. Pleats of various lengths can be prepared by hand or by using machines. Blade-type and rotary-type machines can be used for rapid and accurate pleat creation.

In all the above pressing equipment, it is essential to precisely control the variables so that all the components or finished items are subjected to same conditions to avoid variability. The technological developments for accurate sensing, pressure monitoring and measuring the strength of a cooling vacuum can assist in avoiding the variations. Hence, garment manufacturers should adopt the automatic units to achieve consistent results.

Some garments are manufactured as wash-and-wear or permanent-press garments. These clothes are prepared by special finishing treatment to provide crease-resistant properties. For example, the resin treatment of 100% cotton items can provide crease resistant properties at the expense of loss in strength and abrasion resistance. The manufactured garments can be treated with the speciality chemicals by dipping in the chemical, spraying and vapour phase treatment. In the former two cases, the garment needs to be cured after application of the chemicals.

Incorrect selection of parameters during pressing can lead to shrinkage, colour loss or degradation of the fabric. Hence, all the parameters should be precisely controlled to avoid any damage to the garments. In addition, the accessories used should be able to withstand the processing conditions. In some instances the lack of understanding of the material and the process can cause permanent damage to the batch of garments. Hence, a perfect understanding and training of the operators is essential to avoid such problems. In addition, while specifying the care conditions related to ironing/pressing for a care label, the garment should be tested so that it can withstand the specified conditions (Nayak and Padhye, 2014b).

15.6 Future trends

We have already seen the trend of increased automation in pressing together with increases in the use of computer technology to not only monitor production but also reduce the effect of human error in the pressing operation. It is clear that this trend will continue. Pressing will become more automated, and pressing pressures, temperatures and the moisture content of steam will be critically controlled using appropriate software integrated into a network. Similarly, automation is applied to other garment-finishing operations. Wet finishing and dyeing processes will continue to be modified as the range of fibres and fibre mixtures increases. Furthermore, consumers are continually demanding increased comfort properties from their garments. These properties will be enhanced by the addition of specific chemical finishes to the garment together with differing fibre mixtures.

The concept of nanotechnology is also making its way into garment finishing. New nanotechnology-based concepts such as ease release, quick wick and rare care finishes further improve the functionally of the textile by imparting the various properties such as soil release, anti-pilling effect, water/oil repellency, hygiene effect, easy care and odour-free effect.

15.7 Conclusions

Automation and development in technology, in particular the introduction of microprocessors into the garment manufacturing industry, have had a significant effect on production methods. These changes have greatly increased productivity, improving working conditions as well as product quality. The introduction of more complex finishing machinery will of necessity result in increased training requirements for operators not only in machine operations but also in the maintenance of a safe working environment. The introduction of new fibres such as Sarona® from Invista® will be of beneficial to the garment dyer due to its improved dyeing properties at lower temperatures. Advances in fibre technology will have significant flow-on effects into the finishing of fashion garments, resulting in higher quality, better performance and more economical processing. Introduction of automatic and robot-controlled equipment can result in better garment finishing. Hence, the manufacturers should implement the concept of automation to produce tomorrow's finished garments.

References

Bahners, T., Textor, T., Opwis, K., Schollmeyer, E., 2008. Recent approaches to highly hydrophobic textile surfaces. J. Adhes. Sci. Technol. 22, 285–309.
Bone, J., Collishaw, P., Kelly, T., 1988. Garment dyeing. Rev. Prog. Color. Relat. Top. 18, 37–46.
Cheng, K., Poon, C., Au, C., Tsun, S., 1998. The enzyme washing on silk fabrics. Res. J. Text. Apparel 2.
Deaton, A., 1980. Economics and Consumer Behavior. Cambridge University Press, Cambridge.
Freddi, G., Mossotti, R., Innocenti, R., 2003. Degumming of silk fabric with several proteases. J. Biotechnol. 106, 101–112.
Gao, Y., Cranston, R., 2008. Recent advances in antimicrobial treatments of textiles. Text. Res. J. 78, 60–72.
Gulrajani, M., Agarwal, R., Grover, A., Suri, M., 2000. Degumming of silk with lipase and protease. Indian J. Fibre Text. Res. 25, 69–74.
Harnden, A., Donnelly, M., York, J., 2001. Laboratory-and commercial-scale investigations into the action of cellulase enzymes on Tencel. Color. Technol. 117, 217–224.
Heikinheimo, L., Buchert, J., Miettinen-Oinonen, A., Suominen, P., 2000. Treating denim fabrics with *Trichoderma reesei* cellulases. Text. Res. J. 70, 969–973.

Horrocks, A., 1986. Flame-retardant finishing of textiles. Rev. Prog. Color. Relat. Top. 16, 62–101.
Joshi, M., Wazed Ali, S., Purwar, R., Rajendran, S., 2009. Ecofriendly antimicrobial finishing of textiles using bioactive agents based on natural products. Indian J. Fibre Text. Res. 34, 295–304.
Kang, I.-S., Yang, C.Q., Wei, W., Lickfield, G.C., 1998. Mechanical strength of durable press finished cotton fabrics part I: effects of acid degradation and crosslinking of cellulose by polycarboxylic acids. Text. Res. J. 68, 865–870.
Koo, H., Ueda, M., Wakida, T., Yoshimura, Y., Igarashi, T., 1994. Cellulase treatment of cotton fabrics. Text. Res. J. 64, 70–74.
Nadvi, K., Thoburn, J., Thang, B.T., Ha, N.T.T., Hoa, N.T., Le, D.H., 2004. Challenges to Vietnamese firms in the world garment and textile value chain, and the implications for alleviating poverty. J. Asia Pac. Econ. 9, 249–267.
Nayak, R., Chatterjee, K., Khandual, A., Jajpura, L., 2008. Evaluation of functional finishes-An overview. Man-Made Textiles in India, 51, 130–135.
Nayak, R., Padhye, R., 2014a. Antimicrobial Finishes for Textiles. Elsevier, Cambridge, UK.
Nayak, R., Padhye, R., 2014b. The care of apparel products. Elsevier, Cambridge, UK.
Nostro, P.L., Corrieri, D., Ceccato, M., Baglioni, P., 2001. Enzymatic treatments on Tencel in water and microemulsion. J. Colloid Interface Sci. 236, 270–281.
Partridge, H., 1975. Hosiery and knitted garment dyeing. Rev. Prog. Color. Relat. Top. 6, 56–65.
Ramaswamy, K., Gereffi, G., 2000. India's apparel exports: the challenge of global markets. Dev. Econ. 38, 186–210.
Simoncic, B., Tomsic, B., 2010. Structures of novel antimicrobial agents for textiles-a review. Text. Res. J. 80, 1721–1737.
Tesoro, G.C., Rivlin, J., Moore, D.R., 1972. Flame-retardant finishing of polyester-cellulose blends. Ind. Eng. Chem. Prod. Res. Dev. 11, 164–169.
Ulrich, P.V., Anderson-Connell, L.J., Wu, W., 2003. Consumer co-design of apparel for mass customization. J. Fashion Mark. Manage. 7, 398–412.
Wei, W., Yang, C.Q., 1999. Predicting the performance of durable press finished cotton fabric with infrared spectroscopy. Text. Res. J. 69, 145–151.

Quality control and quality assurance in the apparel industry

C.N. Keist
Western Illinois University, Macomb, IL, USA

16.1 Introduction

Quality control and quality assurance are complex areas of the apparel industry. First off, quality assurance is not quality control, but quality control is an aspect of quality assurance. Quality assurance is the "process of designing, producing, evaluating, and assessing products to determine that they meet the desired quality level for a company's target market" (Kadolph, 2007, p. 6). Quality assurance looks at a product from the first design concept until it is sold to the consumer. Quality control is generally understood as assessing for quality after products have already been manufactured and sorted into acceptable and unacceptable categories. It is costly for companies that do not take a quality assurance method, but only look at quality in terms of quality control. Quality is a multifaceted concept that describes how well a service, process, material, or product possesses desired intangible or physical attributes (Kadolph, 2007).

Several organizations have created standards and specifications to help in assessing consistent quality. Major organizations include the American Association of Textile Chemists and Colorists (AATCC) and the American Society for Testing and Materials (ASTM). These organizations account for the majority of test methods written for the apparel industry. Other organizations include the American Society for Quality (ASQ), American Apparel and Footwear Association (AAFA), Textile Clothing and Technology Corporation (TC^2), American National Standards Institute (ANSI), and the International Organization for Standardization (ISO). These organizations publish industry-wide standards or "commonly agreed on aid for communication and trade; a set of characteristics or procedures that provide a basis for resource and production decisions; a product that meets all specifications and company or product requirements" (Kadolph, 2007, p. 551). These written standards assess fabrics and apparel products usually in terms of characteristics such as pilling, frosting, or color transfer. Individual companies write their own specifications or "a precise statement of a set of requirements to be satisfied by a material, product, system, or service that indicates the procedures for determining whether each of the requirements are satisfied" based on their target market's expectations (Kadolph, 2007, p. 550).

16.2 Quality control in the apparel industry

16.2.1 Preproduction quality control

In preproduction quality control, each component of a garment is tested prior to assembling. Fabric is assessed for major and minor defects as described in Chapter 5. Closures, interlinings, sewing threads, and other design elements are tested for their quality and durability. Fabric with too many defects or closures that do not work properly can be detected prior to construction, which saves time and money in the long run. Fabric, accessories, closures, interlinings, sewing threads, and other design elements are all tested prior to the garment manufacturing in the preproduction quality control phase.

16.2.1.1 Fabric

Fabric quality is of utmost importance to the overall quality of apparel and textile products. Regardless of how well a product is designed or constructed, if the fabric is of poor quality, the product will most likely to fail with the consumer. Most fabric is comprised of fibres that are spun into yarns and then woven or knitted into fabric. Support materials like interlinings usually go from the fibre to the fabric stage. Since fibres are the building blocks of all apparel and textile products, it is important to start with quality fibres regardless if they are natural, manufactured, regenerated, or synthetic. The essential properties of fibre and yarn are beyond the scope of this book. The role of fabric properties in apparel manufacturing has been described in Chapter 3. In addition, various fabric faults and their inspection have been described in Chapter 5. In addition to the inspection parameters described in Chapter 5, fabric can also be assessed in terms of comfort, colorfastness, and durability as described below.

Comfort

Comfort describes "how materials interact with the body and addresses how the body's functional environment can be expanded" (Kadolph, 2007, p. 187). Comfort is studied by looking at fabric in terms of elongation and elasticity, heat retention and conduction, moisture absorbency, water repellency, waterproofing, hand and skin contact, drape, and air permeability (Nayak et al., 2009). Elongation is the fabric's ability to stretch without recovery, whereas elasticity is the fabric's ability to stretch and recover to its original dimension without distortion. Heat retention and conduction of a fabric addresses the way the body reacts to heat. Moisture absorbency is the ability of a fabric to absorb liquid water. Water repellency is a fabric's ability to repel water or other liquids upon initial wetting. This is usually accomplished with a combination of densely woven material and a water-repellent finish. Water will soak through the fabric with extended time and pressure of water. Waterproofing, on the other hand, is a fabric that will not allow water to penetrate through it no matter how much exposure to water, duration, and pressure used. Water repellent and waterproof fabrics are often seen as uncomfortable due to their stiffness and inability to breathe. Hand and skin contact is the way a fabric feels to the touch. Drape is how well a fabric hangs over a body or object. Fabrics with more drapes bend easily around objects and are often seen as more comfortable because the fabric moves with the body. Fabrics with less drape are stiffer and hang away from the body.

Colorfastness

Colorfastness relates to appearance retention and can be described as "how consumers use textile products and includes factors that may cause colorants to change color or migrate from one material to another" (Kadolph, 2007, p. 266). Colorfastness is studied by exposing the fabric to different conditions including acids and alkalis, crocking, environmental conditions, frosting, heat, light, perspiration, or water. Depending on what condition the material is exposed to, colorfastness is analyzed using one of three different approaches: color change, color transfer, or a combination of both change and transfer.

Various dye classes can be susceptible to acids and alkalis in terms of color loss. Crocking is when color from one surface is transferred to another by rubbing. Frosting is color loss on the surface of the fabric and is evaluated by looking at color loss. Frosting can be looked at in combination with abrasion resistant tests or tested on its own. Some dye classes are sensitive to color change when exposed to heat and different light sources. Tests for perspiration and water can be performed together. Fabric samples are soaked in an artificial perspiration and distilled water and placed into the AATCC perspiration tester with swatches of multifibre test fabric.

Durability

Durability evaluates "how various materials used in a product perform when subjected to different conditions" (Kadolph, 2007, p. 152). Durability of a fabric is tested until it fails, and both warp and weft yarns are tested. There are many ways to assess fabric durability, including strength (tensile, tear, and bursting), abrasion, pilling, snagging, and dimensional stability (Nayak and Padhye, 2014). Tensile, tearing, and bursting strength looks at the amount of force used in order for the fabric to rupture. Tensile strength, also known as breaking force, tests for durability when a fabric is placed under tension. The yarns in a fabric (warp or weft), are put under force and stretched (elongation) until they can no longer stretch and then rupture or break. Tearing strength is the force needed to continue a tear that has already been created in a fabric, which is essential for loose-fitting garments. Bursting strength looks at both the warp and weft yarns simultaneously as a diaphragm is inflated under the fabric or a large ball is forced through the fabric.

Abrasion is a progressive loss of fabric caused by rubbing against another surface. It has also been reported to occur through molecular adhesion between surfaces which may remove material. The hard abradant may also plough into the softer fibre surface. The breakage of fibres has been reported to be the most important mechanism causing abrasion damage in fabrics. Abrasion can be of three types: flat or plane, edge, and flex. In flat abrasion, a flat part of the material is abraded, edge abrasion occurs at collars and folds, and flex abrasion rubbing is accompanied by flexing and bending. Abrasion is a series of repeated applications of stress. The selection of suitable yarn and fabric structure can therefore provide high abrasion resistance. Abrasion resistance is the ability of fabric to withstand destruction when a fabric rubs against a surface or other textiles, as when wearing layers or during the cleaning process. Pile retention is a fabric's ability to retain its pile during abrasion. Pile fabrics such as terrycloth and corduroy are susceptible to losing their three-dimensional form when rubbed against another surface as part of the fabric is raised and more exposed to abrasion.

Pilling resistance is the ability for a fabric to withstand forming pills. Pilling is a common problem with apparel products prepared from synthetic fibres. Pills are tiny balls of entangled fibres that stay on the fabric surface. Both woven and knitted fabrics are prone to pilling. The propensity may be related to the type of fibre used in the fabric, the type and structure of the yarn, and the fabric construction. Generally, pills are formed in areas which are especially abraded or rubbed during wear and can be accentuated by laundering and dry-cleaning. The rubbing action causes loose fibres to develop into small spherical bundles anchored to the fabric by few unbroken fibres.

Snagging occurs when yarns are pulled from the surface of the fabric. This is usually a problem with fabrics with longer floats like in satin fabrics or in larger knits, knits with multiple colors, or open-work and lacey knits. Only the appearance of a garment is changed by snagging and its other properties are not affected. Snagging is observed particularly in filament-type fabrics and in extreme cases, a single blemish may render an article unserviceable even though unsightly ladders do not necessarily ensue. Soft twisted yarn and loose fabric structure are prone to snagging which may rupture the yarn and ruin the fabric. Woven fabrics with long floats and fabrics made from bulked continuous filament yarns are susceptible to snagging. Frosting is often a side issue when testing for abrasion resistance. Frosting is the white or discolored areas on a fabric before a hole usually appears. With frosting, the dye of yarns is rubbed off, exposing the insides of a yarn in yarn or fabric/piece-dyed fabrics.

16.2.1.2 Inspection of other accessories

Apparel accessories are inspected in the same manner as other textile and apparel products. Accessories are checked during preproduction, production, and postproduction with a final inspection. Various fashion accessories include closures, interlinings, sewing threads, elastic waistband, and other design elements. These accessories should be able to withstand the care and maintenance procedures devised for the clothing (Nayak and Padhye, 2014). The selection criteria for various accessories are described in Chapter 6. A brief inspection procedure for the accessories is described in the following section.

Closures

Closure strength and durability is extremely important to garment construction and consumer satisfaction. Closures for apparel and textiles products include zippers, buttons, hooks, snap fasteners, drawstrings, hook-and-loop fasteners, and others. Zippers are tested for cross-wise strength of the zipper chain, the holding strength of zipper stop to prevent the slide from coming off, and the gripping strength of the zipper around the teeth. Zippers are also assessed in terms of teeth strength to prevent pulling apart while zipped, prevention of twisting or rippling of the zipper tape, and the zipper's resistance to crushing or breaking. Zippers should be inspected for correct dimensions (i.e., tape width and overall length) and other manufacturing defects (color uniformity of tape, securely locking by slider, smooth slider movement, securely attached top and bottom stops, proper attachment of pull tab with the slider, etc.).

Zippers often fail in the garment as a result of human error during their use, due to: (1) improper attachment during sewing, (2) wrong combination of zipper and garment type, (3) inappropriate garment design, and (4) wrong use by the customer. Zippers may show various problems such as ratcheting, shear, crushed slider, puckering at the attachment, humped and popped zipper, after they are attached to the clothing. Ratcheting (forcibly pulling the slider down the chain by holding the two zipper halves) and shear (relative shifting of one half with respect to the other, with the slider mounted) are the major use related errors, which can permanently damage the zipper. Zippers should be inspected in accordance with the ASTM standards (D 2061, D 2062, D 2057, D 2059, etc.). Buttons are tested for their ability to stay in their buttonhole and impact resistance against creaking, chipping, or breaking during sudden external force. Hook-and-eye or snap fasteners are tested for the amount of force needed to separate the hook and eye or snaps from each other or to tear from the fabric. Hook-and-loop fasteners (also known by the trademark name Velcro) are tested similarly to snap fasteners and measured for the amount of force needed to peel the fasteners apart. Hook-and-loop fasteners are also assessed for the amount of force needed to shift each side of the strip while connected, known as shear strength.

Interlinings

Interlinings, also called interfacing, are generally nonwoven fabrics that add more structure and body to garment components like collars, button plackets, waistbands, and cuffs. Interlinings may be fusible or sew-on. Interlining durability is important for garment construction. Fusible interfacing can become unglued from fabric and shift, creating rippling, puckering, and unevenness. Hence, the fusible interfacing should be tested for their performance for defects such as cracking, bubbling, and delamination during their regular use. Fusible interfacings are susceptible to the adhesive bleeding through causing darker spots on the surface called strike-through. Fusible interlinings are assessed for their ability to stay bonded to the fashion fabric and not shift during wear and cleaning. They are also tested for compatibility and shrinkage. Compatibility indicates good drapability, bulk, and support of the fabric at the attachment point. Shrinkage can cause puckering of the attached point and bubbled appearance. The three parameters such as temperature, pressure, and time should be appropriately selected to avoid improper interlining attachment.

Sewing threads

Sewing thread is the yarn used to combine two or more fabric pieces together in garments, accessories, and other textile products. Thread may be comprised of the same construction and fibre content as the garment, but is often different. Thread encompasses the majority of the stress and strain from movement and needs to be strong and durable. It must resist breaking and be compatible with the rest of the garment in terms of color, care instructions, and construction. Major quality checks for sewing threads include construction (diameter and fineness), strength and elongation, shrinkage, twist, twist balance, and color. The other parameters include sewability (Nayak et al., 2010), imperfections, finish, package density, and winding. Sewability is tested by sewing the thread in the intended fabric at the highest machine speed. The sewn fabric should consist uniform and consistent stitches, which indicates good sewability. Sewing thread should be free from imperfections such as knots, slubs, thick and thin places.

Elastic waistband

Elastic waistbands are tested for fit (as per size) and durability (loss of elasticity). The fit is measured by the force needed to stretch the waistband about $2''$ more than the hip size (as per the size label) and bringing back to the waist size. The force can be measured by a tensile testing machine, which simulates the condition that exists while putting the garment. The durability can be measured by stretching the waistband by 50% and measuring the force needed to stretch it. Then the waistband is laundered 3 times as per a specific standard and again the force is measured to stretch it by 50%. The loss of force in the two cases should be less than 10% for the waistband to be acceptable. If it exceeds 10%, the elastic waistband will be loose due to the loss of elasticity. Alternatively, the durability can be measured by accelerated aging method. The elastic waistband, with marks $10''$ apart is subjected to high temperature (150 °C) for 2 h, followed by cooling to room temperature. The waistband is stretched by 50% and kept stretched for 24 h by a tensile testing machine. Then it is relaxed for 10 min and the distance between the $10''$ marks is measured and the percentage change in the size is calculated. A growth of 8% or higher is not acceptable, whereas any shrinkage is not acceptable as it will cause tighter fit.

Other design elements

Other design elements include beads, sequins, braids, and fringes. They are tested for quality in similar ways as closures. Beads are similar to buttons and are tested for their impact resistance against creaking, chipping, or breaking during sudden external force. Sequins are assessed for their strength and resistance to breaking or tearing. Braids and fringes are checked for their quality in terms of durability from fraying, unraveling, tearing, and ripping.

16.2.2 Quality control during production

Each step in the production process is vital to the overall quality of apparel products. The production of apparel products includes cutting, assembling, pressing and other finishing procedures, and final inspection. Pattern pieces need to be cut with precision and on grain. Cut pattern pieces should be assembled with accuracy and care. Assembled garments are finished and pressed. Poor attention to detail, or carelessness when sewing, could have the domino effect on other components or future assembling. For example, skewed fabric pieces will not fit together easily and sewing is difficult. Poorly sewn garments have popped stitches and loose seams. Poorly pressed garments will not lie on the body correctly and could have permanent wrinkles. The following section describes the quality control of apparels during various production processes.

16.2.2.1 Spreading and cutting defects

Proper care should be taken to avoid any mistakes during spreading, otherwise, it will result in improperly cut components. The major parameters such as ply alignment, ply tension, bowing, and splicing should be done with a great care. Not enough plies to cover the quantity of garment components required should also be taken care. Misaligned plies will result in garment parts getting cut with bits missing in some plies at the edge of the spread. Narrow fabric causes garment parts at the edge of the lay getting cut with bits missing. Incorrect tension of plies, i.e., fabric spread too tight or too loose, will result in parts not fitting in sewing, and finished garments not meeting size tolerances. Not all plies facing in correct direction (whether "one way" as with nap, or "either way" as with some check designs), may create in pattern misalignment or mismatch. This happens when the fabric is not spread face down, face up, or face to face as required.

Garment parts not fully included owing to splicing errors, should be dealt carefully. Spread distorted by the attraction or repulsion of plies caused by excessive static electricity, can result in mismatching checks or stripes. The patterns should be aligned with respect to the fabric grain, or else may not fit or drape properly. Insufficient knife clearance space; mismatching of checks and stripes; missing notches and drill marks; poor line definitions; mixing of wrong-sized components; and too wide markers; are some of the mistakes that should be avoided during spreading.

Cutting is an important phase of the production process. Yards of fabric are laid flat in multiple layers and the marker arranged on top. Precision is needed to cut accurate pieces that will fit together during the assembly process. Cutting defects include frayed edges; fuzzy, ragged, or serrated edges; ply-to-ply fusion; single-edge fusion; pattern imprecision; inappropriate notches; and inappropriate drilling (Mehta, 1992). Frayed edges, fuzzy or serrated edges, scorched or fused edges, are caused by a faulty knife, not sharp enough knife, or knife working at too high a speed. Garment parts are damaged by careless use of knife, perhaps overrunning cutting previous piece. Failure to follow the marker lines results in distorted garment patterns. Top and bottom plies can be of different sizes if the straight knife is allowed to lean, or if a round knife is used on too high spreads. Marker incorrectly positioned on top of the spread can lead to garment parts having bits missing at the edge of a lay. If too tight or too loose, then garment parts are distorted. Notches, which are misplaced, too deep, too shallow, angled, omitted, or wrong type to suit fabric are notch faults. Drill marks, which are misplaced, wrong drill to suit fabric, omitted, not perpendicular through the spread are drilling faults.

16.2.2.2 Defects in assembling

After the pattern pieces have been cut, they are assembled. Many issues and defects can arise during the sewing process. Defects in assembling include defects with both stitches and seams. Possible stitching defects include needle damage, feed damage, skipped stitches, broken stitches, wrong or uneven stitch density, balloon stitches, broken threads, clogged stitches, hangnail, and improperly formed stitches. Possible seam defects include seam grin, seam pucker, incorrect or uneven width, irregular or incorrect shape, insecure back-stitching, twisted seam, mismatched seam, extra material caught in seam, reversed garment part, wrong seam type used, slipping seam, and wrong thread used (Mehta, 1992).

Various types of stitch and seam defects are discussed below:

Needle damage causes the fabric to be damaged by the wrong type or size of the needle during sewing. Feed damage is caused by inappropriate feed system, excessive pressure by foot, higher machine speed, damaged throat plate, and misalignment of feed and foot. Skipped stitches are caused when the sewing thread partially skips and stitching is not performed completely. Broken stitches are caused by the selection of wrong stitch type, too tight thread tension, excessive pressure and machine speed. Uneven stitch density arises when stitching is not performed straight and the machine is not controlled properly. Balloon stitches occur when large or small thread loop projects from the surface of the fabric (face or back) at the knotting point of the thread. Threads can break during sewing due to wrong thread size, needle heat, excessive tension, improper combination of thread and needle, etc. Clogged stitches are caused due to resistive force applied to the garment components during sewing such that there is less feed than the normal. Hangnail arises when several parts of the sewing thread are cut during sewing. Improperly formed stitches can be caused by wrong tension, lack of sewing machine maintenance, and improper machine timings.

Seam grin is the defect when stitch opens due to insufficient sewing thread tension. Seam pucker is the distorted appearance of the sewn garment near the stitches and there are various causes for this. Uneven seam width and incorrect shape (runoffs) are caused by wrongly set guide or improper material handling by the operator. Insecure backstitching arises as subsequent rows of stitch do not cover the first row. Twisted seam is caused by improper fabric alignment, mismatching of notches, and when one fabric ply creeps over the other. Mismatched seam is the problem where transverse seams do not match. Many foreign materials can be caught in the seam; the fabric components may be not positioned in correct side facing up/down, leading to poor appearance. Slipping seam is caused by the slippage of the upper and lower fabric with respect to each other.

In addition to the above faults, there may be other faults related to stitch and seam, such as the wrong type of seam or stitch selection, wrong shade of sewing thread selected, oil marks or stains in the fabric during sewing, blind stitching, which can arise during the production process. Some of the major sewing room problems and possible remedies are discussed in Chapter 12. Furthermore, there are several faults that can arise during the assembling, which are not related to the stitches and seams, some of these faults are discussed below:

Finished garment may not be in proper size due to incorrect patterns, inaccurate marking or cutting, shrinkage or stretch, etc. The garment components may not be symmetrical or wrong size; there may be shade variation due to mixing of batches; wrongly positioned or misaligned components due to incorrect marking or sewing not in the right direction. The accessories such as buttons, zippers are damaged during their attachment or attached inappropriately. Interlinings are positioned incorrectly, twisted, cockling, too tight, or too full; linings may be too tight, twisted, incorrectly pleated, projecting beyond the bottom of the garment, which can lead to fault in the final garment.

16.2.2.3 Defects during pressing and finishing

After garments are constructed, final preparations are completed. These final preparations include pressing garments to help set seams and finish garment shaping. Defects during pressing and finishing include burned garments, water spots, change in original color, flattened surface or nap, creases not correctly formed, fabric of finished garment not smooth, edges stretched or rippled, pockets not smooth, garment not correctly shaped, and shrinkage from moisture and heat (Mehta, 1992). Finishes might also be added during this phase of apparel production. Finishes can be temporary or permanent. Temporary finishes need to be reapplied and include starching and some waterproofing finishes. Permanent finishes change the chemical composition of the fibre, which cannot be changed back to its original form. An example of a permanent finish is mercerization on cotton. During the cleaning process, the effectiveness of finishes can be compromised and diminished.

16.2.3 Final inspection

After materials have been tested for quality and the products have been manufactured, products are tested for their performance requirements, overall appearance, and sizing and fit. Proper sizing and fit can be measured as per the size of the garment or they can be tested by putting the garments in manikins or even live models. They are also checked visually for any faults during the production process. Hence, the quality of stitching, joining of garment components and accessories are inspected. Although each component of a garment is tested individually, in preproduction quality control, products are tested for a final time to assess the compatibility of materials used together and any noticeable fault. Garments are inspected for off-grain fabric, poor or uneven stitching, mismatched plaids or stripes along seams, puckered or extra material caught in seams, and uneven seams along hems, among many other problems that can occur in the apparel industry.

During inspection, some parts of a product are more important than others in terms of allowable defects. Each company defines its own product zones and includes these in their specifications as there is no industry standard. The highest priority zone (could be identified as Zone 1, Zone A, or Zone I) is usually identified as the area that will most likely been seen during a face-to-face conversation, whereas the inside of a garment is not as critical in terms of acceptable defects. Companies will also define what they deem as critical, major, and minor defects. A critical defect results in a flaw that produces an unsafe or hazardous situation like a hole in a latex glove that would compromise the safety of the wearer. A major defect is a flaw that often contributes to product failure or lack of usability for a product. Examples of a major defect could be a broken zipper, broken stitches, or tears in the fabric. A minor defect is a flaw that does not reduce the usability of a product, but still deviates from standards and specifications. Examples of minor defects could be an unclipped thread, untrimmed seam allowance, or slubbed yarns in the fabric.

16.2.3.1 Material interactions

Products are tested for their material interactions or "the way in which materials that are combined in a product act and react when their performance is influenced by the presence of another material" (Kadolph, 2007, p. 384). Even simple textile products, like tablecloths, are comprised of at least two materials: fabric and thread. The way those materials interact with each other is important to the manufacturer and the consumer. Threads can shrink, causing the hem of the tablecloth to scallop. Product performance may increase, stay the same, or decrease based on the material interactions. Material interactions include fabric combinations, thread, closures, and trims. Many of the 1 garments include closures like buttons and zippers.

In many apparel products, more than one fabric is used in the construction of a garment. Collars, cuffs, and button plackets incorporate fusible interfacing for stability; vinyl for chair cushions might have attached foam for comfort; and outdoor gear might have a laminate covering for waterproofing. These fabrics or materials might be successful on their own, but when combined with each other they should be compatible and should not, cause buckling or rippling of the component.

16.2.3.2 Process–material interactions

Process–material interactions are "the effect of a production process on the performance of a material and its effects on the performance of the finished product" (Kadolph, 2007, p. 545). The processes include sewing, fusing, and finishing processes. Many issues can arise when combining both fabric and thread in a garment. Both fabric and thread need to be compatible for optimal process performance; the conditions of the workroom could also be out of balance, and the equipment could not be in ideal working condition or not suited for the material. Many issues can arise when sewing garment pieces including buckling (a tuck in the seam when two fabrics of different lengths are sewn together), yarn distortion, seam grin (individual stitches can be seen in the seam), and seam slippage (incorrect thread and seam type are used for the fabric and yarns start to pull from the seam), among other issues. Issues of fusing occur when inappropriate fusible interfacing, in terms of thickness or permanence of adhesive, is used or is not fused properly. Improperly fused interfacing can cause ridging in components or the adhesive can bleed through to the face of a product, causing unsightly spots. These issues cannot generally be repaired on the consumer side.

16.2.3.3 User–production interactions

User–production interactions are "an examination of the ways in which product and user interact and influence consumer satisfaction with the product" (Kadolph, 2007, p. 555). Products that are finished are used to test these interactions. This includes the design of the product, function, appearance, size or dimension, fit, construction, and packaging. Whole products are inspected for their overall design and whether or not they meet the specified criteria set forth by the company. Seams are checked for overall smoothness, and design motifs, like stripes and plaids, are checked for matching along seams. Products are tested for their function, to see if they are performing the way they are intended to perform. Consumers can participate in wear studies at this point to assess products for their mobility and dexterity. The appearance of a product is assessed visually at what would be considered a conversational distance. The outside and the inside of a garment are examined for conformance to company standards and specifications.

Sizing is not standardized in the apparel industry, but is one of the main concerns for consumers and the manufacturers. Several products that are stated to be the same size are compared in each area of the important dimensions (chest, waist, sleeve length, etc.) for conformance to a company's stated size requirements. Fit is related to size, but the same garment can fit differently on different body types. Live models or mannequins are used to see how the same garment fits on different shapes.

Products are evaluated for their construction during the manufacturing process. Seams are viewed for their integrity and appropriateness for the garment. Packaging is one of the last criteria to be checked. Garments are checked for their adherence to number of items per box, that hang tags are correct and attached in the correct spot, and fibre and care labels are correct and affixed in the correct spot.

16.2.3.4 General requirements for final inspection

The following points should be taken care for the final inspection.

- Work area must be well lighted and the measuring table should be large enough to hold the entire garment spread out flat and buttoned.
- Use a soft fibreglass ruler or a metal ruler that has been calibrated against a rigid steel ruler.
- Cuts should be stored in the auditing storage area to facilitate the access of the boxes for the auditor.
- Sample boxes must be randomly obtained. Cuts that are only partially boxed are not ready for the final statistical audit and should not be audited until all boxes are complete. Samples must be randomly obtained from finished sealed boxes.
- Final Statistical Audits are done following a 4.0 AQL (Acceptable quality level).
- Auditors should establish a routine for inspecting garments in order to eliminate the possibility of overlooking an operation.
- The auditor must be aware of the specifications of the garment.
- Round measurements are made to the nearest 1/8th unless specifications require that it is taken to the 1/16th.
- All operations must be checked in the final audit. Also, tacks, shading, long threads, raw edges, skip stitches, and other defects must be checked.
- Garments with major defects are to be marked by colored tape and set aside for repair.
- Detailed records should be recorded and major defects must be properly recorded with their code.
- Cuts that have not passed a final audit or that have only been partially audited should not be loaded on the truck.
- After inspection, the remainder of the garments in the box must be counted and checked for size. The label on the exterior of the box must reflect what is inside the box.

- Garments that have passed the inspection must be returned to the box in the same manner that they were in when they were taken out. All repairs should be set aside and marked.
- Detailed records of any defects must be recorded.

As an example, general final inspection procedure for a pant and a shirt is discussed below.

Pant inspection procedure
- Lay garment face up and visually check the front for shading, fabric defects, and soil.
- Measure the waist with a metal or fibreglass ruler. Check that the measurement of the waist is the same as the size on the label.
- Check that pockets are functional and have no shaded pieces, missing tacks, and are overall correct.
- Check the placement of the button and that it lines up with the hole. Button and unbutton the garment to ensure that there are no problems with function.
- Check the zipper making sure it is properly placed, the right length and that it is functional (must zip and unzip smoothly).
- Check that the crotch has the correct tacks and no "dog ears". A slight pull should be administered to the crotch area to ensure that all the seams are secure.
- Measure the inseam and verify that it is the same as the size label. Also, measure the inseam to ensure that both legs are the same length.
- Flip garment over and visually inspect the back for shading, fabric defects, and soil.
- Check that the back pockets are properly aligned, have tacks, and are not too open (exposing the inside of the pocket).
- Compare the sobar in the back pocket to the paper ticket and the woven size label to be sure that the garment is correctly labeled.
- Check belt loops for correct size, attachment, and alignment.
- Ensure that the label is properly placed and aligned correctly.
- Turn the pants inside out and inspect all seams and operations.
- Then turn the garment outside in and rebutton, zip, and fold the garment.

Shirt inspection procedure
- Visually inspect the front of the garment for any defect.
- Check that the two sides of the shirt are the same length and evenly meet at the bottom.
- Check that all buttons line up with their button holes and are properly placed. (Also, make sure that the number of buttons is correct and that all of them are securely attached to the garment.)
- Buttons should be checked for function (button and unbutton to ensure that no button holes are too small).
- When checking short-sleeve garments, both arm holes must be checked for size.
- Pockets must be checked for shading, tacks, and placement. Crooked or uneven pockets are unacceptable. Pockets of a patterned fabric must line up according to the print. (A pocket set even slightly off can be very apparent when using a patterned fabric.)
- Garments must be turned inside out and all seams must be checked.
- Three garments of every size must be measured. (Bust, sweep, collar, yoke, cuff, arm hole, natural shoulder, and pockets must all be measured and compared to the specifications of the garment.

Lot failure

If a lot fails, then a 100% inspection must be done. First, 20% must be inspected and those results should be combined with the failure results. If the lot still fails, then continue to inspect 100%.

16.2.4 Developing a sampling plan

Although quality has been incorporated into each product up to this point, products are selected for audits and sorted into acceptable or unacceptable categories prior to shipment to their final destination. There are many types of samples including random, representative, convenience, stratified, constant percentage, and systematic samples. A random sample is where every item has an equal chance of being selected. A representative sample includes a planned variation of items in a ratio that is appropriate. A convenience sample is made up of items that are easier to inspect over others and not random. A stratified sample is selecting a sample when a large lot of similar items exist. A constant percentage sample is sampling with a known constant percentage regardless of lot size to determine the sampling size. A systematic sample consists of items from equal intervals of time or the same location (Kadolph, 2007).

Types of sampling plans include lot-by-lot sampling, lot-by-lot sampling by attribute, skip-lot sampling, continuous production sampling, and arbitrary sampling.

16.2.4.1 Lot-by-lot sampling

A lot-by-lot sampling is the case where samples are taken from each production lot. A lot-by-lot sample by attribute is similar to a lot-by-lot sampling plan, but items are inspected according to their attributes. A single sampling plan occurs when the total size of the sample used, and the total number of sample products inspected, is equal. A double or multiple sampling plan extends the sampling to more products when the first sampling plan barely meet the acceptance levels (Kadolph, 2007).

16.2.4.2 Skip-lot sampling

A skip-lot sampling is where some production lots may not be represented. The number of items sampled is decreased. Skip-lot sampling is used in the apparel industry for factories that produce basic goods or consistently high-quality goods (Kadolph, 2007).

16.2.4.3 Continuous production sampling

A continuous production sample is for products that are staple items and stay consistent for extended periods of time. This practice is common for basic apparel items like pantyhose, where the production line does not stop for long periods of time (Kadolph, 2007).

16.2.4.4 Arbitrary sampling

If a company does not employ a 100% inspection plan, arbitrary sampling is used. With arbitrary sampling, companies determine a set percentage to sample and do not inspect the whole production lot. The most common percentage amount used in arbitrary sampling is 10%. Arbitrary sampling is not always the most sound sampling plan in the apparel industry as sometimes 10% inspection is too small a sample size or too large a sample size based on the production lot (Mehta, 1992). Statistical sampling represents a better sampling plan in the apparel and textile industries.

16.2.4.5 Statistical sampling

Statistical sampling indicates "selecting a sample of units from a lot or shipment of material or product, inspecting the sample for defects, and making a decision as to whether the lot is acceptable or not based on the quantity of the sample" (Mehta, 1992, p. 56). Conducting statistical sampling is often the best choice of sampling plan in the apparel industry by representing the whole of the production lot relatively well.

Associated with statistical sampling of apparels, acceptance quality level (AQL) is the "minimal standard for a satisfactory process or product average" (Kadolph, 2007, p. 434). The detailed information on AQL can be found in the Military Standard 105 (MIL-STD-105). This standard is widely used for acceptance sampling around the world and the corresponding international standard is ISO 2859. Since 100% inspection is not conducted, AQL inevitably allows some defective products through the inspection process, but calculates the maximum number of defective products in a lot before the production lot is rejected. AQL depends on percentage defective or defects per 100 units. Percentage defective is used in the apparel industry for simply made products, components, and materials. It is calculated by taking the number of defective products, multiplying by 100 and dividing by the number of units inspected. Defects per 100 units is a more exact way to find acceptable production lots. Defects per 100 units is calculated by taking the total number of defects, multiplying by 100 and dividing by the number of units inspected (Kadolph, 2007).

Three groups of sampling plans are mentioned in MIL-STD-105 such as single, double, and multiple sampling plans. This chapter will deal with single sampling plan and double sampling plan only as they are generally used for apparel inspection rather than the multiple sampling plan. Although there are three tables in MIL-STD-105 standard, only two tables will be discussed for the sake of simplicity for apparel inspection. The table used to find the code letter for a lot size, is not discussed here. However, the corresponding lot sizes are mentioned in these two tables, Tables 16.1 and 16.2, which describe the single sampling plan and double sampling plan, respectively, for general inspection levels. These tables also indicate the number of samples to be inspected and the number for acceptance or rejection of a lot. Generally, for clothing items 1.5%, 2.5%, 4.0%, 6.5%, and 10% AQL values are used depending on the type and price of the clothing. The following examples will explain the method of using the AQL table.

Table 16.1 Master table for normal inspection (single sampling)

Lot size	Inspect (sample size)	AQL (normal inspection)				
		1.5	2.5	4.0	6.5	10
		Accept/Reject	Accept/Reject	Accept/Reject	Accept/Reject	Accept/Reject
2–8	2	↓	↓	↓	0/1	↓
9–15	3	↓	↓	0/1	↑	↓
16–25	5	↓	0/1	↑	↓	1/2
26–50	8	0/1	↑	↓	1/2	2/3
51–90	13	↑	↓	1/2	2/3	3/4
91–150	20	↓	1/2	2/3	3/4	5/6
151–280	32	1/2	2/3	3/4	5/6	7/8
281–500	50	2/3	3/4	5/6	7/8	10/11
501–1200	80	3/4	5/6	7/8	10/11	14/15
1201–3200	125	5/6	7/8	10/11	14/15	21/22
3201–10,000	200	7/8	10/11	14/15	21/22	↑
10,001–35,000	315	10/11	14/15	21/22	↑	↑
35,001–150,000	500	14/15	21/22	↑	↑	↑
150,001–500,000	800	21/22	↑	↑	↑	↑
500,001 and Over	1250	↑				

↓ = Use first sampling plan below arrow. If sample size equals or exceeds lot or batch size, do 100% inspection. ↑ = Use first sampling plan above arrow. Accept/Reject indicates the numbers for acceptance/rejection, respectively. This table is a part of the Military Standard MIL-STD-105 for garment inspection only. The full table can be viewed in the original standard.
Source: MIL-STD-105E/ BS 6001/ DIN 40080/ ISO 2859.

Table 16.2 Master table for normal inspection (double sampling)

Lot size	Sample	Sample size	Cumulative sample size	1.5 Accept/Reject	2.5 Accept/Reject	4.0 Accept/Reject	6.5 Accept/Reject	10 Accept/Reject
2–8								
9–15	First	2	2	→	→	→	•	→
	Second	2	4	→	→	•	←	→
16–25	First	3	3	→	•	←	→	0/2
	Second	3	6	•	←	→	1/2	1/2
26–50	First	5	5	←	→	0/2	0/2	0/3
	Second	5	10	→	0/2	1/2	1/2	3/4
51–90	First	8	8	0/2	0/3	0/3	0/3	1/4
	Second	8	16	1/2	3/4	3/4	3/4	4/5
91–150	First	13	13	0/3	0/3	1/4	1/4	2/5
	Second	13	26	3/4	4/5	4/5	4/5	6/7
151–280	First	20	20		1/4	2/5	2/5	3/7
	Second	20	40			6/7	6/7	8/9
281–500	First	32	32			2/5	3/7	5/9
	Second	32	64				8/9	12/13

← AQL (normal inspection) →

Quality control and quality assurance in the apparel industry

Lot size		Sample size	Cumulative sample size					
501–1200	First	50	50	1/4	2/5	3/7	5/9	7/11
	Second	50	100	4/5	6/7	8/9	12/13	18/19
1201–3200	First	80	80	2/5	3/7	5/9	7/11	11/16
	Second	80	160	6/7	8/9	12/13	18/19	26/27
3201–10,000	First	125	125	3/7	5/9	7/11	11/16	↑
	Second	125	250	8/9	12/13	18/19	26/27	
10,001–35,000	First	200	200	5/9	7/11	11/16	↑	↑
	Second	200	400	12/13	18/19	26/27		
35,001–150,000	First	315	315	7/11	11/16	↑	↑	↑
	Second	315	630	18/19	26/27			
150,001–500,000	First	500	500	11/16	↑	↑	↑	↑
	Second	500	1000	26/27				
500,001 and Over	First	800	800	↑	↑	↑	↑	↑
	Second	800	1600					

↓ = Use first sampling plan below arrow. If sample size equals or exceeds lot or batch size, do 100% inspection. ↑ = Use first sampling plan above arrow. Accept/Reject indicates the numbers for acceptance/rejection, respectively. • = use corresponding single sampling plan (or alternatively, use double sampling plan below where available). This table is a part of the Military Standard MIL-STD-105 for garment inspection only. The full table can be viewed in the original standard.

Source: MIL-STD-105E/ BS 6001/ DIN 40080/ ISO 2859.

Example 1: In a single sampling if we need to inspect a lot size of 1500 garments, with an AQL of 6.5%, the following steps should be followed:

1. The number of samples to be inspected for a lot size of 1500 = 125 (Table 16.1)
2. The corresponding values at 6.5% AQL for Accept/Reject are 14/15. This indicates, if the number of defective garments is 14 or less, out of the 125 garments inspected, the whole lot (1500) should be accepted. If the number of defective garments is 15 or more out of the 125 garments inspected, the whole lot (1500) should be rejected.

Example 2: In a double sampling if we need to inspect a lot size of 1500 garments, with an AQL of 6.5%, the following steps should be followed:

1. The number of samples to be inspected for a lot size of 1500 = 80 (Table 16.2)
2. The corresponding values at 6.5% AQL for Accept/Reject are 7/11. This indicates, if the number of defective garments is seven or less out of the 80 garments inspected, the whole lot (1500) should be accepted. If the number of defective garments is 8, 9, or 10, take another sample of 80 garments for second inspection (which makes a total of 160 garments for inspection). The Accept/Reject values now for the second inspection are 18/19 (cumulative). This indicates, the total (cumulative) defective garments out of the 160 inspected should be 18 or less defective. If it is 19 or more, the lot is rejected.

16.2.5 Post-production quality evaluation

Post-production quality evaluation in the apparel industry includes wear testing for realistic reactions to everyday scenarios and testing with a simulation study when a consumer's reliability is in question. In wear testing, which is sometimes called product testing, companies provide a small group of consumers with products. Consumers are contracted to wear garments under certain stated guidelines and requirements in order to determine whether they meet the company's intended performance criteria. Consumers report back to the company and identify issues with the product before an entire production lot of garments are produced. Testing with a simulation study is similar to wear testing, but a consumer's safety might be in question. Companies would test items like helmets with a simulation prior to producing an entire production lot, or would test the effectiveness of nonskid shoes on wet surfaces. Appearance retention and care are other aspects of post-production quality evaluation.

16.2.5.1 Appearance retention

Appearance retention is "the degree to which a textile product retains its original appearance during storage, use, and care" (Kadolph, 2007, p. 526). Aspects of appearance retention include wrinkle recovery, storage, and resistance to insects, fungus and bacteria, aging, and dye transfer (Nayak and Padhye, 2014). Wrinkle recovery is a fabric's ability to recover from creases and fabric deformations created during storage and is different from durable press during care of a garment. In addition to wrinkle recovery during storage, products can be susceptible to insects, fungi, bacteria, aging, and dye transfer, which should also be taken care to retain the appearance.

16.2.5.2 Care

Care is "the procedure(s) recommended for returning a soiled item to its clean and as near to new condition as possible" (Kadolph, 2007, p. 529). Care is looked at in terms of home care and dry cleaning. Home care is assessed in terms of colorfastness, dimensional change, appearance retention during cleaning, and durability of finishes. Colorfastness is assessed by looking at both color loss and color transfer. Color loss occurs when dye molecules detach from the fabric; these might attach to another surface, which is called color transfer. Dimensional change is when a product shrinks or grows due to the laundering process. Products can shrink (or grow) in the lengthwise direction, in the cross-wise direction, or in both directions. Appearance retention during cleaning looks at how a product retains its original appearance during the cleaning process. Products abrade against each other or transfer dye between garments. The care labels in a garment describe the appropriate methods of washing or dry-cleaning and ironing to take care of it. The details of care labellings is described in Chapter 17.

16.3 Future trends

What an exciting time for the textile industry with new technologies always on the forefront of the marketplace. Current and future trends in quality control and quality assurance in the apparel industry include, but are not limited to, green or environmentally friendly textile testing, sustainability, and analyzing new horizons in nanotechnology textiles. The "green" movement, or practicing environmentally friendly methods, is not a new trend, but has gained momentum in the last several years. The apparel industry is one of the biggest culprits and promoters of this movement. The entire manufacturing systems from fibre to garment, in apparel production is not environmental friendly.

Traditionally in the textile and apparel industry, fabric inspection has been done by trained inspectors. With any process that is conducted primarily by humans, error is inevitable. Physically inspecting fabric can become tedious and is repetitive in nature, which causes eye fatigue. Textile companies are now installing machines that inspect fabric prior to shipment. With automated fabric inspection, 100% of the fabric is inspected. Although the initial cost is much greater than using skilled fabric inspectors, the cost is reduced as more fabric defects are identified prior to shipment and manufacturing. Widths of fabric are run through automated fabric-inspections machines that use light and cameras to look for fabric defects including slubbed yarns, color discrepancies, holes, etc. Complex algorithms are used to identify the number of defects in the fabric (Chan et al., 1998; Banumathi and Nasire, 2012).

Merriam-Webster (an on-line dictionary) defines sustainability as "able to be used without being completely used up or destroyed." Cotton Incorporation stated that the industry is replacing practices in order for denim jeans to be more environmentally friendly. These practices include replacing sodium hypochlorite with enzymes for bleaching, using laser etching to replace harsh chemicals for sanding and spraying, dry ice blasting in place of sand blasting, eliminating washing and abrading, digital printing on denim to create different effects, and energy audits in textile mills (Musante, 2013a).

Smart textiles or "e-textiles" incorporate electronics, lights, circuits, and other nontraditional components into fabrics and garments. For example, designers incorporate fibre-optic lights that make their clothing glow. New standards and specifications will have to be written to account for these new e-textiles. One major concern for these futuristic fibres is durability, specifically related to care. Another concern is whether the e-textile will detract from the overall appearance or hand of the garment (Musante, 2013b).

Safety "addresses the physical risks to which the user of a textile product is exposed" (Kadolph, 2007, p. 215). A major safety issue is flammability, which is mandatory to address for children's sleepwear, mattresses, and other products that might come into contact with a heat source like firefighter uniforms. Voluntary procedures include thermal protection, chemical resistance, and impact resistance. Chemical resistance would be important for people working in a factory, laboratory, or medical environment. Impact resistance would be used for safety gear such as helmets and knee or elbow pads. Health is the "interaction of physical, mental, emotional, and social aspects of an individual" (Kadolph, 2007, p. 223). Health issues include allergens and irritants, ultraviolet protection, and biological resistance. Ultraviolet protection is a new and emerging area of apparel where finishes are added to garments and provide a barrier from the sun's harmful rays similar to the effects of wearing sunscreen.

16.4 Conclusions

Quality assurance and quality control are the important, unique, and complex area of the textile, apparel, and accessories industries. Quality assurance is not quality control, but quality control is an aspect of quality assurance. Quality needs to be incorporated into every aspect of a product from the original design concept to the marketing and selling of a product. Many organizations, such as AATCC and ASTM, focus on setting standards for the industry. Individual companies look at those standards and set their own specifications based on their target market.

Quality is assessed in the apparel industry during preproduction, production, and post-production operations. In the preproduction phase, fabric, accessories, closures, interlinings, sewing threads, and other design elements are tested prior to the construction of garments. The production of apparel products includes cutting, assembling, pressing and other finishing procedures, and final inspection. At each step of the production phase, garments are assessed for quality. Companies develop various sampling plans to inspect products during the final inspection. Post-production quality evaluation includes wear testing and testing with a simulation.

Quality assurance and quality control are evolving in apparel industry with technological advances, similar to most other industries in the world. Fabric inspection is now an automated process, and can find defects in fabrics quicker, more accurately, and save money in the long run. Current and future trends in quality assurance and quality control in the apparel industry include, but are not limited to, green or environmentally friendly textile testing, sustainability, and analyzing new nanotechnology applications in textiles. Incorporating quality assurance into an apparel company program can improve product quality and aid in increasing customer satisfaction.

16.5 Sources of further information and advice

For further information on the topic of quality assurance and quality control in the apparel industry, consult the following materials:

> American Apparel & Footwear Association 2014. Available from: http://www.wewear.org/ (14.07.14).
> American Association of Textile Chemists and Colorists 2014. Available from: http://www.aatcc.org/ (14.07.14).
> American National Standards Institute 2014. Available from: http://www.ansi.org/ (14.07.14).
> American Society for Quality 2014. Available from: http://asq.org/index.aspx (14.07.14).
> American Society for Testing and Materials 2014. Available from: http://www.astm.org/ (14.07.14).
> Das, S., 2009. Quality Characterisation of Apparel, Woodhead Publishing India Pvt Limited.
> International Organization for Standardization 2014. Available from: http://www.iso.org/iso/home.html (14.07.14).
> Mehta, P., 2004. An Introduction to Quality Assurance for the Retailers, iUniverse, Inc., New York.
> Stojanovic, R., Mitropulos, P., Koulamas, C., Karayiannis, Y., Koubias, S. & Papadopoulos, G., 2001. Real-time vision-based system for textile fabric inspection. Real-Time Imaging, pp. 1–12. Available from: http://www.idealibrary.com.
> Textile and Clothing Technology Corp 2014. Available from: http://www.tc2.com/ (14.07.14).
> Textile Learner 2001. Quality control system in garments industry. Textilelearner Blogspot. Available from: http://textilelearner.blogspot.com/2011/08/quality-control-system-in-garments_2589.html (14.07.14).

References

Banumathi, P., Nasira, G.M., 2012. Fabric inspection system using artificial neural networks. Int. J. Comput. Eng. Sci. 2 (5), 20–27.

Chan, C., Liu, H., Kwan, T., Pang, G., 1998. Automation Technology for Fabric Inspection System. Available from: http://www.researchgate.net/publication/228577577_Automation_technology_for_fabric_inspection_system.

Kadolph, S.K., 2007. Quality Assurance for Textiles & Apparel, second ed. Fairchild Publications, New York.

Mehta, P.V., 1992. An Introduction to Quality Control for the Apparel Industry. ASQC Quality Press, Milwaukee, WI.

Musante, G., 2013a. The greening of cotton. AATCC News, 19 November. Available from: http://aatcc.informz.net/aatcc/archives/archive.html (02.03.14).

Musante, G., 2013b. Wash and wear out? AATCC News, 8 October. Available from: http://aatcc.informz.net/aatcc/archives/archive.html (02.03.14).

Nayak, R., Padhye, R., 2014. The care of apparel products, in Textiles and fashion: Materials, design and technology. Elsevier, United Kingdom, pp. 799–822.

Nayak, R., Punj, S.K., Chatterjee, K.N., Behera, B.K., 2009. Comfort properties of suiting fabrics. Indian J. Fibre. Text. Res. 34, 122–128.

Nayak, R., Padhye, R., Gon, D.P., 2010. Sewing performance of stretch denim. J. Text. Apparel, Technol. Manage. 6 (3), 1–9.

Rana, N., 2012. Fabric inspection systems for apparel industry. Indian Text. J. Available from: http://www.indiantextilejournal.com/articles/FAdetails.asp?id=4664 (15.07.14).

Care labelling of clothing

17

R. Nayak, R. Padhye
School of Fashion and Textiles, RMIT University, Melbourne, VIC, Australia

17.1 Introduction

Apparel and textiles are soiled during their normal use. From an economic point of view, these items must be cleaned and refurbished for reuse without substantially altering their functional and aesthetic properties. It is essential that the various processes to which the apparel is subjected should maintain and restore the desirable and functional properties. This is a joint responsibility of the textile and apparel industries, the textile care organizations and the consumers.

The Federal Trade Commission (FTC) in United States promulgated a trade rule on care labelling of wearing textiles and certain piece goods in 1972 (Chatterjee et al., 2006). The rule requires that apparel items should have a permanent care label that provides information about their regular care instructions. The purpose of the rule is to give the consumer accurate care information to extend the useful life of garments (Davis, 1987).

Care symbols provide all the necessary information on washing, bleaching, ironing, dry-cleaning and tumble drying (Mehta and Bhardwaj, 1998). The consumer usually does not have the experience or technical knowledge to decide which care treatment is suitable. So it is the responsibility of the apparel manufacturers to provide the necessary care information for the products. All the textile wearing apparel used to cover or protect the body, and all piece goods sold for making home-sewn apparel, are covered in care labelling apart from shoes, belts, hats, neckties, non-woven garments and one-time garments.

Without care information, the consumers will face trouble in deciding on the appropriate conditions for care treatment of apparels (Mupfumira and Jinga, 2014). Care labels should not be considered as a guaranty or a quality mark of the product. The following people or groups are covered in care labelling:

- Manufacturers of textiles and apparels.
- Manufacturers of piece goods sold at a retail price to consumers for making wearing apparel.
- Importers of wearing apparel and piece goods for making wearing apparel.
- Any organization that directs or controls the manufacture and/or import and export of textile wearing apparel or piece goods for making wearing apparel.

17.2 Requirements of care labelling

According to the FTC rule, anyone dealing with apparels or other textile products must establish a suitable basis for care information and it should be sufficient to keep the garment safe during its use (Kadolph, 2007). Various care labelling systems are followed worldwide, which may differ in terms of the symbols or the wordings that convey the message. Whatever the system may be, it should follow the following set of guidelines (Davis, 1987):

- All the symbols used in the care labelling system should be placed directly on the article or on a label, which shall be affixed in a permanent manner to the article.
- The symbols may be produced by weaving, printing or other processes.
- Care labels should be made of a suitable material with resistance to the care treatment indicated in the label at least equal to that of the article on which they are placed.
- Labels and symbols should be large enough so that they are easily visible and readable.
- All the symbols should be used in the prescribed order, and they denote the maximum permissible treatment.
- The consumers should easily understand the symbols irrespective of the language.
- The care symbols are applicable to whole of the garment including trimmings, zippers, linings, buttons, etc. unless otherwise mentioned by separate labels.
- The care symbols selected should give instructions for the most severe process or treatment the garment can withstand while being maintained in a serviceable condition without causing a significant loss of its properties.
- The label, with the symbols and words on it, should be legible throughout the useful life of the garment.
- The machines used for washing and drying should be able to provide the conditions mentioned in the care label.
- The care labels should not be visible from outside.
- They should not be inconvenient or cause irritation to the wearer.
- They should be easily visible and not hidden, which would otherwise lead to difficulties in conveying information.
- The labels for a particular style should be positioned at one place for the whole lot, either at the back, top or middle.
- If not readily seen due to packaging, care information must be repeated on the outside of the package or on a hangtag attached to the product.
- It is always not possible to have all the information in one label due to the type of the garment, material and fashion requirements. In these cases it is permissible to go for the second label.
- When a garment consists of two or more parts and is always sold as a unit, only one care label can be used if the care instructions are the same for all the pieces. The label should be attached to the major piece of the suit. If the suit pieces require different care instructions or are designed to be sold separately, like coordinates, then each item must have its own care label.

The care labels should be used for a wide range of products such as apparel textiles, household textiles, home furnishings, resin-coated fabrics, piece goods made from textiles, suede skins, leathers, and furs (Nayak et al., 2014). Care labelling used in clothing should provide the consumers enough information on: (1) care instructions for clothing and other textile products, (2) prior knowledge of care and maintenance costs of the materials such as dry-cleaning, (3) the processes and conditions to avoid in order to maximize the useful life of clothing and textile products and (4) possible damages that can happen such as dyes running out (e.g. wash separately) during care and maintenance.

The information provided in the care labels also affects the purchase decision of the consumers (Davis, 1987; Koester and May, 1985; Heisey, 1990; Holmlund et al., 2011).

Care labelling of clothing 429

During clothing purchase, the information on the fibre content (Huddleston et al., 1993) and the cost involved with the care procedures (Huddleston et al., 1993) are the most sought in addition to the price, physical characteristics (colour, size and style) and brand name.

17.3 Definition of care label

According to ASTM D 3136-96, a care label is a label or other affixed instructions that report how a product should be taken care. Care instructions are a series of directions that describe practices that should refurbish a product without adverse effects and warn against any part of the directions that one could reasonably be expected to use that may harm the item. The FTC definition states that 'Care label means a permanent label or tag, containing regular care information and instruction, that is attached or affixed in some manner that will not become separated from the product and will remain legible during the useful life of the product'.

The care label informs sales personnel and consumers of the appropriate care and treatment of the textile and the other materials used in its production. Correct labelling and careful compliance with the information given on the care label help to ensure a long life for the textile items. Care labels help to prevent irreversible damage to the textile article during its care processes. The care labels generally contain the following information, although the statutory provisions may vary from country to country:

- Care symbols
- Fibre content (percent of each fibre)
- Size
- Country of origin
- Further information, such as eco labels, etc.

17.3.1 Different processes described by care labels

Care labels describe useful information on the processes used for care and maintenance of clothing items, which include laundering or washing, bleaching, tumble drying, ironing and dry-cleaning.

17.3.1.1 Laundering

A process intended to remove soil or stains by treatment (washing) with an aqueous detergent solution (and possibly bleach) and normally including subsequent rinsing, extracting and drying. The process may be further divided as hand washing, home laundering and commercial laundering.

- Hand washing: The gentlest form of home laundering, using hand manipulation without the use of machine or device such as a scrubbing board.
- Home laundering: A process, by which textile products or parts thereof may be washed, bleached, dried and pressed by non-professional use.

- Commercial laundering: A process, by which textile products or specimens may be washed, bleached, rinsed, dried and pressed typically at higher temperatures, higher pH and longer times than used for home laundering.

The laundering process may include various operations in relevant combinations such as soaking, pre-washing and proper washing carried out usually with heating, mechanical action and in the presence of detergents or other products and rinsing. Water extraction, that is spinning or wringing, is performed during and/or at the end of the operations mentioned above.

17.3.1.2 Bleaching

Bleaching helps to remove stains on white clothes and retain their brightness. It can remove the colour when used on coloured clothes. The bleaching agents can be classified as chlorine bleach or non-chlorine bleach.

1. **Chlorine bleach:** A process carried out in an aqueous medium before, during or after washing processes, requiring the use of chlorine-based bleaching agents for the purpose of removing stains and/or improving whiteness.
2. **Non-chlorine bleach:** Bleach that does not release the hypochlorite ion in solution, for example sodium perborate, sodium percarbonate, etc.

17.3.1.3 Tumble-drying

A process carried out on a textile article after washing, with the intention of removing residual water by treatment with hot air in a rotating drum.

17.3.1.4 Ironing or pressing

Ironing is a method that uses a heated iron (with or without the presence of steam) to smooth or retain the shape, by the application of heat, moisture, and pressure.

17.3.1.5 Dry-cleaning

Dry-cleaning is the process of cleaning textile articles by means of organic solvents (e.g. petroleum, perchloroethylene and fluorocarbon). This process consists of cleaning, rinsing, spinning and drying.

17.3.2 Terminologies used in care labelling

Besides the above processes used for care labelling, various other terminologies are related to care labelling, and the consumers should understand them thoroughly, which are described below.

1. **Detergent:** A cleaning agent containing one or more surfactants as the active ingredient(s).
2. **Soap:** A cleaning agent usually consisting of sodium or potassium salts of fatty acids.
3. **Bleach** (in care of textiles): A product for brightening and aiding the removal of soils and stains from textile materials by oxidation that is inclusive of both chlorine and non-chlorine products.

4. **Cleaning agent:** A chemical compound or formulation of several compounds that loosens, disperses, dissolves or emulsifies soil to facilitate its removal by mechanical action.
5. **Consumer care:** Cleaning and maintenance procedures as customarily undertaken by the ultimate user.
6. **Professional care:** Cleaning and maintenance procedures requiring the services of a person specially trained or skilled in these.
7. **Refurbish:** To brighten or refresh up and restore the wearability or use by cleaning such as dry-cleaning, laundering or steam cleaning.
8. **Stain removal:** A cleaning procedure for localized areas with cleaning agents and mechanical action specific to the removal of foreign substances present in a clothing.
9. **Solvent relative humidity:** The humidity of air over dry-cleaning bath and in equilibrium with the solvent and small amount of water.

The care labels for various types of clothing should be positioned at appropriate places in a particular clothing style (see Table 17.1). However, individual manufacturers can slightly vary these positions, but should follow the instructions described in the 'Requirements of Care Labelling' section earlier in the chapter.

17.4 Care labelling systems

At present, there is no universal care labelling system. In the United States, the Wool Products Labelling Act (1938), the Fur Products Labelling Act (1951), the Flammable Fabrics Act (1958) and Rule on Care Labelling (1972) are in force. The JIS (Japan Industrial Standard) for care labelling came into force in 1962. Similarly, in Korea, the rule on Quality Labelling came into force in 1969, and the use of symbols for care labelling of apparel products was published in 1972.

The ASTM system is accepted in North American Free Trade Agreement (NAFTA) countries. The International Organization for Standardization (ISO or GINETEX) system is accepted in most of Europe and Asia. Japan has its own system, JIS. Negotiations are under way to harmonize the two major systems (ASTM and GINETEX) into a universal labelling system for care procedures. An international labelling system can facilitate global trade by avoiding technical or standards barriers. The major systems followed worldwide are ASTM, ISO (GINETEX), British (Home Laundering Consultative Council System, or HLCC), Canadian, Dutch and Japanese care labelling systems. Although there are some variations in the symbols, the five basic symbols used in many of these systems (e.g. ISO) are given in Figure 17.1.

The technical committee (TC-38) of the ISO handles all types of textile standards through several subcommittees. Subcommittee SC-11 is concerned with developing standards for care labelling, with the primary objective of developing

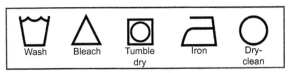

Figure 17.1 Five symbols used in the ISO system for care labelling.

Table 17.1 Positioning of care labels in various garments

Type of garment	Position of label	Type of garment	Position of label
Clothing for men and boys			
Coats, formal jackets, overalls	On the left inside breast pocket; if there is not one, on the lining or the facing, on the left side.	Shirts	At the back or near the neck; if possible on the side seam above the hem.
Sports jacket	On the left inside breast pocket; if there is not one, on the lining or the facing.	Ties	On inside.
Trousers	At the top centre of the right rear pocket; if there is not one, on the waistband at the back.	Pre-packed shorts	At the back on the inside, in the middle of the waistband.
Ski-pants and trousers to be worn with a belt (knitted)	On the waistband at the back; if there is not one, at the top of the back centre seam.	Pre-packed vests Swimming trunks	At the top centre back. At the top of the left side seam.
Clothing for women and girls			
Coats, suits	On the lower front facing. If there is no facing or if, after making up, it is not suitable for carrying a label, at the top centre back.	Overalls and jumpers	At the top centre back (or, if the material is transparent or the overalls have no neck, in the left side seam above the hem).
Dresses	When fashion permits, at the top centre back; otherwise on the left side seam, above the hem.	Pinafore dresses	At the top centre back (with size indication).

Care labelling of clothing

Blouses	At the bottom, on the left side seam, above the hem.	Underwear	At the top centre back. Exception: For cami-knickers, in the middle of the side seam.
Skirts, trousers	At the waistband, at the top centre back.	Aprons	At the joint between the body of the apron and the left tie.
Corsetry			
Brassieres, short or long	At the back left, at the lower edge of the garment.	Non-stretch corselettes, woven corselettes (non-stretch)	At the back left, at the lower edge of the garment.
Stretch girdles short or long, stretch panty girdles short or long, stretch corselettes	At the top centre back.	Athletic support (non-stretch), suspender belt (non-stretch)	At the back left, at the lower edge of the garment.
Tights	At the centre back.	Stockings	On the packing.
Women's swimwear			
One-piece	At the top of the left side seam.	Two-pieces Top piece Bottom piece	At the top of the left side seam. At the top of the left side seam.
Clothing for men, boys, women and girls			
Pullovers, knitted waistcoats, knitted jackets, anoraks, track suits, tops and bottoms, nightwear for men, women and children	At the top centre back.	Reversible anoraks, Dressing gowns, housecoats, bathrobes Baby clothes (Excluding nappies)	In the pocket. At the neck, beside the size marking. For articles with side seams, on the left one. For articles without side seams, on the left shoulder seam. For all-in-one rompers, on the top outside hem.

Continued

Table 17.1 Continued

Type of garment	Position of label	Type of garment	Position of label
Other articles			
Table and bed linen (white or coloured) not to be soiled (i.e. easy-care articles)	At the corner and on the underside, in the hem.	Shawls and neckerchiefs Finished curtains	At the corner. On the draw tape.
Hand and bath towels, not to be soiled	At the corner or, if possible, on the hanging tab.	Woven and knitted gloves	On the left glove.
Ribbon sold by length, pre-packed	On the packing.	Hand knitting yarn	On the band.
Patterned covers, wool covers (blankets)	At the corner, with the marking.	Hand embroidery yarn, crochet yarn, hand knitting wool	On the band.

an international symbol system. Manufacturers and retailers follow the ASTM standard (ASTM D 3938 – Determining or confirming care instructions for apparel and other textile consumer products) to ensure correct information is included on care labels. The other standards dealing with care labelling include ASTM D 3136 (Standard terminology relating to care labels for textile and leather products other than textile floor coverings and upholstery); ASTM D 6322 (International test methods associated with textile care procedures) and ASTM D 5489 (Standard guide for care symbols for care instructions).

The most recent amendment to the rule states that the manufacturers can use a set of basic care label symbols developed by ASTM instead of using words. These symbols are graphic images that function like universal symbols on highway signs, and which do not need to be translated into a variety of languages. The intention of using symbols is that an individual without any previous experience should be able to interpret the symbols correctly and follow the actions suggested by it. However, it is often hard for the consumers to understand the symbols correctly. Following the FTC rules, products sold in the United States can use text only, symbols only, or both text and symbols. Products that are destined for multiple countries should adopt the symbols-only format to avoid the need to label in multiple languages. Consumers with a high need for cognition prefer labels that present care information in text format, while those with a lower need prefer the information in symbol format.

Care labels that are easily understood by consumers increase their confidence in caring for the apparel and reduce their perceptions of risk concerning the purchase of the item. The care instructions can be passed to the consumers with text only, symbols only and a combination of text and symbol (Yan et al., 2008). The manufacturers of clothing and other textiles prefer to use symbols on the care label, as the care symbols are globally recognizable and do not need to be translated into other languages. A majority of consumers prefer care labels that contain text and symbols, as these skills are taught and reinforced from an early age.

There are various care labelling systems followed around the world. The main systems followed are:

- International Care Labelling System
- ASTM Care Labelling System
- Canadian Care Labelling System
- British Care Labelling System
- Japanese Care Labelling System
- Australian Care Labelling System

17.4.1 International (ISO) care labelling system

The ISO system, commonly known as GINETEX (International Association for Textile Care Labelling) for care labelling, was established in 1963 in Paris following several international symposiums for textile care labelling at the end of the 1950s (GINETEX, 2014; ISO, 1991). A large number of national organizations are members of GINETEX. The countries that are members of GINETEX are Austria, Belgium, Brazil, the Czech Republic, Denmark, England, Finland, France, Germany, Greece, Italy, Lithuania, Netherlands, Portugal, Slovakia, Slovenia, Spain, Switzerland, Tunisia, Turkey and the United Kingdom. GINETEX has the following objectives:

- To define symbols for textile care at an international level
- To define the regulations for the use of care symbols
- To promote the use of clothing care symbols
- To acquire all markings and all rights relative to the symbols
- To register the symbols, both national and international
- To insure protection for all marks and symbols as adopted in all member countries of GINETEX
- To conclude all agreements liable to the promotion of the above-mentioned objectives
- To take all measures and carry out all actions in order to promote the above objectives, either directly or indirectly

An internationally applicable care labelling system based on symbols for textile materials has been devised by GINETEX. The care labelling system provides the correct information on the care instruction of textile products to consumers, retailers and textile manufacturing companies. The care labels describe various processes the clothing item can tolerate to avoid any irreversible damage to the product. The symbols or pictograms used in most countries are registered trademarks of GINETEX.

The international system of care labelling symbols is defined by GINETEX. In addition, GINETEX promotes and coordinates its technical background on an international level. The care labelling system takes into account any new technical and ecological developments together with changes in consumer practices. The symbols used in the GINETEX system represent the garment can withstand the process, and a cross indicates the process is not possible for the garment. The five symbols shown in Figure 17.1 are used in this system.

The washtub represents the washing or laundering process, which may contain some number inside. This number indicates the maximum permissible temperature of the water in degrees centigrade. Both the washtub and the number indicate that machine washing is possible. A hand in the washtub indicates only hand wash is possible. If there is an underline beneath the washtub, this indicates a milder treatment is in order. Numbers above the washtub indicate different washing programs and these are not always identical with those actually used in washing machines. There may be some additional indications that are not followed everywhere. The letters 'CL' inside the triangle indicates that chlorine bleaching is possible. The dots (1, 2 or 3) inside the iron symbol indicate the maximum temperature at which ironing can be done. The letters A, P or F inside the circle indicate the dry-cleaning process with the solvent to be used (A, P and F indicate any solvent, any solvent except trichloroethylene and petroleum solvent, respectively.) A circle inside a square indicates the particular garment can be tumble-dried.

17.4.2 ASTM care labelling system

In the ASTM System there are five basic symbols: washtub, triangle, square, iron and circle indicating the process of washing, bleaching, drying, ironing or pressing and dry-cleaning, respectively (ASTM, 1998). The prohibitive symbol 'X' may be used only when evidence can be provided that the care procedure on which it is superimposed would adversely change the dimensions, hand, appearance or performance of the textile. The washtub with a water wave represents the washing process in home laundering or commercial laundering process. Additional symbols inside the washtub represent the washing temperature and the hand washing process. The water temperature in the hand washing process may be 40 °C. Additional symbols below the washtub indicate the permanent press cycle (one underline, minus sign or bar) and delicate-gentle washing cycle (two underlines, minus sign, or bar). The full details of the ASTM System can be found at http://www.textileaffairs.com/docs/acsguide-050608.pdf.

The triangle represents the bleaching process, and an additional symbol inside the triangle indicates the type(s) of bleach to be used. The square indicates the drying process. Additional symbols inside the square represent the type of drying process to use including tumble dry, line dry, dip dry, dry flat and dry in shade. Additional symbols below the square indicate the permanent press cycle (one underline, minus sign or bar) and the delicate-gentle cycle (two underlines, minus sign or bar). Permanent press and gentle/delicate cycle instructions may be reported in words along with symbols instructions for tumble-drying and the dryer heat setting. The dots are used to represent the dryer temperature: three dots (high), two dots (medium), one dot (low), no dots (any heat), and a solid circle (no heat/air).

The hand iron represents the hand-ironing process and the pressing process. Dots inside the iron represent the temperature setting. The maximum temperatures by different number of dots are three dots (200 °C), two dots (150 °C) and one dot (110 °C). The warning symbol, crossed-out steam lines under the iron, or words, may be used to report the warning 'do not steam'. The circle represents the dry-cleaning process. A letter inside the circle represents the type of solvent that can be used. Additional symbols may be used with the circle to furnish additional information concerning the dry-cleaning process. The iron symbol may be used with the dry-cleaning symbol to represent how to restore the item by ironing after wearing.

17.4.3 Canadian care labelling system

Canadian Care Labelling System consists of five basic symbols that are illustrated in the conventional traffic-light colours. If any message is not conveyed by the care labelling symbols, words in English and French may be used. The five symbols must appear in the following order on the care labels: washing, bleaching, drying, ironing and dry-cleaning. The symbols and the care labelling system are described in Table 17.2.

Table 17.2 **Symbols and processes used in Canadian care labelling system**

Symbol	Instructions
Washing (a washtub without water wave represents washing process)	
70°C	Green washtub: machine wash in hot water (not exceeding 70 °C) at a normal setting.
50°C	Green washtub: machine wash in warm water (not exceeding 50 °C) at a normal setting.
50°C	Orange washtub: machine wash in warm water (not exceeding 50 °C) at a gentle setting (reduced agitation).
40°C	Orange washtub: machine wash in lukewarm water (not exceeding 40 °C) at a gentle setting (reduced agitation).
(hand)	Orange washtub: hand wash gently in lukewarm water (not exceeding 40 °C).
(hand)	Yellow washtub: hand wash gently in cool water (not exceeding 30 °C).
(crossed out)	Red washtub: do not wash.
Bleaching (triangle symbol indicates bleaching process)	
△ Cl	Orange triangle: use chlorine bleach with care. Follow package directions.
△ Cl (crossed out)	Red triangle: do not use chlorine bleach.
Drying (square symbol indicates drying)	
◯	Green square: tumble dry at medium to high temperature and remove article from machine as soon as it is dry. Avoid over-drying.
◯	Orange square: tumble dry at low temperature and remove article from machine as soon as it is dry. Avoid over-drying.
⎕	Green square: hang to dry after removing excess water.
⦀	Green square: 'drip' dry hang soaking wet.
—	Orange square: dry on flat surface after extracting excess water.

Care labelling of clothing

Table 17.2 Continued

Symbol	Instructions
Ironing (the iron with a closed handle represents ironing process.)	
[200°C iron] or [iron with •••]	Green iron: iron at a high temperature (not exceeding 200 °C – recommended for cotton and linen).
[150°C iron] or [iron with ••]	Orange iron: iron at a medium temperature (not exceeding 150 °C – recommended for nylon and polyester).
[110°C iron] or [iron with •]	Orange iron: iron at a low temperature (not exceeding 110 °C – recommended for acrylic).
[iron crossed out]	Red iron: do not iron or press.
Dry-cleaning (the circle indicates dry-cleaning)	
○	Green circle: dry-clean.
○	Orange circle: dry-clean, tumble at a low safe temperature.
⊗	Red circle: do not dry-clean.

17.4.4 British care labelling system

The British Care Labelling System uses graphic symbols and numbers to provide information on care labels. The care symbols should appear in the order of washing, bleaching, drying, ironing and dry-cleaning. The symbols used for various processes with necessary explanation are given in Table 17.3.

17.4.5 Japanese care labelling system

The Japanese Care Labelling System uses basic symbols that are different from other systems for care labelling. The symbols and the care labelling system are shown in Table 17.4.

Table 17.3 **Symbols and processes used in British care labelling system**

Symbol	Process
Washing instructions	
1 / 95°	Maximum temperature of wash: 95 °C. Mechanical action: maximum. Rinsing: normal. Spinning or wringing: normal.
2 / 60°	Maximum temperature of wash: 60 °C. Mechanical action: maximum. Rinsing: normal. Spinning or wringing: normal.
3 / 60°	Maximum temperature of wash: 60 °C. Mechanical action: medium. Rinsing with gradual cooling before spinning. Water extraction with care.[a]
4 / 50°	Maximum temperature of wash: 50 °C. Mechanical action: medium. Rinsing action: medium. Rinsing with gradual cooling before spinning. Water extraction with care.[a]
5 / 40°	Maximum temperature of wash: 40 °C. Mechanical action: maximum. Rinsing with gradual cooling before spinning. Spinning or wringing: normal.
6 / 40°	Maximum temperature of wash: 40 °C. Mechanical action: maximum. Rinsing with gradual cooling before spinning. Water extraction with care.[a]
7 / 40°	Maximum temperature of wash: 40 °C. Mechanical action: minimum. Rinsing: normal. Rinsing or machine wringing: normal, do not wring by hand.
8 / 30°	Maximum temperature of wash: 30 °C. Mechanical action: minimum. Rinsing: normal. Water extraction with care.[a]
9 / 95°	Maximum temperature of wash: 95 °C. Mechanical action: minimum. Drip dry.
(hand symbol)	Hand wash, do not machine wash. Maximum temperature of wash: 40 °C. Wash time: short. Wash, rinse and gently squeeze by hand – do not wring by hand.
(crossed tub)	Do not wash.
Bleaching instructions (only for chlorine-based bleach)	
△	Bleaching (chlorine-based).
⟁ (crossed)	Do not use chlorine-based bleach.

Table 17.3 Continued

Symbol	Process
Drying instructions	
▢	Tumble-dry.
⊠	Do not tumble-dry.
Ironing instructions	
🔥 (three dots)	Iron at a maximum sole-plate temperature of 200 °C (hot iron).
🔥 (two dots)	Iron at a maximum sole-plate temperature of 150 °C (warm iron).
🔥 (one dot)	Iron at a maximum sole-plate temperature of 110 °C (cool iron).
🔥 (crossed)	Do not iron.
Dry-cleaning instructions	
Ⓐ	Articles that are normal for dry-cleaning in all solvents normally used for dry-cleaning.
Ⓟ	Articles that are normal for dry-cleaning in tetrachloroethylene, trichlorofluoromethane (solvent 11), hydrocarbons (white spirit)[a], trichlorotrifluoroethane (solvent 113), using the normal dry-cleaning procedures without restriction.
Ⓟ (underlined)	Articles dry-cleanable in the solvents given in the preceding paragraph but are sensitive to some dry-cleaning procedures for which there is a strict limitation on the addition of water during cleaning and/or certain restrictions concerning mechanical action and/or drying temperature.
Ⓕ	Articles that are normal for dry-cleaning in hydrocarbons (white spirit)[a] and trichlorotrifluoroethane (solvent 113) using the normal dry-cleaning procedures without restriction.
Ⓕ (underlined)	Articles dry-cleanable in the solvents given in the preceding paragraph, but which are sensitive to some dry-cleaning procedures and for which there is a strict limitation on the addition of water during cleaning and/or certain restrictions concerning mechanical action and/or drying temperature.
⊠	Do not dry-clean.

[a] Indicates reduced spinning, or wringing at reduced pressure.

Table 17.4 **Symbols and processes used in Japanese care labelling system**

Symbol	Instructions
Washing instructions	
95	Machine wash in water temperature of 95 °C or less. No other restrictions.
60	Machine wash in water temperature of 60 °C or less. No other restrictions.
40	Machine wash in water temperature of 40 °C or less. No other restrictions.
40	Machine wash at delicate cycle or hand wash in water temperature of 40 °C or less.
30	Machine wash at delicate cycle or hand wash in water temperature of 30 °C or less.
30	Hand wash in water temperature of 30 °C or less.
	Do not wash (not washable).
Bleaching instructions	
	Use chlorine bleach.
	Do not use chlorine bleach.
Ironing instructions	
	May be ironed directly at 180–210 °C.
	May be ironed directly at 140–160 °C.
	May be ironed directly at 80–120 °C.
	Do not iron.
	May be ironed at 180–210 °C if a cloth is placed between iron and garment.

Table 17.4 Continued

Symbol	Instructions
Dry-cleaning instructions	
(symbol)	Dry-clean, use any dry-cleaning agent.
(symbol)	Dry-clean, use only a petroleum-based agent.
(symbol)	Do not dry-clean.
Wringing instructions	
(symbol)	Wring softly by hand or spin dry by machine quickly.
(symbol)	Do not wring by hand.
Drying instructions	
(symbol)	Hang dry.
(symbol)	Hang dry in shade.
(symbol)	Lay flat to dry.
(symbol)	Lay flat to dry in shade.

Taken from Japanese Care Label Standard.

17.4.6 *Australian/New Zealand care labelling system*

The joint systems AS/NZS 1957:1998, Textiles – Care Labelling; and AS/NZS 2622: 1996, Textile Products – Fibre Content Labelling with variations and additions, are followed in Australia and New Zealand. The joint standards specify that care instructions must be permanently attached to articles, written in English and appropriate for the care of the article. The instructions can be clarified by additional symbols if needed, but symbols alone are not sufficient. The symbols are used for washing, bleaching, drying, ironing and dry-cleaning instructions. In addition, professional wet cleaning is also explained by the use of one symbol 'circle' and the letter 'W' inside it, which indicates normal wet cleaning process. A line under the circle indicates mild process, whereas two lines indicate very mild process. A cross over the circle containing 'W' indicates no professional wet cleaning.

The washtub symbol indicates the washing process. The number inside the washtub indicates the maximum temperature for washing. A single line underneath the washtub indicates mild treatment or washing cycle using the permanent press setting, whereas two lines underneath the washtub indicate gentle or delicate cycle. A cross indicates that the process is not appropriate for the clothing.

The bleaching, drying, ironing, dry-cleaning and professional wet cleaning processes are indicated by the use of symbols. The bleaching process is indicated by a triangle symbol, whereas a cross 'X' on the triangle indicates do not bleach. The drying process is indicated by either with a circle or circle inside a square. The dots inside the circle indicates the amount of heat (three dots: high heat, two dots: medium heat and one dot: low heat). Similarly, air drying is indicated by various symbols using squares and lines. The iron symbol indicates ironing, with the dots inside indicating the acceptable temperature range.

The dry-cleaning process is indicated by a circle, and the letters within the circle provide information on the solvents to be used during the dry-cleaning process, which is needed by professional textile cleaners. The letters A, P or F inside the circle indicate any solvent, any solvent except trichloroethylene and petroleum solvent, respectively. The line below the circle indicates limitations in the dry-cleaning process, which may be related to mechanical action, addition of moisture and/or drying temperature. A cross on the circle indicates "do not dry-clean".

17.5 Future trends

Care labels always play an important role in the appropriate care and maintenance of many textile products. The durability, aesthetic values and dimensions of these items can be altered if the processes, process conditions and chemicals needed for care and maintenance are wrongly selected. Hence, the manufacturers should always include the right parameters in the care instructions and the consumers should follow them.

The main difficulties associated with care labels are: (1) some labels indicate procedures that are far more restrictive than necessary, (2) some instructions make no sense or are difficult to understand and (3) some abrasive and coarse labels cause skin irritation. These problems can be avoided by the manufacturers with necessary action. The conditions essential for a clothing care should always be clearly demonstrated with a universal language or symbol. The selection of soft material for preparing the labels or directly printing the instructions at some inner part of a textile item can avoid the problems of skin irritation.

The use of Internet selling of various products has grown tremendously. Purchasing clothing over the Internet poses risks concerning size, fit, aesthetics, feel of the fabrics and reading information on care and content labels (Park and Stoel, 2002; Kim and Lennon, 2010; Xu and Paulins, 2005).

In addition, to the essential information, the manufacturers can include additional information such as environmental labels, guarantees, finishing information and sweatshop label. Environmental labels provide information on environmentally friendly, ozone-friendly, biodegradability and recycling. Environmental labels can assist consumers in the selection of products with lower impact on the environment. With the guarantees in the clothing, consumers are assured on the quality of the product. The information on textile finish can provide additional information to the consumers. The sweatshop label provides information on employees working conditions during the textile production process.

The care labels used in garments are either printed or woven, which store limited information. Recently, electronic care labels, which use radio frequency identification (RFID) technology for storing information electronically on a garment, are paving their way (Nayak et al., 2007). These labels can store sufficient amount of care information in an electronic chip, which can be read by suitable devices or readers. However, the feasibility of using this technology in care labelling is a challenge due to its high cost, health risks and other technical challenges (Nayak et al., 2007).

The consumers should also properly understand the meaning of the care instructions before any care and maintenance process. A survey found that many people do not fully understand care label information and select more vigorous cleaning methods than those recommended. Some respondents indicated that they thought bleaching was acceptable, though the instructions warned against it. Similarly, 'line dry' was interpreted incorrectly. Educational programmes are therefore necessary to maximize the number of consumers correctly interpreting the labels. Standardizing information on care labels can also minimize misunderstanding. It is essential for the manufacturers to always include the care label with right care instructions that will be an integral part of the clothing for the useful life of a product. However, several studies have demonstrated no existence of a direct relationship between information provided and information used.

17.6 Conclusions

Consumers do not have the experience and technical knowledge to decide which care treatment is suitable for a product. Thus, care labelling, which is the responsibility of garment makers, helps the consumers maintain the apparel's aesthetic value and durability. The manufacturer is responsible for proper labelling of textile fibre products when they are ready for sale or delivery to the consumer. The importer is responsible for proper labelling of imported textile products. Custom merchants and tailors are responsible for showing properly labelled bolts, samples and swatches to customers. Domestic manufacturers must attach care labels to finished products before they sell them.

For consumers, care symbols make sense when they can understand and follow the instructions. Symbols should provide the same information to everyone without language barriers. Use of symbols allows for smaller and more comfortable care labels, and symbols are easy to understand. Smaller labels also cost less, and this could translate into consumer savings. For manufacturers, care symbols make even more sense. When harmonized with other countries, symbols will allow participation in a global marketplace where symbols will clearly communicate the same information in all countries. Smaller labels cost less to buy or manufacture and also cost less to inventory. Total inventory can be further reduced by eliminating the need for different labels for different countries. Therefore, all manufacturers should attach care labelling instructions to the garment for the benefit of the consumers and for upkeep of their brand.

Due to globalization and liberalization processes, it becomes more relevant for the garment manufacturers to use care labelling. It is also very important for garment manufacturers to conduct an awareness campaign regarding the utility of care labels for the consumers. However, while applying the care labels, it is very important for garment manufacturers to understand the requirement criteria of care labels.

References

ASTM, 1998. ASTM D 5489 — 98, Standard Guide for Care Symbols for Care Instructions on Textile Products.

Chatterjee, K.N., Nayak, R.K., Bhattacharya, S., Kansal, N.R., 2006. Care labeling of apparels. Indian Text. J. 116 (12), 55—60.

Davis, L.L., 1987. Consumer use of label information in ratings of clothing quality and clothing fashionability. Clothing Text. Res. J. 6 (1), 8—14.

GINETEX - the International Association for Textile Care Labeling, 2014. Available from: http://www.ginetex.net/ginetex.

Heisey, F.L., 1990. Perceived quality and predicted price: use of the minimum information environment in evaluating apparel. Clothing Text. Res. J. 8 (4), 22—28.

Holmlund, M., Hagman, A., Polsa, P., 2011. An exploration of how mature women buy clothing: empirical insights and a model. J. Fash. Mark. Manage. 15 (1), 108—122.

Huddleston, P., Cassill, N.L., Hamilton, L.K., 1993. Apparel selection criteria as predictors of brand orientation. Clothing Text. Res. J. 12 (1), 51—56.

ISO, 1991. ISO 3758: 91 Textiles — Care Labeling Code Using Symbols.

Kadolph, S.J., 2007. Quality Assurance for Textiles and Apparel, second ed. Fairchild Publications, Inc, New York.

Koester, A.W., May, J.K., 1985. Profiles of adolescents' clothing practices: purchase, daily selection, and care. Adolescence 20 (77), 97—113.

Kim, J.-H., Lennon, S.J., 2010. Information available on a web site: effects on consumers' shopping outcomes. J. Fash. Mark. Manage. 14 (2), 247—262.

Mehta, P.V., Bhardwaj, S.K., 1998. Managing Quality in the Apparel Industry. New Age International, New Delhi.

Mupfumira, I.M., Jinga, N., 2014. Clothing Care Manual. Strategic Book Publishing Rights Agency.

Nayak, R., Padhye, R., 2014. The Care of Apparel Products. Elsevier, UK, pp. 799—822.

Nayak, R., Chatterjee, K.N., Khurana, G.K., Khandual, A., 2007. RFID: tagging the new era. Man-Made Text. in India 50 (5), 174—177.

Park, J.H., Stoel, L., 2002. Apparel shopping on the internet: information availability on US apparel merchant web sites. J. Fash. Mark. Manage. 6 (2), 158—176.

Xu, Y., Paulins, V.A., 2005. College students' attitudes toward shopping online for apparel products: exploring a rural versus urban campus. J. Fash. Mark. Manage. 9 (4), 420—433.

Yan, R.-N., Yurchisin, J., Watchravesringkan, K., 2008. Use of care labels: linking need for cognition with consumer confidence and perceived risk. J. Fash. Mark. Manage. 12 (4), 532—544.

Garment costing

A. Singh[1], K. Nijhar[2]
[1]RMIT University, Melbourne, VIC, Australia; [2]Yum Productions Pty. Ltd, Melbourne, VIC, Australia

18.1 Introduction

Globalization, which has shortened the world's distance, impacted the garment industry a lot because it is a massive labour-dependent and tech-savvy industry. Although the world has come closer spatially, the garment industry's supply chain has gone multi-dimensional, which practically means that it is almost impossible to incorporate all the processes under one roof, so various companies opt to outsource some of the activities/processes. The process outsourcing may be in the same or different country, depending on the lead time requirement. As globalization has opened new markets to stimulate trade, especially in garment industry, companies are moving their operations to low-cost countries to compete with their counterparts on cost. The impact of international outsourcing on the practice of costing and labour cost is significant (Miller, 2013). Due to deficient resources (resource scarcity) and technological know-how, each country is not competent in producing the garments up to the same standards and matching the cost by their counterparts. The focus of this chapter will remain on the supply-chain cost of the garment industry, in addition to the garment costing.

This chapter outlines the cost component and driving factors in the garment industry. In the first section, it introduces the conceptual framework of cost classification, and in the second section the basic breakdown of the cost components to produce the garment are discussed. Finally, the last section explains the price at the retail store and the contributing factors such as mark-up, mark-down and margin and how these contributing factors impact the final cost of garments and profitability of a business. As the supply chain of the garment industry is quite diverse, so each component of the supply chain contributes towards the final cost of the garment. The simple supply chain in the garment industry starts from the fibre extraction or fibre manufacturing, in case of manufactured fibre, and runs through to various processes such as yarn manufacturing, weaving, dyeing/printing, stentering, cutting, sewing, packing and many more. Each constituent involves the cost directly or indirectly, which is counted to the cost of the garment. The intent of this chapter is to discuss the costs that are involved directly or indirectly to manufacture the garments, from the retail side mainly.

18.2 Costing need

What is cost? Why it is important to understand? How do we evaluate the cost? These are the most basic and essential questions to be asked while dealing with the subject of costing.

According to Langfield-Smith et al. (2008), cost refers to any resources that are given up to achieve a goal. So, the cost is measured in monetary terms in accounting to show up on a balance sheet. The cost can also be defined as an economic value, spent to produce a product. The cost has a direct relationship with the economic activities used to add the value to a product from the raw material through to the finished commodity.

The survival of any business depends upon how profitable it is. So, the need to understand the cost of a product, primarily, is to cover the expenses incurred by business. There are always two perspectives on the cost; one is viewed from the customer's perspective and the other from the supplier. The supplier's perspective is often termed as the 'cost of a garment', which means how much is spent to manufacture a garment. When a garment is manufactured in an industry it is viewed as an asset, or an inventory; and when the garment is sold, the cost of making the garment is classified as a cost of goods sold expense. If the revenue generated is more than the expenses incurred, the company is said to be profitable; otherwise the company shows losses on their balance sheet.

From the customer's perspective, the cost of the garment, which is often termed the 'price of a garment', is the price paid by the customer to own a garment, which includes the mark-up and margin. The received money is distributed upstream to the supply chain accordingly. The supplier is always willing to reduce the cost of production operations and the whole supply chain in order to impact the profitability positively. So, practically the price of a garment is always higher than the cost to make the garment, for the sake of generating profit. The price of a garment is decided based on what revenue a company wants to generate so that after paying expenses a company can earn a reasonable profit. Therefore, it is an elementary requirement to know the cost and factors impacting cost by the manufacturer in order to set a price for the finished garment.

There are various drivers to decide the price of a garment. If the product is to be sold directly from the manufacturer without any intermediary, then the price may be kept lower; but if an intermediary is involved, the price needs to be set higher because the intermediary will take its profit share as well. In the garment industry, it is not pragmatic to distribute the garments straight from the manufacturer's hands unless it is totally an online business. Even in online business there has to be distribution cost involved, which acts as an expense; an example may be as seen on screen (ASOS). In the garment industry, the driver of the price is not the manufacturer but the retailer, so acceptance of the pricing decision will depend on the elementary costing done by the manufacturer.

Due to globalization, usually the products are not manufactured in the same place where they are sold. Earlier the dominant market of garment manufacturing was the United Kingdom and Europe, but due to elevated labour costs, there has been a shift of production from these regions to developing countries like India, China, Bangladesh, etc. The manufactures in these countries are trying to be cost competitive so as to win the orders from big companies and make permanent ties in order to attract larger orders. Therefore, to compete in rigorous market, the cost is an essential element to be known apart from availability of resources and access to the latest technology. If a company fails to estimate an accurate cost to make the garments, then it might result in redundant business or loss of profitability. Although the expected outcome of outsourcing is a positive impact on the garment supply-chain cost, it is also associated with hidden costs. The hidden costs may include transportation cost, lead time, quality costs, communication, travel costs, administration costs, etc. (Hergeth, 2002). Lowson (2002) also identified a hidden cost in offshore sourcing, and to explain this he compared domestic and offshore apparel sourcing. He created a model to measure the actual cost of a garment being imported from low-cost countries and quantified the relationship between the major impacted components in sourcing, which are lead time, inventory, communication, service level and supplier performance.

18.3 Cost classification

Any activity performed within the garment industry supply chain incurs cost, and understanding the behaviour of each cost is very important to plan and manage the cost. There are number of ways in which the cost can be classified. The main consideration in this chapter will be on the basis of behaviour and traceability.

18.3.1 Cost classification on the basis of behaviour

The cost is classified as fixed or variable according to the behavioural aspect. The total cost in garment manufacturing on the basis of behaviour is thus the sum of fixed cost and variable cost.

Fixed cost remains unchanged, irrespective of the number of garments produced, for a fixed period. The term 'fixed' should not be thought that the fixed cost never changes. An example of fixed cost may be the rent of a company, which is fixed for a certain time-frame. Conventional economics defines **variable cost** as the cost that changes with short-term fluctuations in the output (Cooper and Kaplan, 1988). Modern management accounting defines variable cost as the cost that varies depending on the number of garments produced, in garment context; for example, the raw material used. So the total variable cost tends to increase in proportion to the output (garments produced). The variable cost is also referred to as marginal cost. For example, to produce a shirt, approximately 1 m of a fabric may be needed; but to produce 100 shirts, 100 m of fabric will be required. The variable cost will increase in direct proportion to the number of units produced unless a volume discount is offered.

18.3.2 Cost classification on the basis of traceability

Cost can be classified as direct or indirect, depending on their traceability.

Direct cost is a cost that can be traced in relation to a produced product. The direct cost in the garment industry may be the direct cost of material (cost of fabric, cost of trims, etc.) and direct cost of labour. **Indirect cost** is a cost that is unidentifiable; e.g. the job of a design manager is to oversee the designs of the garments, so how much time he is investing on checking each design is difficult to quantify, so the salary of a design manager in this situation may be treated as an indirect cost.

The major concern in the garment industry is to control costs due to the fast-paced and competitive market. The Asian market has dominated in garment production because of achieving economies, but few top-end brands are still manufactured in Europe and other parts of world irrespective of being expensive. The major proportion of the garment cost is the material cost, followed by overhead (indirect cost) and labour. These three costs summarize the cost of a garment, where materials cost is the raw material used, to be value added into the finished garment; labour cost is the cost of the operators who help in manufacturing the garment; and finally, the overhead cost is the cost to run the business.

18.4 Cost elements

As discussed previously, cost is classified as direct or indirect on the basis of traceability. These costs can be further elaborated. The direct cost elements are as follows.

18.4.1 Direct material cost

The direct material cost is the cost of material that after value addition is converted into a finished garment. The direct material cost is approximately 45–60% of the garment's cost, depending on the types of the garment and the automated sophistication used to produce the garment (Jeffrey and Evans, 2011). So, raw material is still a major cost driver. Even a little reduction in direct material cost significantly impacts the total cost of a garment. The major reason to select an appropriate supplier, using supplier selection criteria, is to bring the total cost down. The fabric, as a raw material for the garments, is a major cost component, so optimum utilization of fabric is vital at the planning, laying out and cutting stages. An example of direct material cost in apparel manufacturing may include yarn, fabric, zippers, buttons, fabric fillings, hooks, labels, etc. The price paid to bring the material from the supplier to the manufacturing unit is also accounted as a direct material cost, which is transportation cost and also called 'inbound logistics cost'.

18.4.2 Direct labour cost

Labour cost is the most essential component in the overall cost of garment. According to Fiallos (2010), labour cost can be divided into direct labour and indirect labour, where direct labour is skilled and indirect is unskilled labour. The term 'direct labour cost' in this chapter includes both types of labour costs defined above, which are involved in making a garment. Thus, direct labour cost means the wages given to the workers who are directly involved in manufacturing of the garments and the cost that can be traced per garment. So, the wages cost may include the salaries for cutters, sewers, press operators and packing operators as well. 'Labour on cost' is an additional cost that is paid to the employees managing payroll tax, employer contribution to superannuation funds, workers insurance, etc. (Langfield-Smith et al., 2008). Some companies treat these costs as manufacturing overheads.

Garment costing

18.4.3 Direct expenses

The proportion of **direct expenses** is not very significant, but this is the cost involved in assisting garment manufacturing. An example would be a company purchases any license to run a machine or process or any royalty paid by the company to another company as a part of business. Also, any small contract work involved, such as embroidery, special finish to a garment, etc., is treated as direct expense.

18.4.4 Indirect cost or overhead

Another element of cost is indirect cost, the cost that cannot be traced to a garment in an economic way. Indirect cost is the cost to run a company, which is otherwise referred to **manufacturing overhead** or simply **overhead**. Manufacturing overhead covers all other cost not associated with direct material and labour cost. The overhead contributes around 30% to the total cost of the garment (Jeffrey and Evans, 2011). The cost to move the garment batches from one process to another may be regarded as an indirect cost. Other overhead costs may include machinery depreciation, factory insurance, electricity, overtime and the support department. The support departments are not directly involved in production but assist in production. **Idle time** is also an overhead when machine and labour are not being utilized due to machine breakdown, scheduled maintenance or machine set-up time. Apart from manufacturing overhead there is **administrative overhead**, **selling overhead** and money spent in **research and development**.

18.5 Measures of efficiency

It is important that retailers make a decent profit to cover operational expenses and gain actual profit. In order to be profitable, a company needs to fulfill its duties and tasks efficiently. The objectives on which the efficiency of retailer can be measured are the following:

- **Period sales**: The sales are compared with the previous year's sales data and are always expected to be more than last year's sales. Primarily the sales for a period are set on the basis of last year sales in terms of dollars, units sold or sales per unit area. Every year an increase or decrease in sales is calculated. An increase in sales indicates success and profitability.

$$\% \text{ Increase or Decrease} = \frac{(\text{New period sales} - \text{Last period sales})}{\text{Last period sales}} \times 100$$

- **Inventory/stock**: Stock is second measure of success and efficiency. Inventory needs to be replenished, but should be in an optimum quantity. More stock than required increases in-house cost and reflects loss, as the stock may not be able to be sold. Less stock means a fear of losing customers, which is also a loss as customers may decide not to come back next time. To measure stock efficiency, inventory turnover ratio is calculated, which means how quickly the product is sold.
- **Margin**: Another measure is the initial mark-up, as in second, third and subsequent mark-ups the profitability starts going down.

- **Discounts**: Miscalculation of stock will bring the company to a position of reducing the price of merchandise. Discounts are losses for a company.

18.6 Profitability

As a retailer, merchandiser or buyer, the major responsibility is to earn profit. Profits are essential to any business in order to achieve the following (Easterling and Ellen, 2012):

- Cover expenses in the business.
- Buy more inventory and replenish stock.
- Be able to expand the business.
- Provide a return on investment.

18.6.1 Profit and loss statement

The profit and loss (P&L) statement is also referred to as income statement. In general, it provides information on profit gained by the company (Guilding, 2010). The P&L statement is a skeletal statement consisting of three factors: net sales, which is sales volume; cost of merchandise sold, which is also called cost of goods sold; and operating expenses.

Net sales denotes actual sales in dollars after deducting any returns or allowances made.

Sales return is the money returned when customers bring the garments back due to some defect or displeasure.

Also, **allowances** are made sometimes to satisfy the customer or to make the customer happy; these are treated as discounts. A **trade allowance** is given to the retailer by vendors to display their product. Before deducting any returns or any allowances, the total amount received after selling the garments is called **gross sales**.

$$\text{Net sales} = \text{Gross sales} - (\text{Merchandise returns} + \text{Discounts or Allowances})$$

These are three main factors, but a detailed P&L statement shows five factors, (Net sales, cost of goods sold, gross profit, operating expenses and net profit) which are shown in Example 18.1.

Example 18.1. Example of a profit and loss statement

P&L factors		Value in $	%
Net sales		$50,000	100
− (Minus)	Cost of goods sold	$30,000	60
= (**Equals**)	**Gross profit (margin)**	**$20,000**	**40**
− (Minus)	Operating expenses	$12,500	25
= (**Equals**)	**Net profit**	$7500	15

The final P&L statement is developed in detail, covering each transaction and information regarding stock levels. If there is an opening stock that is carried forward from the last period, it is added in the sales for the operating period too, and the closing balance is deducted from the total inventory to get the gross sales. An example of a P&L statement with opening and closing stock is given in Example 18.2. A children's wear department has an opening inventory worth $25,000 and the billed cost in the sales period is $50,000. The freight inward is worth $2000, and the closing inventory cost is $15,000. What is the gross sales?

Example 18.2. Example of gross sale calculation

	Particulars	Value in $
	Opening inventory (Stock)	$25,000
+	Billed cost in the sales period	$50,000
+	Freight inward	$2000
=	**Total merchandise handled in the sales period**	**$77,000**
−	Closing inventory (Stock)	$15,000
=	**Gross cost of the goods sold**	**$52,000**

- **Billed cost in the sales period** means new merchandise purchased in the running/operating period of sales.
- **Freight inward** means the cost of transported merchandise to the premises.
- **Closing inventory** means the stock left at the end of the operating period.
- **Operating expenses** are the expenses incurred by a business to run a store/business, except the cost of merchandise. The expenses to run a store includes wages for employees, salaries of other staff, rent, advertising, information technology used, fittings and fixtures, insurance, bad debts, overhead and many more. Controlling the expenses will result in a direct positive effect on profit. Recently, the wages and energy costs have increased, so retail stores must control their expenses to avoid squeezing profits.

In the aforementioned example the cost of the goods sold is affected by a number of factors. First, the transportation cost to bring the material is included in the merchandise cost. Second, once the goods are brought to the premises, they might need some alterations, so the expenses spent on alterations/modifications are also added in the merchandise cost. Third, the company may get some discount on buying, so all these discounts are deducted from the cost to get the actual cost of merchandise. There are various discount types, such as a quantity discount, also known as a patronage discount, which is a discount given on bulk orders as a company achieves economies of scale in producing bulk consignments. Seasonal discounts are given to the retailer if the merchandise is purchased out of season; for example, if woollen jumpers are bought in the summer season to stock the garments for next season. Cash discounts are given if the invoice is paid within a specified time or paid in cash on buying or during delivery. A trade discount is given on the listed price, which is normally higher than the retail price.

18.7 Garment sales element analysis

Once the garment is manufactured it is subjected to be sold in the market. The market may be the same as where it was manufactured or in another part of the world. Nowadays, due to price competitiveness, normally the garments are manufactured in one part of the world and sold somewhere else. So after the garment is manufactured in a company, various other types of costs are involved or added before being sold to the final customer. This section will deal with the cost components after the garment is manufactured and is ready to be sold. First, it will start with defining and explaining various terms and key concepts in garment retailing and conclude with an expanded version of a garment costing sheet. In addition, this section will touch upon base costing from a customer's perspective.

To understand profit and loss, there needs to be a clear distinction between cost and price (Kunz and Glock, 2000). Cost is an amount that a retailer pays to own a product, which will be in turn sold to gain profitability. This is also termed as 'cost of goods sold'. Price is an amount asked for a product. The retailer adds its own profit and sells the garment to the final customer at a price known as the retail price or selling price. Retailing is the last step in the supply chain of the garment industry, unless a garment is returned due to faulty workmanship or a change of mind by the customer. Retailing involves all the operations from obtaining the garments from manufacturing to wholesalers and directing them to the final customers. The garments are sold through physical retail stores and online as well. The only difference between these two modes of selling is that the former needs cost to set up operations, in turn increasing the final cost of the garments; but in the latter case, due to exclusion of an intermediary, the cost is brought down.

18.7.1 Factors affecting pricing

Since pricing impacts directly on profitability, so it is decided carefully considering many factors. If the product is new and first in the market, then the company may achieve a first-mover advantage and set the retail price higher, as opposed to when the product becomes generic the companies will have to reduce the prices. The factors affecting pricing are given below.

- **Brand recognition**: Popular brands can be sold at a higher price due to brand acknowledgement and popularity.
- **Nature of garment**: The quality of the merchandise will decide its price and mark-downs.
- **Market pressure**: Competition in the market is a major driving force behind the price decision. One company out of two, selling the same product will not be able to set the price higher; otherwise they will be out of the market.
- **Supply and demand**: In the case of low demand and high supply, the prices are to be lowered, and vice versa.
- **Other costs**: Inventory control, storage cost, pilferage cost, handling cost, alteration cost, delivering cost, etc. will also be an input in deciding the final price of a merchandise.

Garment costing

18.7.2 Retail price and relationship with mark-up

In the aim to achieve greater profit, the company will decide the retail price of garments for which the garment will be sold after adding mark-up. So the mark-up is the amount added to the buying price to achieve the retail price. Mark-up will decide the gross margin or profit achieved. How to set a merchandise retail price and add mark-up is a skillful task demanding experience. When the price of a garment is decided, the main consideration is to check the saleability of the garment at the set retail price. The retail price is set high enough to cover the expenses and the cost of merchandise with an additional profit. Due to fierce competition in the garment industry, the price cannot be set too high so that garments are left unsold.

Mark-up = Retail price − Buying price (Cost price)

Retail price = Buying price (Cost price) + Mark-up

Buying price (Cost price) = Retail price − Mark-up

Retailers find percentages more useful and meaningful than dollar figures because percentages allow the company to compare their figures with those of their competitors. The comparison of margin and mark-up percentage gives a company their position in market. By and large, margin, mark-up and gross profit, when expressed in dollars, have the same meaning, which is the difference between net sales and the cost of goods sold. But when these terms are expressed in percentages, the meaning is different. Mark-up percentage is a function of buying cost, whereas margin percentage is a function of retail price. The margin percentage has more meaning to garment stores, since expenses and profits are figured as a percentage of net sales.

18.7.3 Calculating mark-up percentage and margin percentage

To calculate mark-up percentage, divide dollar mark-up by the cost price, where dollar mark-up is the difference between retail price and cost price (see Example 18.3).

$$\text{Mark-up (\%)} = \frac{(\text{Retail price} - \text{Cost price})}{\text{Cost price}} \times 100$$

To calculate margin percentage, divide dollar mark-up by the retail price, where dollar mark-up is the difference between retail price and cost price.

$$\text{Margin (\%)} = \frac{(\text{Retail price} - \text{Cost price})}{\text{Retail price}} \times 100$$

Example 18.3 calculates the mark-up percentage and margin percentage for a casual dress if the cost price is $60 and the retail price is $120.

Example 18.3. Calculating mark-up and margin percentage.

Answer:

$$\text{Mark-up} = \$ \text{ Mark-up} \div \$ \text{ Cost Price}$$
$$= (\$120 - \$60) \div \$60$$
$$= \$60 \div \$60$$
$$= \$1$$
$$\text{in \%} = 100\%$$

$$\text{Margin} = \$ \text{ Mark-up} \div \$ \text{ Retail Price}$$
$$= (\$120 - \$60) \div \$120$$
$$= \$60 \div \$120$$
$$= \$0.5$$
$$\text{in \%} = 50\%$$

Note 1: $ Mark-up in Example 18.3 is the difference between retail price and cost price.

Note 2: Mark-up percentage can be more than 100%, but margin percentage never exceeds 100%.

In calculating margin percentage, the retail price is kept to 100%, which in turn is addition of cost price and margin percentage.

18.7.4 Basic margin calculations

Margin is calculated on retail price, keeping retail price as 100% (see Example 18.4). It is a very useful tool to compare the performance of two similar types of companies. A higher profit margin depicts a more profitable company. The decision of estimating retail price is the customer's perception about the product.

Example 18.4. Retail price of a shirt that cost $70 and has a 55% margin

Particulars	Cost ($)	Percentage
Retail price	To be calculated	100%
Cost price	$70	To be calculated
Margin	To be calculated	55%

Garment costing

The answer is:

$$\text{Cost }(\%) = \text{Retail }(\%) - \text{Margin }(\%)$$
$$= 100\% - 55\% = 45\%$$

$$\text{Retail price} = \frac{\text{Cost} \times \text{Retail }(\%)}{\text{Cost }(\%)}$$
$$= \frac{70 \times 100(\%)}{45(\%)}$$
$$= \frac{70 \times 100}{45}$$
$$= \frac{7000}{45} = \$155.55$$

The final answer of Example 18.4:

Particulars	Cost ($)	Percentage
Retail price	$155.55	100%
Cost price	$70	45%
Margin	$85.55	55%

Formulas:

$$\text{Cost }(\%) = \text{Retail }(\%) - \text{Margin }(\%)$$

$$\text{Retail }(\$) = \frac{\text{Cost }(\$)}{\text{Cost }(\%)}$$

$$\text{Cost }(\$) = \text{Retail }(\$) \times \text{Cost }(\%)$$

18.7.4.1 Calculating cost when retail and margin % are known

See Example 18.5.

Example 18.5. Determine the cost of a jacket that retails for $200 and has a margin of 60%

Particulars	Cost ($)	Percentage
Retail price	$200	100%
Cost price	To be calculated	To be calculated
Margin	To be calculated	60%

The answer is:

$$\text{Cost } (\%) = \text{Retail } (\%) - \text{Margin } (\%)$$
$$= 100\% - 60\% = 40\%$$
$$\text{Cost } (\$) = \text{Retail } (\$) \times \text{Cost } (\%)$$
$$= \$200 \times 40\%$$
$$= \$80$$
$$\text{Margin } (\$) = \text{Retail } (\$) - \text{Cost } (\$)$$
$$= \$200 - \$80$$
$$= \$120$$

The final answer of Example 18.5 is as below:

Particulars	Cost ($)	Percentage
Retail price	$200	100%
Cost price	$80	40%
Margin	$120	60%

18.7.4.2 Calculating mark-up and margin for multiple products (averaging mark-up)

So far the mark-up and margin, both in dollars and percentages, have been calculated for one type of merchandise. For a number of different products, the applied mark-up percentage cannot be the same. Even for the same categories of the product line, the mark-up applied may be different; for example, a low margin percentage may be applied on promotional products and a higher margin can be obtained on distinctive products. So, the company will decide the overall margin percentage to be achieved and that basis company will choose low or high margin placed on different types of merchandise. Due to rapidly changing fashion, the buyers have lesser control over the margin of regular items but can compensate their overall margin through high margins decided for the promotional products.

Average margin percentage cannot be calculated by averaging margins on individual items. Average margin is calculated based on total cost and total retail of all the items. If the quantity is the same, then individual percentages can be averaged; otherwise not. The example of average margin (%) is shown in Example 18.6.

Garment costing

Example 18.6. Find the margin % if 15 shirts are bought at $50 price each and sold at $80, and 20 pairs of jeans costing $65 each retailed at $100

Answer:

Particulars	Quantity	Cost per item	Total Cost
Shirts	15	$50	15 × $50 = $750
Jeans	20	$65	20 × $65 = $1300
Total			$750 + $1300 = $2050

Particulars	Quantity	Retail per item	Total retail price
Shirts	15	$80	15 × $80 = $1200
Jeans	20	$100	20 × $100 = $2000
Total			$1200 + $2000 = $3200

$$\text{Margin } (\%) = \frac{(\text{Retail price} - \text{Cost price})}{\text{Retail price}} \times 100$$

$$= \frac{(\$3200 - \$2050)}{\$3200} \times 100$$

$$= \$1150/\$3200 \times 100$$

$$= 0.36 \times 100 = 36\%$$

18.7.4.3 Averaging cost when retail price and margin percentage are known

In Example 18.7, a company plans to buy 500 pairs of jeans for Christmas Eve and wants them to retail at $80. The company has already placed an order for 200 pairs of jeans at $50 each. What maximum amount can they pay for the rest of the jeans if they plan to achieve a margin of 45%?

Example 18.7. Cost averaging from retail price and margin percentage.

(1) Total retail price (500 pairs of jeans @ $80)

$$= 500 \times \$80$$

$$= \$40,000$$

(2) Total cost price

$$\text{Cost}(\%) = \text{Retail}(\%) - \text{Margin}(\%)$$
$$= 100\% - 45\% = 55\%$$
$$\text{Cost}(\$) = \text{Retail}(\$) \times \text{Cost}(\%)$$
$$= \$40,000 \times 55\%$$
$$= \$22,000$$

(3) Cost already paid for 200 pairs of jeans @ $50

$$\text{Cost}(\$) = 200 \times \$50$$
$$= \$10,000$$

(4) Final balance to be paid to buy remaining pairs of jeans

$$\text{Remaining cost}(\$) = \$22,000 - \$10,000$$
$$= \$12,000$$

(5) Number of pairs of jeans remaining to be bought = 500 − 200 = 300
(6) Unit cost to be paid for the remaining jeans = Remaining cost ($)/Remaining pairs to be bought

$$= \$12,000/300$$
$$= \$40 \text{ each}$$

In Example 18.8, the retail price of pairs of slacks needs to be $50,000 for a sales period. The first margin is 52%. The purchase cost of 300 pairs of slacks is $50 each and sold at a price of $110. What is the margin percentage required on rest of the slacks?

Example 18.8. Margin percentage from total cost (%)

(1) Total cost price

$$\text{Cost}(\%) = \text{Retail}(\%) - \text{Margin}(\%)$$
$$= 100\% - 52\% = 48\%$$
$$\text{Cost}(\$) = \text{Retail}(\$) \times \text{Cost}(\%)$$
$$= \$50,000 \times 48\%$$
$$= \$24,000$$

Garment costing

(2) Cost already paid for 300 pairs of slacks @ $50 each

$$\text{Cost (\$)} = 300 \times \$50$$
$$= \$15,000$$

(3) Cost of remaining purchases of pairs of slacks = Total cost price − Cost already paid

$$= \$24,000 - \$15,000$$
$$= \$9000$$

(4) Retail price of 300 pairs of slacks @ $110 each = $300 \times \$110 = \$33,000$
(5) Retail price of remaining purchases = Total retail price − Retail price of pairs of slacks already bought

$$= \$50,000 - \$33,000$$
$$= \$17,000$$

(6) Remaining Margin (%) = Cost of remaining pairs of slacks/Retail price of remaining purchases × 100%

$$= (\$9000/\$17,000) \times 100\%$$
$$= 52.9\%$$

18.7.5 Initial mark-up

Initial mark-up is very important in garment industry, and is widely used. Initial mark-up is placed on the garments upon receiving them into the store. It can either be calculated in dollars or in percentage. It is important to bear in mind that this initial mark-up should be sufficient to cover reductions, expenses, etc. It is marked to obtain the original retail price of the garment. As it is hard to achieve 100% forecast accuracy in the garment industry, there will be a situation when more garments are ordered than are sold, so some percentage of garments will be put on sales or the prices will be marked down. The discounted price means the retail prices now are less than the original retail price. So to have a decent profit in business, a company will need to put the initial mark-up high enough to average out the mark-downs, which is given by the formula:

Initial mark-up = (expenses + profit + reductions)/(net sales + reductions)

18.8 Mark-downs

The vibrant nature of the garment industry pushes companies to adjust the prices either by increasing or decreasing from the original retail price. Normally, an increase in price does not happen in the garment industry, which means reductions in the retail price that was previously set higher. All the garments cannot be sold at the original retail price set by the company, and the downward adjustment of the retail price is known as markdown. For example, a jumper sold at retail price of $150 and was now sold at $85, perhaps due to being in the warehouse for too long. This lowering of the retail price from $150 to $85 is called mark-down. In this case, the mark-down is of $65. So the company will be gaining lesser profit by selling the garments at this price. The markdown percentage is a function of net sales. Mark-downs are also a promotion tool to increase sales. There are numerous purposes of mark-downs, which are as follows:

- To encourage the sales of the garments.
- Buying error, such as purchasing the wrong styles, designs, etc., which cannot be sold and need to be marked down.
- Miscalculation at the time of ordering.
- If the demand is low and supply is more.
- If the products are not displayed properly and out of the reach of customers.
- To compete in the market for the same category.
- To provide money to the business to buy more products for next season.
- To create seasonal sales, which may include winter ending and summer coming in, or any occasion such as Christmas, birthdays, etc.
- To sell slow-moving items because slow-moving items increase the inventory cost.
- A high number of returns due to quality issues or other reasons.

The major cause of mark-down is buying more quantity than required. In this fast-fashion world, it is very difficult to estimate the exact number of garments to be sold, so estimation has to be done, which is known as forecasting. Forecasting is a strategic decision on choice of the product and quantity to be ordered. There are a number of forecasting methods, such as averaging past data, exponential smoothing, regression models, causal or econometric models, time series analysis, etc. Although forecasting is a scientific discipline, 100% accurate forecasting is not possible as customer's demand can change at any time (Barbee, 1993). Other causes may include:

- Launching the product at the wrong time may adversely affect sales.
- Longer lead time means the product is not hitting the market on time, and in turn may not be able to be sold.
- Poor shelving of the garment may destroy it.
- Careless handling of the garments.

It is very important to decide the timing and amount of the mark-down, which is an experienced person's job. When the stock is not being sold, that means customers are not interested anymore in that style, so mark-down may help in getting the customer's interest back. In addition, the right price is to be decided for the mark-down garments so as to clear the stock quickly. The amount depends upon factors such as quantity in hand and the first mark-up. If the first mark-up the company has gained a reasonable profit, then the retail price can be set significantly lower, but if the initial mark-up was not high enough, then mark-down should be decided carefully. At the same point in time, the mark-down should not be large enough to make the customer suspicious that there is something wrong in the stock; also, very small mark-downs are unnoticed by the customer, so makes no difference.

Example 18.9 describes the method of calculating mark-down.

Example 18.9. Calculating mark-down.

A retail store advertises a mark-down of 15% on its pullovers. The store is left with 250 pullovers, which were retailed at $100. What is the total $ mark-down?

Answer:

Planned mark-down = 15%
Retail price = $100
$ Mark-down per pullover = $15 (i.e, 15% × 100)
Total mark-down = $15 × 250 pieces = $3750

18.9 Managing cost through inventory control

To control the operation of an organisation, it is very important to keep records. Inventory includes all the items held by a company, ranging from raw material to finished goods (Hogan et al., 2011). Controlling inventory impacts significantly on the cost. The company should order enough stock, which would not be left unsold. Both over-ordering and under-ordering will influence the profitability, where in former case ordering more than required will leave the stock unsold and in the latter case the company may lose the customer. Hence, it is imperative to order an optimum amount. The main concern in the garment industry is how much to order so that the inventory cost is the least. The role of a successful merchandiser is to order enough to fulfil the demand and keep the investment as low as possible. This gives rise to the concept of stock turnover, which is discussed below.

Stock turnover or inventory turnover ratio measures the firm's performance in managing inventory (Cooper and Kaplan, 1988). It compares the cost of the goods sold in a year with an average inventory held throughout the year. According to Kunz and Glock (2000), stock turnover gives an indication of how an inventory is moving through. If the inventory is moving quickly, then stock turnover is higher, which is an ideal situation for a company. Every company wants to achieve higher inventory turnover ratio because it directly impacts on the capital investment. In the case of higher stock turnover, the capital investment by the company will be lower, to gain the same profitability. The reason for lesser investment is buying lesser quantity, selling the stock quickly and rotating the same money in the business rather than spending a lot of money at the start and then holding the stock for a long time. Especially in the garment industry, it is advisable to have a quick turnover because in this fast-fashion world, the same style does not stay on the market for long. The Zara model of fast fashion is a perfect example, where their inventory turnovers are too high compared to their competitors. Although each industry plans to increase their stock turnover, it depends upon the company type and cannot be compared with 'inter-industry'; for example, stock turnover in the garment industry cannot be compared with the dairy industry as the dairy industry has a high inventory turnover.

In managing inventory in the garment industry, two questions are very important to answer:

1. How much to order?
2. When to order or reorder?

The garment industry needs to keep total cost of inventory low, which comprises total ordering cost and total carrying cost. The total cost of inventory is least on a quantity known as the economic order quantity or EOQ (Hergeth, 2002). In dependent demand, where the quantity to be sold per year is known, it is important to order that much stock per order to keep inventory cost under control. The EOQ is calculated by the formula:

$$Q^* = \sqrt{\frac{2DK}{h}}$$

where

Q = Economic order quantity
D = Annual demand
K = Fixed cost per order
h = Annual holding cost per item.

18.10 Apparel costing sheet analysis

The process of approximation of the total cost to produce a garment, from a buyer's perspective, including raw materials, labour, and other expenses, can be defined as costing (Brown and Rice, 2001). A costing sheet is used to trace the costing components. It is quite useful and handy to understand the percentage share of cost for each constituent of the apparel supply chain, including material used. The allocation of the cost depends upon multiple variables. Due to competitive manufacturing in Asian countries, a large proportion of the world's garments are manufactured in China, India, Bangladesh, Pakistan, etc. The 'cost to the factory' where the garment is manufactured include types of fabrics used, dyeing cost, trims and accessories used (such as labels, hangers, threads, fusing, buttons, zippers, etc.), cutting cost, stitching cost, trimming cost, packaging cost, company overhead, labour cost, administration cost, etc. A reasonable mark-up is added on the finished garment to cover expenses incurred by the manufacturer and to earn profit.

Once the garment is manufactured, road/rail transportation cost is added to deliver the garments to a port of loading. At the port of loading, the cost of freight forwarders and stevedoring is added. Normally, the garments are transported by sea unless the lead time requirement is very tight that demands air transport. The landed cost of the garment in the buyer's country includes the FOB (free on board) price, shipping cost, clearance cost, custom duties and maritime insurance in the case of sea freight. The cleared garments from customs are road/rail transported to the buyer's warehouse, where the cost of inventory holding is added onto the garment price. Margin is added to this final price (which is cost paid until the warehouse), which largely covers any mark-downs, expenses like salaries, sales promotion, rent, administration cost, insurance, taxes, etc., and finally the profit for the store. A detailed example of a costing sheet for a garment is given in Table 18.1.

Table 18.1 **Costing sheet**

Garment costing sheet			
Product name	Jacket		
Style no	AAA	Date	12/05/2014
Buyer	BBB	Country	Australia
Size range	XL	Season	Autumn/winter 2014
Product description		Selling price	$297.66
Colour	Black		
Fabric costing	**Particulars**	**Details**	**Amount (AUD)**
Fabric material	Metre per unit	Cost per metre ($)	Sub-total ($)
Fabric 1 Viscose	2	$9.00	$18.00
Fabric 2 Lawn	1.5	$4.00	$6.00
Fabric 3 Interfacing	0.78	$1.29	$1.01
Total fabric cost			**$25.01**
Average dyeing cost			$1.50
Dyed fabric cost			**$26.51**
Trims and accessories			
	Quantity	**Price/unit**	**Amount (AUD)**
Labels	2	$0.20	$0.40
Hangtags	1	$0.10	$0.10
Tickets	2	$0.10	$0.20
Tag	0	$0.00	$0.00
Thread	10 m	$0.05	$0.50
Fusing	0	$0.00	$0.00
Zipper	0	$0.00	$0.00
Buttons	8	$0.50	$4.00
Total trim and accessories			**$5.20**
Packing material			
Polybag	1	$0.20	$0.20
Paper	0	$0.00	$0.00
Board	0	$0.00	$0.00
Hanger	1	$0.05	$0.05
Hang tag	1	$0.02	$0.02
Carton	1/15	$0.50	$0.03
Total packing material			**$0.30**

Continued

Table 18.1 Continued

Garment costing sheet			
Product name	Jacket		
CMTP charges			
Stitching		$15.00	
Cutting		$4.00	
Finishing		$4.00	
Packaging		$2.00	
Embellishment		$6.00	
Trims		$4.00	
Total CMTP charges			$35.00
Sub total (fabric cost + trims + packing material + CMTP charges)			67.01
Overhead cost	12%		$8.04
Total factory cost			**$75.05**
Mark-up	25%		$18.76
Total wholesale price			**$93.81**
Transport to port of loading		$1.00	
FOB (Free on board)			$94.81
Insurance of FOB	1%		$0.95
Custom duty on FOB	10%		$9.48
Shipping		$1.00	
Clearance		$0.50	
Transport to warehouse		$1.50	
Cost paid till warehouse			**$108.24**
Mark-up	150%		
Margin	60%		$162.36
Retail price exclusive of GST			$270.60
GST	10%		$27.06
Retail price inclusive of GST			**$297.66**

CMTP = Cut, Make, Trim and Pack
FOB = Free on Board
GST = Goods and Service Tax

18.11 Conclusions

Costing includes all the activities related to purchase of raw materials and accessories, fabrics, processing and finishing of fabrics, sewing and packing of garments, transport and conveyance, shipping, overheads, banking charges and commissions, etc. There are always fluctuations in the costs of raw materials and accessories; charges of weaving or knitting; processing, finishing, sewing and packing; charges of transportation; and conveyance. Hence, it is essential to have knowledge updates about the latest prices, procedures, quality systems, market prices and availability, transportation and freight charges. The volatile nature and rigorous competition in the global garment manufacturing industry drive all the companies to minimize their cost by controlling inventory, accurate forecasting and low mark-downs. It must be remembered that the quality depends on price; and price depends on quality. Each product will have different price according to its quality. While the manufacturers and retailers decide the retail price of a garment, factors such as the average customer's buying level, quality and quantity and payment terms, should be taken into consideration.

References

Barbee, G., 1993. The ABCs of costing. Bobbin 34, 64.
Brown, P.K., Rice, J., 2001. Ready-to-Wear Apparel Analysis. Prentice-Hall.
Cooper, R., Kaplan, R.S., 1988. Measure costs right: make the right decisions. Harvard Bus. Rev. 66 (5), 96–103.
Easterling, C.R., Ellen, L., 2012. Merchandising Math for Retailing. Addison-Wesley Longman Limited.
Fiallos, M., 2010. Developing a Cost Model for Sourcing Products for Different Distribution Channels.
Guilding, C., 2010. Profit and Loss Statement.
Hergeth, H., 2002. Target costing in the textile complex. J. Text. Apparel, Technol. Manage. 2 (4).
Hogan, C., et al., 2011. A framework for supply chains: logistics operations in the Asia-Pacific region. MHD Supply Chain Sol. 41 (3), 73.
Jeffrey, M., Evans, N., 2011. Costing for the Fashion Industry. Berg.
Kunz, G.I., Glock, R.E., 2000. Apparel Manufacturing: Sewn Product Analysis.
Langfield-Smith, K., et al., 2008. Management Accounting: Information for Creating and Managing Value. McGraw-Hill Higher Education.
Lowson, R., 2002. Apparel sourcing offshore: an optimal operational strategy? J. Text. Inst. 93 (3), 15–24.
Miller, D., 2013. Towards Sustainable Labour Costing in the Global Apparel Industry: Some Evidence from UK Fashion Retail.

Index

Note: Page numbers followed by "f" and "t" indicate figures and tables respectively.

A

Abrasion resistance, 60, 310, 407
 as advantage of adhesive bonding, 342t
 as polyurethane property, 344t–345t, 354t
Acid washing, 396
Acrylics, 344t–345t
Active Tunnel Infusion (ATI™) dyeing, 48
Adhesion theories, 343t
Adhesives, 338
 adhesive bonding
  adhesion, 341
  advantages and disadvantages of, 342t
  description, 339t
  examples of applications, 356t
 bonding process, 346–349
  basic steps in, 348f
 classification of, 346t
 heat-activated adhesives, 349–352
 liquid adhesives, 352–355
  reactive liquids, 355
  solvent-based systems, 353
  spray equipment, types of, 352t
  waterborne systems, 353–355
 materials, 341–346
 selection of, 340f
 textile adhesives
  advantages and disadvantages of, 347t
  from base polymers, 344t–345t
  properties of hot-melt textile adhesives, 351t
  surface treatments for improving adhesion, 349t
Administrative overhead, 451
Advanced thermal-welding processes, 363–370
 dielectric welding, 368–370
  advantages and disadvantages of, 370t
 laser-assisted welding, 365–368
  types of, 368t
 ultrasonic welding, 364–365
  equipment used in, 364f
  fabrics joined using, 366t–367t
AentezVogue (tool for production planning), 105
AGMS (tool for production planning), 105
Agreement on textiles and clothing (ATC), 4
AIMS (tool for production planning), 105
Alternatives to sewing
 adhesive bonding and welding, advantages and disadvantages of, 339t, 355–356. *See also* Adhesives; Thermal-welding processes
 examples of applications, 356t
 competitive methods of joining garments, 337–338, 337f
 selecting joining process, 338–341
American Apparel and Footwear Association (AAFA), 43, 405
American Apparel Manufacturers Association (AAMA), 313–314
American Association of Textile Chemists and Colorists (AATCC), 34, 405, 424
American National Standards Institute (ANSI), 405
American Society for Quality (ASQ), 405
 four-point system, 119, 119t
American Society for Testing and Materials (ASTM), 34, 405
 system, in North American Free Trade Agreement (NAFTA) countries, 431
 care labelling system, 437
Antimicrobial treatment, 390
Apparel costing sheet analysis, 464
 costing sheet, 465t–466t

Apparel industry, 1–18
  challenges. *See* Apparel manufacturing challenges
  future trends, 13–15
    Fashion Like campaign, 14
  global scenario of. *See* Apparel production
  organisations, role of
    academic institutions, 12
    governing bodies, 12–13
    industries, 13
  production planning in, 81–83
  production systems, 83–88. *See also* Production systems
    types of, 83–88
  sewing
    art of joining materials, 1–3
    technology of making clothes, 3–4
  technological advancements, 1, 8
Apparel Made for You (AM4U), 48–49
  integrated mini-factory (IMF) of, 48–49, 48f
Apparel manufacturing challenges
  consumer choice and demographic variability, 10
    online scheme used, 9f
  in design and production, 11
  ecological challenges, 11–12
  related to supply chain, 10–11
  stiff competition, 11
Apparel product development, 29–32. *See also* Product development (PD)
  future trends, 44–49. *See also* Product development; future trends
  measures for, 43–44
    initial forecast accuracy and forecast accuracy, 44
    product-development cycle time, 44
    sample adoption ratio, 43
    seasons per year, 43–44
    SKU planning frequency, 44
  product-development technologies, 32–33
    digital color communication, 32–33
    technology solutions, 32
  variations, demand-led product development, 29–32
    fast-fashion product development, 29–32. *See also* Fast-fashion product development

Apparel product life-cycle management (PLM), 41–43
  PLM system, 41–43
Apparel product standards and specifications, 33
  product technical package, 34–41
    design specifications, 35
    general style information, 41
    performance specifications, 41
    product specifications, 35
  quality assurance through, 34
Apparel production
  in China, 4–5
  global exporters of textiles and clothing, 5f
  leading exporters of clothing, 6t
  leading importers of clothing, 7t
  3D-printed fashion materials, 9f
Apparel+ (tool for production planning), 105
ApparelMagic (tool for production planning), 105
Australian/New Zealand care labelling system, 443–444

B

Backpack bag, 160
Belts, 151, 176–178
  manufacturing design process, 176–178, 177f
  types of, 176
Bias binding, 269, 270f
Bias tape, 133
Bill of materials (BOM), 35
Bio-polishing, 390–391
BMSVision Cyclops (fabric inspection system), 116f, 117
Braids, 133
Briefcase, 159
British care labelling system, 439
  symbols and processes used in, 440t–441t
BS 3870 standards, 265
Bundle system, 84–85
Buttonhole tape, 133
Buttons, 131, 141–145
  application of, 141
  attaching mechanisms, 143
    styles of attaching, 143
  importance of, 143
  sizes of, 144
  types of, 142–143, 142f, 147
    fabric buttons, 144

Index

C

Canadian care labelling system, 437
  symbols and processes used in, 438t–439t
Care labelling of clothing, 427
  definition of, 429–431
  different processes described by, 429–430
    bleaching, 430
    dry-cleaning, 430
    ironing or pressing, 430
    laundering, 429–430
    tumble-drying, 430
  future trends, 444–445
  positioning of, 432t–434t
  requirements of, 427–429
  system of, 431–444
    ASTM care labelling system, 437
    Australian/New Zealand care labelling system, 443–444
    British care labelling system, 439
    Canadian care labelling system, 437
    ISO care labelling system, 435–436
    Japanese care labelling system, 439
  terminologies used in, 430–431
Cell phone case, 161
Checkbook case, 160
ClearWeld™ process, 368
  waterproof laminated fabrics welded using, 369f
Closures, 138–145, 408–409
  buttons. *See* Buttons
  hooks, 144–145
  velcro, 144–145
  zippers. *See* Zippers
Clothing-manufacturing process, 59
  buckling and formability, 70–71
  control system, 67–69
    inspection of fabric properties, 68–69
    ranges of rejection for tailoring, 68t
    tailoring process control chart, 67f
  sewability, 71–79. *See also* Sewability
    quality parameters for sewability assessment, 71–79
  tailorability, 69–70
Clutch, 159
Coin purse, 160
Colorfastness, 407
Comfort, 406
Compressional behavior of fabrics, 76–77
  lateral compression, 76
  longitudinal compression, 76
Computer-aided design (CAD) pattern tools, 215–218, 216f
  architects in, 218f
  jacket pattern constructed in, 217f
Computerised sewing machines, 291–295
  first generation of, 293
  fourth generation of, 295
  second generation of, 293–294
    control panel in, 294f
  third generation of, 294–295
    step programming problem, 295f
Consumer Product and Safety Commission (CPSC), 34
Conventional welding process. *See also* Thermal-welding processes
  direct thermal welding, 356–357
  heated-tool welding, 359–363
  hot-air welding, 358–359
    hot-air wedge-welding equipment, 359f
    parameters affecting quality of, 358t
  hot-wedge welding, 360
    fabrics joined using, 361t–362t
Cords, 133
Cosmetic bag, 160
Cosmotex spreader Apollo 100, 226–227, 226f
Credit card holder, 161
Cross-body bag, 157
Cut process planning, 221–223
  automated cut planning process, 221–223
    establishing marker processing time, 222
    performing marker calculations, 222
    processing of manufacturing reports, 222–223
    running of different planning scenarios, 222
    spreading planning, 222
Cutting of textile materials
  automated cutting of textile materials, 234–240
    automated laser cutting systems, 239
    automated ultrasound cutting systems, 239
    automated water-jet cutting systems, 240
    knife cutting systems and their main parts, 234–237, 235f–236f

Cutting of textile materials (*Continued*)
   multiply cutting cutters, 238–239, 238f–239f
   multipurpose cutters, 240
   single-ply cutters, 237
  fusing of cut textile components, 240–242
   fusing presses. *See* Fusing presses
   optimal fusing parameters, 242
  manual cutting equipment, 232–234, 232f–233f
  manual cutting process, 230–234
   characteristics of, 231–232, 231f–233f
Cutting process, final work operations, 242–244
  numbering of cut components, 243
  quality control of cut components, 242–243
   fabric quality, 242
   notches and drill marks, 243
   size and shape, 242–243
  recutting of faulty components, 243
  sorting and bundling of cut components, 243–244
   formation of, 243–244
   grouping of, 244
Cycle sewing machines, 289–291
  button-sewing machines, 289–290, 290f
  buttonhole-sewing machines, 290–291
  pattern-tacking machines, 291, 292f

D

Decorative needlework, 2
Demand-led product development, 29–32, 55
  fast-fashion product development, 29–32. *See also* Fast-fashion product development
Denim garment finishing, 396–397
Designated supplier, 112–113
Digital demand replenishment (DDR), 48
Direct cost, 449
Direct material cost, 450
Discounts, 452
DMDHEU (di-methylol dihydroxy ethylene urea), 389
Doctor's bag, 159
Drawstring bag, 157
Duffel bag, 160
Durability, 407–408

E

East–west bag, 158
Elastic tape, 133
Embroidery, 2, 130f, 134
  inspection of embroidery yarn, 149
Enzyme washing, 390–391
Euler's formula, 70
Eyeglass case, 161

F

Fabric inspection, 115–120
  fabric grading, 117–120
  four-point system for, 119, 119t
   defect calculation, 120, 121t
   various defects, 122t–123t
  future trends, 121–126
  knitted fabric defects, 120, 125f
  woven fabric defects, 120, 124f
Fabric properties and performance
  dimensional and physical properties, 59–60
  mechanical and other miscellaneous properties, 60
  performance, 60–61
Fabric sourcing, 109–128
  future trends, 121–126
  materials testing, 113
   testing required for approval, 114t
  merchandise planning, 111
  minimum order and limitations, 113
  product specification and production capacity, 111–112
  sales forecast, 110–111
  source identification/supplier identification, 112–113
   designated supplier, 112–113
   nondesignated supplier, 113
  sourcing responsibilities, 110
  third party accreditation, 114–115
Fabric spreading and cutting, 221–246
  cut process planning, 221–223. *See also* Cut process planning
   automated cut planning process, 221–223
  cutting of textile materials. *See* Cutting of textile materials
  future trends, 244

Index

spreading modes and their applications, 228–230
   face to face in a single direction, 229, 230f
   face to face in both direction, 230, 230f
   face up in a single direction, 229, 229f
   face up in both directions, 229, 229f
spreading of textile materials. See also Spreading of textile materials
   automated spreading, 224–228
   manual spreading, 223–224
   semiautomated and fully automated spreading processes, 228
Fanny pack, 160
Fashion accessories, 150–181
   in children wear, 145–148
      design developments, 145–146
      embellishment, 147–148
      safety issues, 146–148
   types of, 130–135
      basic accessories, 130f, 131–132
      belts. See Belts
      decorative accessories, 131f, 132–135
      footwear. See Footwear
      handbags. See Handbags
      headwear. See Headwear
      neckwear. See Neckwear
Fast-fashion product development, 29–32
   design and development for small production runs, 30
   design approval, 30–31
   raw material readiness, 31
   supply-chain relationships, 31–32
Fasteners
   metal, 146
   zipper, 146
Fixed cost, 449
Flame lamination, 359, 360f
Flame-retardant (FR) finish, 390
Flammable Fabrics Act (1958), 431
Fold-over bag, 159
Footwear, 151–156
   components of, 153–155, 153f
   sketching techniques, 156, 156f
   styles of, 153–155
Four-point system, 119, 119t
Fralix, Dr Mike (apparel industry consultant), 45f

Fringes, 134
Fur Products Labelling Act (1951), 431
Fusing presses
   continuous work process, 241
   and their characteristics, 240–241

G

Galloons, 134
Garment costing, 447
   classification, 449–450
      on the basis of behaviour, 449
      on the basis of traceability, 449–450
   cost elements, 450–451
      direct expenses, 451
      direct labour cost, 450
      direct material cost, 450
      indirect cost or overhead, 451
   costing need, 447–449
   measures of efficiency, 451–452
   profitability, 452–453
      profit and loss (P&L) statement, 452–453
Garment dyeing, 391–392
Garment make-up process
   and fabric properties, 61–63
      garment-manufacturing process, 61f
      lay-up and cutting, 62
      pattern grading, 61–62
      pressing, 63
      seaming and sewing, 62–63
   and low-stress mechanical properties, 63–67
      bending properties, 64
      of fabrics, 65t–66t
      shear properties, 64
      tensile properties, 64
Garment prototyping, 207
Garment sales elements analysis, 454–461
   basic margin calculations, 456–461
      averaging cost, 459–461
      calculating cost, 457–458
      calculating mark-up and margin, 458–459
      calculating mark-up percent and margin percent, 455–456
      example of, 456
      factors affecting pricing, 454

Garment sales elements analysis (*Continued*)
  initial mark-up, 461
  markdowns, 461–463
    calculating markdown, 463
  retail price and relationship with mark-up, 455
Garment size charts, 193–194
  adapted pictogram for woman's jacket, 193f
  bra size chart for under-band sizes, 193f, 194t
  development of
    intimate apparel, 196–197
    measurements used in bra sizing, 196f
    outerwear apparel, 194–195
  sizing and fit systems, 197–200
    fitting models and target markets, 199
  for women's trousers, 193t, 194f
Garment sizes. *See also* Garment size charts
  benefits and challenges of standardisation of, 199–200
Garment sizing and fit, 187–188
  geometry of human form
    golden ratio, 188
    ideal human body, 188–189, 188f
  human figure and body proportions, 189–193
Garment-finishing techniques, 387–388
  drying of garments, 392
  for functionality, 388–392
    antimicrobial treatment, 390
    enzyme washing or bio-polishing, 390–391
    flame-retardant (FR) finish, 390
    garment dyeing, 391–392
    permanent crease, 389
    sequence of operations, 388
    ultraviolet (UV) protection, 392
    water/oil repellent treatment, 389–390
    wrinkle-free treatments, 389
  future trends, 401–402
  garment dyeing, 387–388
  ready-made garments (RMGs), 387
GCS (tool for production planning), 105
General Sewing Data (GSD), 91

Gimp threads, 134
GINETEX system, 431, 435–436. *See also* International Organization for Standardization (ISO) system
Gloves, 172–176
  components, 173, 173f
  driving gloves, 176
  fingerless gloves, 175
  leather gloves, 176
  materials used for making, 172–173
  measurements, 173, 174f
  mittens, 176
  specific-use, 174–175
  styles, 173–176
Grafis CAD system, 209
Graniteville system, 118, 118t
Gross sales, 452
  calculation, example of, 453

H
Handbags, 151, 156–166
  components of, 162–163, 162f
  design and product development, 161, 161f
  design process, 163–166, 167f
  production of, 167f
  styles of, 157f–158f, 158–159
  tips for choosing right bags, 166
  variants of, 157–166, 158f
Handkerchiefs, 181, 182f
Hand-sewing, 2–3
Headwear, 168–172
  caps, 169–172, 170f
    styles of, 167f, 169–172
  hats, 168–169, 168f
    designing and production of, 169
    styles of, 169, 169f
Heat-activated adhesives, 349–352
  advantages and disadvantages for, 350t
Hobo, 159
Hong Kong style, 270
Hot Pop Factory (company), 8
Human resource management, 104–105

I
Idle time, 451
Inbound logistics cost, 450
Indirect cost, 449

Index

Industrial sewing machines, 275–280
　bed type classification, 276t
　cylinder bed, 277–278, 278f
　feed-off arm, 278–279, 278f
　feed types in, 280–289
　　bottom feed and alternating drop feed, 287
　　classification, 282t
　　differential drop feed, 284–285, 285f
　　drop feed (bottom feed), 281, 283f
　　needle feed (compound feed), 282–284, 283f
　　unison feed, 284, 284f
　　variable top and bottom differential feed, 286–287, 287f
　　variable top and bottom feed, 285–286, 286f
　　X-feed, 287–289, 288f
　feed up the arm, 279–280
　　side bed, 280f
　flat bed, 276, 277f
　post bed, 279, 279f
　raised bed, 276–277, 277f
Integrated mini-factory (IMF), 48–49
Interlacing, 247, 248f
Interlinings, 132, 409
Interlooping, 247, 248f
International Organization for Standardization (ISO) system, 34, 405
　care labelling system, 435–436
　five symbols used, for care labeling, 431f
　ISO 4916 standards, 265
Intralooping, 247, 248f
Inventory, 451
　managing cost through inventory control, 463–464
　　stock turnover, 463
Inventory management, 97–99
　average days of inventory in hand, 98
　incidence of out of stock, 99
　inventory accuracy, 98
　inventory carrying cost, 99

J
Japanese care labelling system, 439
　symbols and processes used in, 442t–443t
Jewelry, 150

K
Kaizen system, 14–15
Key holder, 161
Knitwear finishing, 393–395
　dry finishing, 393
　wet finishing, 393–395
　　Smith drums used for, 394f–395f
　solvent finishing, 395

L
Lace, 134
Lean manufacturing, 14–15
Lean production, manufacturing performance improvement through, 99–103
　defects, 102
　　causes of, 102
　excess inventory, 101
　　causes of, 101–102
　motion, 102
　　waste, causes of, 102
　over processing, 101
　　waste, causes for, 101
　overproduction, 100
　　causes of, 100
　transportation, 101
　　waste, causes of, 101
　waiting, 100
　　causes of, 101
Light-curing adhesives, 346
Linings, 132
Liquid adhesives, 352–355. *See also* Adhesives
　reactive liquids, 355
　solvent-based systems, 353
　spray equipment, types of, 352t
　waterborne systems, 353–355
　　characteristics of, 354t

M
Manufacturing overhead, 451
Margin, 451
Material feeding and associated problems, 324–325
Material interactions, 414
MerchanNet (tool for production planning), 105
Messenger bag, 160

Millennium (tool for production planning), 105
Modular system, 86–88
  advantages of, 87
  disadvantages of, 87
Moiré ribbons, 134
Multi-fibre agreement (MFA), 5–8
  and competition, 11
  freedom of free trade, 12
MyShape (online shopping tool), 8

N
N41 (tool for production planning), 105
Neckwear, 178–181
  bows, 178, 179f
  mufflers, 181, 182f
  neckties, 178, 178f
  scarves, 178, 179f
  shawls, 180, 180f
  stoles, 180–181, 181f
Net sales, 452
New Balance (retailer), 8
No-interval coherently phased product development (NICPPD) model, 22–23
  six phases of, 24t
Nondesignated supplier, 113
North–south bag, 157

O
OnePlace PLM and Workflow (tool for production planning), 105
Overlock stitches, 269, 269f

P
Passementerie, 134
Pattern construction, 205–220
  body, material and design, 208–213, 209f
    anthropometric data, 212
    input variables, 210t
    measurements, 210–211
    size standards, 212–213
    virtual fit simulation, 211f
  modes, 206–208
    pattern design considerations, 206
    patternmaking contemporary supply chain influences, 207
    traditional and contemporary approaches, 207–208
  tools, 213–218
    CAD pattern tools, 215–218, 216f–218f
    fit model mannequin with measurement table, 214f
    future technology tools, 218
    manual pattern tools, 213–214, 215f
Period sales, 451
PFAFF 8320 (hot-wedge welding system), 360
Phase-changing materials (PCMs), 13
Pile retention, 407
Pilling, 407
Piping, 134
Pocket squares, 181
Point of measurements (POM), 35
Polyamide, 344t–345t
Polyester, 344t–345t
Polyolefins, 344t–345t
Polyurethane (PU), 344t–345t
Polyvinyl chloride, 344t–345t
Pompons, 135
Pouch, 159
Predetermined Time Standard (PTS) code, 91
Pressing, 397–401
  factors affecting pressing, 398
  pressing equipment, 398–401
    BRI-1200/101 jacket front finisher, 399–400, 399f
    BRI-1400/101 pressing robot, 398, 399f
    to suit various shapes of a garment, 400
    UADD B58 jacket press, 400
    Veit 8326 shirt finisher, 399–400, 400f
    Veit 8900 shirt press, 399–400, 400f
Process–material interactions, 414
Product data management (PDM), 29
Product development (PD), 21. *See also* Apparel product development
  calendar, 27–29, 55
    multi-season, 28f
  process, 23–27
    design concepts, line concept through research, 23–25
    design development, line development, 25–26
    line planning and research, 23
    line presentation and marketing, 26
    production planning, pre-production and optimization, 26–27

Index

process models, 22–23
　　no-interval coherently phased product development (NICPPD) model, 22–23
Product development, future trends
　Apparel Made for You (AM4U), 48–49
　consultant's view
　　digital fabric performance, 46
　　digital printing, 46
　　digital product development, 45–46
　　PD tools, 46–47
　technology companies' view
　　compressing PD cycle time, 47
　　fit customization and virtual human modeling, 47
　　mobile product development, 47
　tools and technologies, 49–55
　　Human Solutions Anthroscan-based avatar, 50f
　　Human Solutions depth-sensor-based portable body-scanner, 51f
　　Optitex 3D virtual-prototyping solution, 52f
　　Tukatech realistic view, virtual dress-drape simulation on virtual-fit model, 54f
Product life-cycle management (PLM), 29, 34
Production planning and control in apparel industry, 88–94
　new tools developed in, 105
　outline of apparel production cycle, 89f
　overcoming hindrances, 89–90
　primary roles of, 81–83
　production control, 90–94
　　attainable standard, 92
　　corrective measures, 93
　　dispatching, 92
　　flexible manufacturing strategy, 93
　　follow-up, 93
　　inspection, 93
　　mass customization, 93–94
　　production strategies in garment industry, 93
　　Six Sigma, 102–103
　　value-added manufacturing strategy, 93

Production systems, 83–88
　types of, 83–88
　　bundle system, 84–85. *See also* Progressive bundle system
　　modular system, 86–88
　　progressive bundle system, 84–85
　　unit production system, 85–86
Progressive bundle system, 84–85
　advantages of, 85
　comparison with unit production system, 86
　disadvantages of, 85
Puckering, 72
Purchase-activated manufacturing (PAM), 48
Purse, 151, 157

Q

Quality control and assurance, in apparel industry, 405
　developing sampling plan, 417–422
　　arbitrary sampling, 418
　　continuous production sampling, 417
　　lot-by-lot sampling, 417
　　skip-lot sampling, 417
　future trends, 423–424
　post-production quality evaluation, 422–423
　　appearance retention, 422
　　care, 423
　preproduction quality control, 406–410
　　fabric, 406–408
　　inspection of other accessories, 408–410
　during production, 410–413
　　cutting defects, 411
　　defects during pressing/finishing, 413
　　defects in assembling, 411–413
　　final inspection, 413–417
　　statistical sampling, 418–422
Quick True Cost (QTC), 46

R

Ready-made garments (RMGs), 3–4
Ribbed tape, 133
Ribbons, 132
Rickracks, 135

Rosettes, 134
Ruffles, 135
Rule on Care Labelling (1972), 431

S

Sales return, 452
Satchel, 159
Seam efficiency, 71–72
Seaming tape, 133
Seamless products
　advantages of, 379–380
　berets, 376
　　half hose or socks, 376
　　protective gloves, 376
　disadvantages, 380–381
　future development, 382
　medical textiles, 377
　pantyhose, 376–377
　upholstery, 377
　upper-body garments/apparel, 377
Seamless technique, 373–376
　course shaping, 375
　products. *See* Seamless products
　raw materials, 377
　seamless knitting
　　formation of courses in, 373
　　historical events of, 374t
　　machines, 377–379
　wale shaping, 375–376
　　casting-off (knitting off), 376
　　running-on (picking up), 376
　　tubular knitting, 375
Selling overhead, 451
Sewability, 71–79
　quality parameters for sewability
　　assessment, 71–79
　　compressional behavior of fabrics and
　　　pucker, 76–77. *See also*
　　　Compressional behavior of fabrics
　　correlation coefficients, 77
　　initial pucker, 74
　　process of pucker formation, 74–75
　　process of seam slippage, 79
　　seam pucker, 72–77
　　seam pucker model, 73–74, 73f
　　seam slippage, 77–79
　　seam strength, 71–72
　　sewing thread properties and pucker,
　　　75–76

　　subsequent pucker, 74–75
　　understanding seam-slippage
　　　mechanisms, 78–79
Sewing, 1–3
Sewing automats, 302–305
　fabric components
　　loading of, 303
　　sewing of, 303
　unloading of sewn components,
　　303–305
　large part stacker, 304f
　shirt pocket attaching workstation, 305,
　　305f
　small part stacker, 304f
Sewing defects caused by needles, 322–324
　cloth points, examples of, 323f
Sewing equipment, 275
　computerised sewing machines, 291–295.
　　See also Computerised sewing
　　machines
　cycle sewing machines, 289–291
　　button-sewing machines, 289–290,
　　　290f
　　buttonhole-sewing machines, 290–291
　　pattern-tacking machines, 291, 292f
　future trends, 312–313
　　modular sewing head, 313f
　industrial sewing machines. *See* Industrial
　　sewing machines
Sewing needles, 306–310
　cloth point types, 309t
　measurement description of, 308t
　needle point types, 307–308
　needle selection, 310
　needle sizes, 308–310
　needle systems, 306
　types of, 307f
Sewing-room problems and solutions,
　317–336
　differential fabric stretch/feed pucker,
　　317–319, 321f
　drop-feed system, 319f
　dimensional change
　　of fabric, 320–321
　　of sewing thread/tension pucker,
　　　319–320, 320f
　future trends, 328–334
　needle thread-force waveform, 332f, 333
　pattern mismatch and seam pucker, 322

Index

presser-foot displacement waveforms, 333, 333f
seam pucker and surface distortions, 317–322
  in fine woven fabric, 318f
  seam undulation in stretch fabric, 318f
  seam pucker due to differential fabric dimensional instability, 321
  sewing defects, causes and possible remedies, 329t–330t
  structural jamming, 321–322
Sewing thread properties, 75–76
  diameter, 76
  fiber density, 76
  initial tensile modulus, 75
  yarn diameter compressibility, 76
Sewing threads, 310–312, 409–410
  different thread packages, 311f
  thread count and numbering system, 312
  thread performance, 311
Shape-memory polymers (SMPs), 13
Shelton WebSPECTOR, 115, 116f, 117
Shoes and boots. *See also* Footwear
  design and product development/manufacturing, 151–153
  of men, 154–155
  types of, 140f
  of women, 155
Shoulder bag, 157
Simparel (tool for production planning), 105
Singer, Isaac Merritt, 3
Size chart, 35
Small leather goods, 155f, 160–161
Socks, 151
Soutaches, 133
Spreading of textile materials
  automated spreading, 224–228
    spreading machine and its main parts, 225–227
  manual spreading, 223–224
    characteristics of, 223–224
    manual transportation of fabric ply, 224f
  semiautomated and fully automated spreading processes, 228
Stamped tape, 133
Standard minute value (SMV), 91

Stitch formation, problems in, 325–327
  skipped stitch, 326–327, 327f
  stitch distortion, 326, 327f
  stitch imbalance, 325–326, 326f
Stitch types
  on blazer, 259f
  on boilersuits, 261f
  on dress, 260f
  on shirts, 260f
  on sport jacket, 261f
  on suits, 259f
  on underwear, 262f
Stitches, 247–248
  future trends, 270–271
  seam types, 258–267
    bound seam, 263, 264f
    classes 1-8, 265–267, 266f–268f
    edge finishing, 265, 265f
    flat seam, 263–264, 264f
    lapped seam, 262–263, 263f
    on necklines and shirts, 267, 268f
    ornamental stitching, 264–265, 264f
    superimposed seam, 261–262, 262f
  seam-neatening, 269–270, 269f–270f
  stitch classes, 248–258, 249t
    Class 100, chain stitches, 249–251, 250f–251f
    Class 200, hand stitches, 251–252, 251f–252f
    Class 300, lock stitches, 252–253, 252f–253f
    Class 400, multi-thread chain stitches, 253–255, 253f–255f
    Class 500, over-edge chain stitches, 255–256, 256f
    Class 600, cover-seam chain stitches, 257–258, 257f–258f
Stitches per inch (SPI), 33, 35
Stock, 451
  stock turnover, 463
Stock keeping units (SKUs), 21
  SKU planning frequency, 44
Stockings, 138f, 151–156
Stone washing. *See* Acid washing
Styrene-butadiene, 344t–345t
Suitcase, 160
Supply chain management, in apparel industry, 94–96
  challenges related to, 10–11

Supply chain management, in apparel industry (*Continued*)
  network for manufacturer, 97f
  responsiveness of, 94–95
  root cause analysis of three echelons of
    cutting, 95–96
    finishing, 96
    sewing, 96
  typical apparel manufacturer, 96f
Supply-chain relationships, 41–43
Supporting materials, 136–137
  linings and interlinings, 136
    applications of, 136–137
    definitions, 136
  shoulder pads, 137, 138f
  wadding/batting, 137

T
Taffeta ribbons, 134
Tassels, 135
Tech-pack, 25–27, 34–35
  bill of materials, 37f
  branding and packaging information, 38f
  fit specifications, 39f
  garment assembly, 40f
  style history, 42f
  style summary, 36f
Ten-point system, 120t, 410
Textile adhesives. *See also* Adhesives
  advantages and disadvantages of, 347t
  from base polymers, 344t–345t
  properties of hot-melt textile adhesives, 351t
  surface treatments for improving adhesion, 349t
Textile and Clothing Technology Corp (TC^2), 405
Thermal-welding processes, 338
  description, 339t
  direct thermal welding, 356–357
  examples of applications, 356t
  heated-tool welding, 359–363
  hot-air welding, 358–359
  laser-assisted welding, 365–368
  ultrasonic welding, 364–365
Thickness strain, 72–73

Thimonnier, Barthélemy, 3
Thread breakage, 328
Three dimensional (3D) body scanning, 200
3D printing, 32, 46
Tote bag, 159
  recycled tote bag, 159
Trade allowance, 452
Travel bag, 160
Travel-related bags, 159–160
Trims and accessories
  evaluation of quality, 148–150
  inspection and testing of
    buckles, 150
    buttons, 149
    embroidery yarn, 149
    fusible interlinings, 150
    labels, 150
    lace and tape, 150
    snap fasteners, 150
    zippers, 149

U
Under-bed trimmer (UBT) sewing machines. *See* Computerised sewing machines
Unit production system, 85–86
  advantages of, 86
  comparison with progressive bundle system, 86
  disadvantages of, 86
User–production interactions, 414–415

V
Velvet ribbons, 134
Vietsoft ERP (tool for production planning), 105
Vinyl acetate, 344t–345t
Virtual human modeling, 47

W
Wallet, 161
Waste management, 103–104
Water/oil repellent treatment, 389–390
Watson, Ken (fashion supply chain consultant), 29f
Welted tape, 133
Wool Products Labelling Act (1938), 431

Work aids, 275, 295–302
  machine options, 296–299
    downturn feller with flatlock sewing, 300f
    hinged presser foot, 297f
    left-compensating presser foot, 299f
    right-compensating presser foot, 298f
    solid presser foot, 297f
  separate devices, 300–302
    condensed stitch in overlocks, 300f
World on a hanger (tool for production planning), 105
Wrinkle-free treatments, 389
Wristlet, 161

X
X-Burner technology, 397, 397f

Y
Yarn bending, 75
Yarn rotation, 75
Yarn sliding, 75
Young's modulus, 70

Z
Zedonk (tool for production planning), 105
Zellweger Uster Fabriscan (fabric inspection system), 116f, 117
Zero waste pattern cutting, 207–209
Zigzag stitches, 269, 270f
Zippers, 132, 138–141
  classification, 140–141
  components of, 138–139, 139f
  length, allowable tolerance of, 140
  quality of, 139
  tips for selecting, 139–140
  types of, 140f
Zulu weavers, 2